Fundamentals of Energy Engineering

Fundamentals of Energy Engineering

ALBERT THUMANN, P.E., C.E.M.

THE FAIRMONT PRESS, INC.
P.O. Box 14227, Atlanta, Georgia 30324

PRENTICE-HALL, INC.
Englewood Cliffs, New Jersey 07632

This Prentice-Hall, Inc., edition published 1984.

Printed in the United States of America

10 9 8 7 6 5 4 3 2 1

Library of Congress Catalog Card No: 83-20518
Cataloging in Publication Data:
    Fundamentals of energy engineering.
    Bibliography: p. 439
    Includes index.
    1. Power (Mechanics)   2. Power resources.   I. Title.
TJ163.9T54 1984   621.042   83-20518
ISBN 0-13-338327-X

ISBN (Prentice-Hall edition): 0-13-338327-X

Prentice-Hall International, Inc., London
Prentice-Hall of Australia Pt. Limited, Sydney
Editora Prentice-Hall do Brasil, Ltda., Rio de Janeiro
Prentice-Hall Canada Inc., Toronto
Prentice-Hall of India Private Limited, New Delhi
Prentice-Hall of Japan, Inc., Tokyo
Prentice-Hall of Southeast Asia Pte. Ltd., Singapore
Whitehall Books Limited, Wellington, New Zealand

Contents

1

Energy Situation

INTRODUCTION

Energy engineering is a profession which applies scientific knowledge for the improvement of the overall use of energy. It combines the skills of engineering with the knowledge of energy problems. The energy engineer must be able to identify problems in the use of energy, creatively design solutions, and implement the process.

In order to develop economic and socially acceptable ways to use available resources for the benefit of mankind, the energy engineer searches for a better way to combine a broad base of knowledge with experience.

Energy engineering requires a "system approach" and is multi-disciplinary in nature. Thus an energy engineer must have both engineering skills in, for example, electrical, mechanical and process engineering, as well as good management knowledge.

Problems, such as the high cost of energy, the depletion of resources, and the degradation of the environment are not transitory, but will plague mankind for decades to come. The energy engineering profession is addressing these and other problems in a systematic and cost-effective manner.

1

ENERGY CONVERSION FACTORS

In order to communicate energy engineering goals and to analyze the literature in the field, it is important to understand the conversion factors used in energy engineering and how they are applied.

Each fuel has a heating value, expressed in terms of the British thermal unit, Btu. The Btu is the heat required to raise the temperature of one pound of water $1°F$. Table 1-1 illustrates the heating values of various fuels. To compare efficiencies of various fuels, it is best to convert fuel usage in terms of Btu's. Table 1-2 illustrates conversions used in energy engineering calculations.

When comparing cost of fuels, the term "cents or dollars per therm" (100,000 Btu) is commonly used.

The chemical energy stored in fuel is sometimes expressed as Higher Heating Value (HHV) or Lower Heating Value (LHV).

Table 1-1. Heating Values for Various Fuels

Fuel	*Average Heating Value*
Fuel Oil	
Kerosene	134,000 Btu/gal.
No. 2 Burner Fuel Oil	140,000 Btu/gal.
No. 4 Heavy Fuel Oil	144,000 Btu/gal.
No. 5 Heavy Fuel Oil	150,000 Btu/gal.
No. 6 Heavy Fuel Oil 2.7% sulfur	152,000 Btu/gal.
No. 6 Heavy Fuel Oil 0.3% sulfur	143,800 Btu/gal.
Coal	
Anthracite	13,900 Btu/lb
Bituminous	14,000 Btu/lb
Sub-bituminous	12,600 Btu/lb
Lignite	11,000 Btu/lb
Gas	
Natural	1,000 Btu/cu ft
Liquefied butane	103,300 Btu/gal.
Liquefied propane	91,600 Btu/gal.

Source: Brick & Clay Record, October 1972.

2

Table 1-2. List of Conversion Factors

1 U.S. barrel	=	42 U.S. gallons
1 atmosphere	=	14.7 pounds per square inch absolute (psia)
1 atmosphere	=	760 mm (29.92 in.) mercury with density of 13.6 grams per cubic centimeter
1 pound per square inch	=	2.04 inches head of mercury
	=	2.31 feet head of water
1 inch head of water	=	5.20 pounds per square foot
1 foot head of water	=	0.433 pound per square inch
1 British thermal unit (Btu)	=	heat required to raise the temperatrure of 1 pound of water by 1°F
1 therm	=	100,000 Btu
1 kilowatt (Kw)	=	1.341 horsepower (hp)
1 kilowatt-hour (Kwh)	=	1.34 horsepower-hour
1 horsepower (hp)	=	0.746 kilowatt (Kw)
1 horsepower-hour	=	0.746 kilowatt hour (Kwh)
1 horsepower-hour	=	2545 Btu
1 kilowatt-hour (Kwh)	=	3412 Btu
To generate 1 kilowatt-hour (Kwh) requires 10,000 Btu of fuel burned by average utility		
1 ton of refrigeration	=	12,000 Btu per hr
1 ton of refrigeration requires about 1 Kw (or 1.341 hp) in commercial air conditioning		
1 standard cubic foot is at standard conditions of 60°F and 14.7 psia		
1 degree day	=	65°F minus mean temperature of the day, °F
1 year	=	8760 hours
1 year	=	365 days
1 MBtu	=	1 million Btu
1 Kw	=	1000 watts
1 trillion barrels	=	1×10^{12} barrels
1 KSCF	=	1000 standard cubic feet
1 Quad	=	one quadrillion (10^{15}) Btu's

Note: In these conversions, inches and feet of water are measured at 62°F (16.7°C), and inches and millimeters of mercury at 32°F (0°C).

The higher heating value is obtained by burning a small sample of fuel in an oxygen environment and recording the heat transferred to the water sample surrounding it. This test includes the latent heat of vaporization of the condensed vapor.

The lower heating value subtracts the latent heat of vaporization since this energy is usually unavailable in practice.

European countries usually use lower heating values for fuels while in the United States higher heating values are used. Thus a heating device tested in Europe will be approximately 10% more efficient than if tested in the United States due to the standard in heating values assumed.

Example Problem 1-1

The supply of fuel oil is projected at 11.5 million barrels per day. What is the supply in Btu per year, assuming an average fuel oil value of 140,000 Btu/gallon?

Answer

Fuel supply = 140,000 Btu/gallon $\times$ 11.5 $\times$ 10^6 barrels/day $\times$ 42 gallons/barrel $\times$ 365 days/year = 24.7 $\times$ 10^{15} Btu/year.

Example Problem 1-2

The warehouse of the plant is required to be minimally heated at night. Two methods of heating are being considered. One method is electric heaters. The second is to operate the oil-fired boiler (using No. 2 fuel oil) which keeps the radiators warm all night. Electricity costs, excluding demand charges, is 10 cents per kilowatt hour; No. 2 fuel oil costs $1.00 per gallon. Assume the boiler is 70% efficient and exclude the costs of running pumps and fans to distribute heat. What is the relative cost to heat the building using electricity or the oil-fired boiler?

Answer

To compare fuel costs, a common base of $ per therm will be used.

Electricity Cost

$$1 \text{ Kwh} = 3412 \text{ Btu}$$

$$\text{Cost} = \frac{100,000}{3412} \times .10 = \$2.93/\text{therm}$$

4

No. 2 Fuel Oil

No. 2 fuel oil $= 140,000$ Btu/gallon

$$\text{Cost} = \frac{100,000}{140,000} \times \$1.00 = 71 \text{ cents/therm}$$

Taking into account an efficiency of combustion of 0.7, the relative cost becomes:

$$\text{Cost} = \frac{71}{0.7} = \$1.02/\text{therm}$$

ENERGY CONSUMPTION WORLDWIDE

Industrial nations are competing in a world-wide market place. Energy consumption of the industrial sectors in each country gives a good comparative indicator of how each nation uses energy.

In order to compare energy consumption, energy per Gross Domestic Product (GDP) is used instead of Gross National Product (GNP) due to the fact that income originating in overseas investments which do not use domestic energy, is excluded.

Figure 1-1 illustrates how various nations compare in energy use. From this figure it is apparent that the United States uses more energy for a given output than the United Kingdom, West Germany, Sweden, Netherlands, France or Italy.

WORLD ENERGY OUTLOOK

The Organization for Economic Co-operation and Development (OECD) was established under a Convention signed in Paris on December 14, 1960. Members of OECD are Australia, Belgium, Canada, Finland, France, Federal Republic of Germany, Greece, Iceland, Ireland, Norway, Portugal, Spain, Sweden, Switzerland, Turkey, the United Kingdom and the United States.

In the 1982 World Energy Outlook published by OECD[*] several trends were discussed. Several conclusions to be drawn are:

---

[*]*Source:* World Energy Outlook, OECD 1982, International Energy Agency, 2 Rue André-Pascal, 75775 Paris Cedex 16, France.

Energy consumption (tons oil equivalent) per million U.S. dollars of gross domestic product

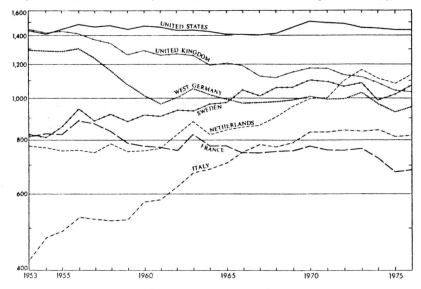

Source: EPRI EA 1471, Aug. 1980

Figure 1-1. Energy Intensity Among Industrial Societies

1. Energy and oil markets will remain stable in mid-1980's but will become increasingly tight thereafter.

2. Energy demand will accelerate between .6 to .8% per year in the first half of the 1980's to as much as 1.7 to 2.6% per year during the 1990's.

3. Electric consumption will grow at a faster rate of 2-3% up to the 1990's and 3-4% thereafter.

4. OECD countries are expected to remain highly dependent on imported oil.

5. Natural gas is unlikely to grow beyond 20% of its present share of total OECD energy use. To maintain the present share of gas use imports of natural gas will have to increase five to sixfold.

6. Coal is expected to grow in use over oil due to its price advantage.

7. Nuclear by the year 2000 is expected to grow 10–11% for OECD total energy.

In the Executive Summary of the World Energy Outlook the finding concluded:

> In sum, while some structural change away from oil can be expected in the energy economy as a result of the price mechanism, the basic vulnerability of the world economy to oil supply disruptions is far from being eliminated. On the contrary, with energy demand pressures growing, there is a continuing risk that a closely balanced demand and supply equilibrium could once again be precipitated by political events in the Middle East.

ENERGY CONSUMPTION DATA SOURCE

The U.S. Department of Energy, Energy Information Administration, Office of Energy Data Operations, publishes the "Monthly Energy Review" based on data from DOE, other government agencies and private establishments. The data which follows is from this source unless otherwise referenced.

U.S. Energy Production

Energy production for 1980 totaled 64.8 quadrillion Btu, a 1.4% increase compared to production for 1979. This increase amounted to 1.2% when measured as a daily rate (a measure which removes the influence of leap year). Increases in production occurred for petroleum and coal. Petroleum production was up 0.3% and coal 6.7% (all measured as daily rates). Natural gas production decreased by 2.1%. All other forms of energy production combined were down by 1.3%, primarily due to a decline in electricity production by nuclear plants.

7

U.S. Energy production, classified by energy type is presented in Figure 1-2.

Yearly

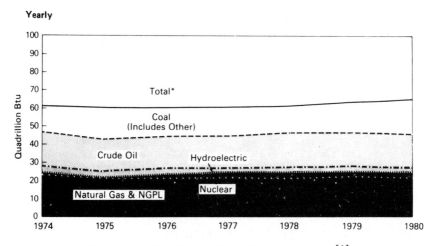

Figure 1-2. U.S. Production of Energy by Type[1]

U.S. Energy Consumption

Total U.S. energy consumption in 1980 dropped to 76.3 quadrillion Btu, 3.4% below 1979 and a 2.4% decrease from the 1978 consumption level. The consumption by energy type is presented in Figure 1-3. Figure 1-4 shows the U.S. energy consumption by end-use sector. Energy consumption per GNP dollar is presented in Figure 1-5.

The Residential and Commercial sector consumption was 27.3 quadrillion Btu in 1980, a 0.5% decrease from the amount consumed the previous year and a 2.9% decrease from the amount consumed in 1978. The Residential and Commercial sector consumed 35.8% of the total consumption for 1980, up from the sector's 34.8% share in 1979. The consumption of energy by the Commercial and Residential sector is presented in Table 1-3.

The Industrial sector consumption was 30.3 quadrillion Btu in 1980, down 4.0% from 1979, but up 3.2% from the consumption

8

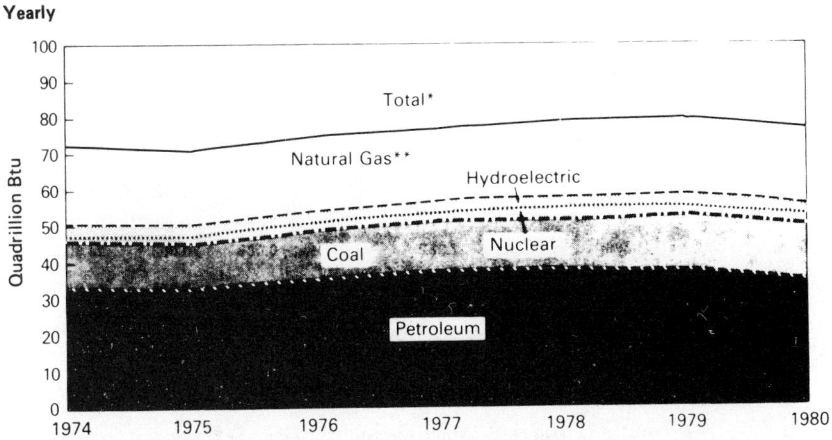

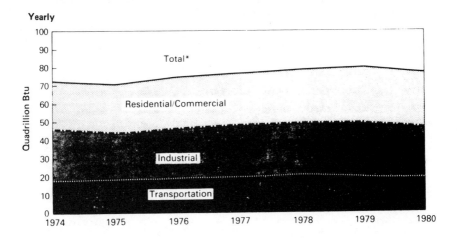

Yearly

Figure 1-3. U.S. Consumption of Energy by Type[1]

Figure 1-4. U.S. Consumption of Energy by End-Use Sector[1]

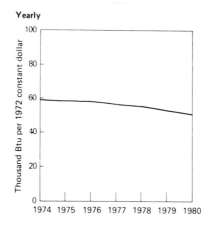

Figure 1-5. Energy Consumption per GNP Dollar[1]

Table 1-3. Consumption of Energy by the Residential
and Commercial Sector*[1]

	Coal	Natural Gas (Dry)	Petroleum	Electricity Sales	Electrical Energy Losses†	Total Energy Consumed
			Quadrillion (10[15]) Btu			
1973	0.291	7.626	6.741	3.495	8.460	26.613
1974	0.293	7.518	6.141	3.475	8.548	25.974
1975	0.239	7.581	5.792	3.588	8.814	26.014
1976	0.227	7.866	6.302	3.729	9.089	27.213
1977	0.226	7.461	6.245	3.936	9.702	27.569
1978	0.250	7.624	6.268	4.100	9.918	28.159
1979	0.210	7.891	5.027	4.184	10.150	27.462
1980	0.194	7.637	4.389	4.354	10.762	27.337

Notes:

* The Residential and Commercial Sector consists of housing units, nonmanufacturing business establishments (e.g., wholesale and retail businesses), health and educational institutions, and government office buildings. Notes on the methodology used for sector calculations are provided in the Notes and Sources of Reference 1.

† Proportion of total electrical energy losses incurred in the generation and transmission of electricity plus plant use and unaccounted for that are attributed to this sector.

level in 1978. The Industrial sector consumed 39.7% of the 1980 total, as compared to the 40.0% share in 1979. The consumption of energy by the Industrial sector is presented in Table 1-4.

The Transportation sector consumption was 18.6 quadrillion Btu in 1980, down 6.6% from 1979 and down 9.6% from the consumption level in 1978. This sector consumed 24.4% of the 1980 total, as compared to a 25.3% share in 1979. Petroleum energy consumption of the Transportation sector is presented in Table 1-5.

The Electric Utilities consumption was an estimated 24.8 quadrillion Btu of energy in 1980, 2.5% higher than in the previous year, and 6.0% higher than the energy consumed in 1978. Coal contributed 48.8% of the energy consumed by Electric Utilities in 1980, while natural gas contributed 15.3%, hydroelectric power 12.5%, petroleum 12.1%, nuclear power 10.9%, and geothermal, wood and waste 0.5%. The consumption of energy by the Electric Utilities is presented in Table 1-6.

Imports and Exports

Net imports of energy for 1980 totaled 12.0 quadrillion Btu, 28.3% below the 1979 level. This decrease amounted to 28.5% when measured as a daily rate. By energy source, the decreases in net imports were petroleum, 21.8%; natural gas 21.7%; and electricity and coal coke combined, 37.1% (daily rates). Net exports of coal for 1980 were 41.0% higher than the level for 1979.

The annual value of U.S. energy imports between 1974 and 1980 is presented in Figure 1-6. The annual production and consumption data is also presented for comparison.

The total annual U.S. energy imports and exports for 7 years is presented in Figures 1-7 and 1-8. This data indicates both a decrease in imports and increase in exports in 1980.

The U.S. dependence on petroleum imports is shown in Figure 1-9.

11

Table 1-4. Consumption of Energy by the Industrial Sector*[1]

	Coal	Natural Gas (Dry)	Petroleum	Hydroelectric	Net Coke Imports†	Electricity Sales	Electrical Energy Losses**	Total Energy Consumed
				Quadrillion (10^{15}) Btu				
1973	4.350	10.397	6.683	0.035	(0.008)	2.341	5.676	29.474
1974	4.057	10.012	6.506	0.033	0.059	2.337	5.751	28.755
1975	3.801	8.531	6.160	0.032	0.014	2.304	5.669	26.512
1976	3.792	8.768	6.951	0.033	0.000	2.525	6.162	28.230
1977	3.494	8.642	7.692	0.033	0.015	2.635	6.513	29.024
1978	3.462	8.540	7.840	0.032	0.131	2.732	6.637	29.373
1979	3.641	8.554	9.401	0.034	0.066	2.873	6.983	31.551
1980	3.354	8.407	8.876	0.033	(0.037)	2.781	6.886	30.300

Notes:

* The Industrial Sector is made up of construction, manufacturing, agriculture, and mining establishments. Notes on the methodology used for sector calculations are provided in the Notes and Sources of Reference 1.

† Net Imports=imports minus exports. Parentheses indicate exports are greater than imports.

** Proportion of total electrical energy losses incurred in the generation and transmission of electricity plus plant use and unaccounted for that are attributed to this sector.

Table 1-5. Petroleum Energy Consumption by the Transportation Sector[1]
(in Quadrillion Btu)

Year	10^{15} *Btu*
1973	17.74
1974	17.31
1975	17.55
1976	18.47
1977	19.16
1978	20.04
1979	19.30
1980	17.98

Table 1-6. Consumption of Energy by the Electric Utilities[1]

	Coal*	Natural Gas (Dry)	Petroleum	Hydroelectric Power[†]	Nuclear Electric Power	Other**	Total Energy Consumed
			Quadrillion (10^{15}) *Btu*				
1973	8.655	3.746	3.671	2.975	0.910	0.046	20.004
1974	8.524	3.518	3.499	3.276	1.272	0.056	20.144
1975	8.783	3.241	3.231	3.187	1.900	0.072	20.414
1976	9.714	3.153	3.454	3.032	2.111	0.081	21.544
1977	10.245	3.285	4.028	2.482	2.702	0.082	22.825
1978	10.134	3.297	3.813	3.132	2.977	0.068	23.421
1979	11.258	3.610	3.392	3.132	2.748	0.089	24.229
1980	12.125	3.789	2.998	3.093	2.704	0.114	24.823

Notes:

* Includes bituminous coal, lignite, and anthracite.

[†] Includes net imports of electricity.

** Includes geothermal power and electricity produced from wood and waste.

13

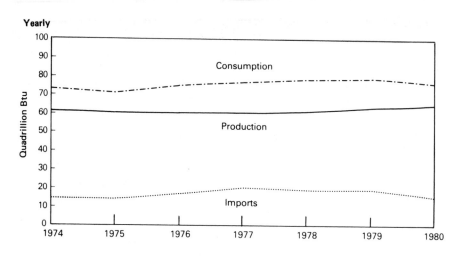

Figure 1-6. Summary of U.S. Energy Consumption
Production and Imports[1]

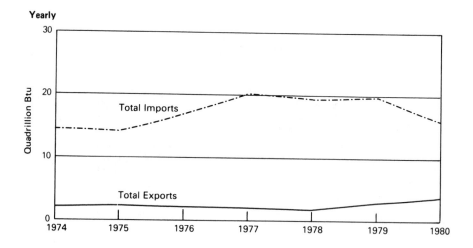

Figure 1-7. U.S. Energy Imports and Exports[1]

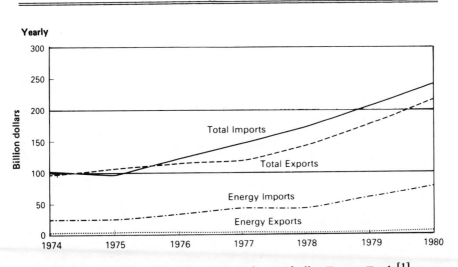

Figure 1-8. U.S. Merchandise Trade Value, Including Energy Trade[1]

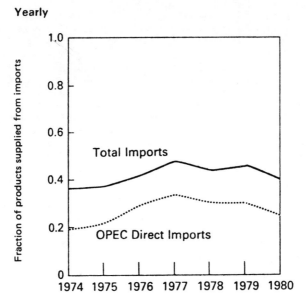

Figure 1-9. U.S. Dependence on Petroleum Imports[1]

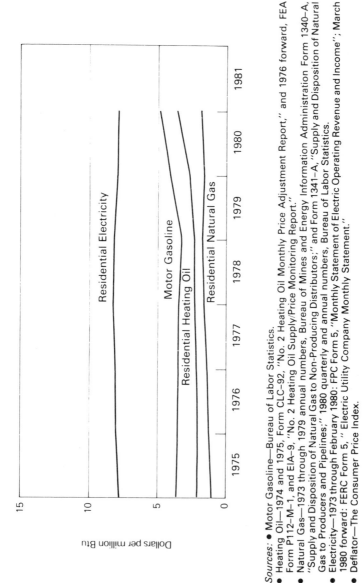

Figure 1-10. Average Cost of Fuels to End Users (1972 Constant Dollars)

Sources: • Motor Gasoline—Bureau of Labor Statistics.
• Heating Oil—1974 and 1975, Form CLC–92, "No. 2 Heating Oil Monthly Price Adjustment Report," and 1976 forward, FEA Form P112–M–1, and EIA–9, "No. 2 Heating Oil Supply/Price Monitoring Report."
• Natural Gas—1973 through 1979 annual numbers, Bureau of Mines and Energy Information Administration Form 1340–A, "Supply and Disposition of Natural Gas to Non-Producing Distributors;" and Form 1341–A, "Supply and Disposition of Natural Gas to Producers and Pipelines;" 1980 quarterly and annual numbers, Bureau of Labor Statistics.
• Electricity—1973 through February 1980: FPC Form 5, "Monthly Statement of Electric Operating Revenue and Income"; March 1980 forward: FERC Form 5, " Electric Utility Company Monthly Statement."
• Deflator—The Consumer Price Index.

Energy Costs

Figure 1-10 presents the average cost of Fuels to end users, in 1972 constant dollars per Btu, based on U.S. Department of Energy data.

Petroleum prices are summarized for five years in Tables 1-7 and 1-8. Table 1-9 summarizes natural gas prices for the period 1973 to 1980. Electricity prices are presented in Table 1-10 for the same years. It is found that electric utility rate structures vary significantly throughout the country. Table 1-11 presents typical electrical utility bills for various major cities throughout the United States.

Table 1-7. Petroleum Price Summary[1]

	Actual Domestic Average Wellhead Price	Refiner Acquisition Cost of Crude Oil			No. 6 Residual Oil Price—Average	
		Domestic	Imported	Composite	Wholesale	Retail
	Dollars per barrel					
1976	8.19	8.84	13.48	10.89	10.72	11.49
1977	8.57	9.55	14.53	11.96	11.96	13.23
1978	9.00	10.61	14.57	12.46	11.51	12.75
1979	12.64	14.27	21.67	17.72	17.66	18.67
1980	21.57	24.23	33.89	28.07	22.40	25.46

Table 1-8. Petroleum Price Summary[1]

	No. 2 Diesel Price—Average		No. 2 Heating Oil Price—Average		Gasoline Price Average All Grades Retail	Propane Price Average Wholesale	Butane Price Average Wholesale
	Wholesale	Retail	Wholesale	Retail			
	Cents per Gallon						
1976	31.9	34.7	32.6	40.6	NA	20.6	21.9
1977	36.1	39.3	36.9	46.0	NA	25.0	25.4
1978	37.1	40.2	38.7	49.4	65.2	24.0	23.0
1979	58.2	62.4	53.0	65.6	88.2	29.5	45.8
1980	80.6	86.8	82.2	97.7	122.1	42.1	61.0

17

Table 1-9. Natural Gas Prices[1]

	Average Wellhead Value	Delivered to Electric Plant	Average Residential Heating
	Cents per thousand cubic feet		
1973	21.6	35.0	108.2
1974	30.4	49.0	125.3
1975	44.5	76.9	154.2
1976	58.0	105.9	184.6
1977	79.0	133.4	226.4
1978	90.5	147.9	262.6
1979	117.8	180.3	323.1
1980	149.1	NA	391.5

Steam Costs

In 1981, industrial purchasers of steam were typically billed rates ranging from $13 per 1000 pounds (Consolidated Edison of New York) to $6-8 per 1000 pounds (Southern California Edison). Ten years earlier, typical prices were $1.50 to $2.00 per 1000 pounds.

Oil Drilling Activities

Oil drilling activities have sharply increased, spurred by the ten-fold escalation in crude oil prices between the energy crisis of 1973 and 1980. Some categories of domestic oil doubled in price in 1980 due to the suspension of federal price controls.

Figure 1-11 shows drilling activity for the years 1973–80. It is expected that the number of oil and gas wells in 1980 will break the 1956 record of 58,160. The Hughes Tool Co. of Houston, which acts as the industry's statistician in such matters, now predicts that an average of 2,800 drilling rigs will be operating each week during 1980. That would be 29% more than in 1979 and would surpass the record of 2,636 in 1955.

18

Table 1-10. Electricity Prices[1]

	Cost of Fossil Fuels Delivered to Steam-Electric Utility Plants				Average Retail Electricity Prices				
	Coal	Residual Oil	Natural Gas	All Fossil Fuels	Residential	Commercial	Industrial	Other	Total
	Cents per million Btu				Cents per kilowatt-hour				
1973	40.5	78.8	33.8	47.5	2.54	2.41	1.25	2.10	1.96
1974	71.0	191.0	48.1	90.9	3.10	3.04	1.69	2.75	2.49
1975	81.4	201.4	75.4	103.0	3.51	3.45	2.07	3.08	2.92
1976	84.8	195.9	103.4	110.4	3.73	3.69	2.21	3.27	3.09
1977	94.7	220.4	130.0	127.7	4.05	4.09	2.50	3.51	3.42
1978	111.6	212.3	143.8	139.3	4.31	4.36	2.79	3.62	3.69
1979	122.4	299.7	175.4	162.1	4.64	4.68	3.05	3.96	3.99
1980	NA	NA	NA	NA	5.36	5.48	3.69	4.76	4.73

19

Table 1-11. January 1982 Typical Electricity Bills ($) in Selected Cities*

	Commercial	Industrial		Commercial	Industrial
ALABAMA			MINNESOTA		
Birmingham	643	9,820	Minneapolis	517	8,329
Huntsville	509	8.681	Rochester	499	8,869
ARIZONA			MISSISSIPPI		
Phoenix	622	8,869	Jackson	681	10,451
ARKANSAS			MISSOURI		
Little Rock	498	6,965	Kansas City	769	10,122
CALIFORNIA			St. Louis	573	7,084
San Francisco	943	18,004	Independence	531	7,699
Sacramento	314	3,640	NEBRASKA		
Los Angeles	679	12,732	Lincoln	432	6,992
COLORADO			Omaha	415	6,286
Denver	693	9,209	NEW JERSEY		
Colorado Springs	339	5,613	Newark	929	13,655
CONNECTICUT			NEW YORK		
Hartford	976	14,159	New York	1,366	22,122
Bridgeport	1,111	18,538	Buffalo	717	10,371
FLORIDA			OHIO		
Miami	676	12,218	Cleveland	697	11,988
Orlando	603	10,004	Canton	562	8,087
GEORGIA			OKLAHOMA		
Atlanta	798	9,883	Oklahoma City	546	7,774
IDAHO			OREGON		
Boise	365	5,179	Portland	403	7,111
ILLINOIS			PENNSYLVANIA		
Chicago	834	12,289	Scranton	692	9,040
Springfield	640	10,385	Philadelphia	930	13,255
INDIANA			SOUTH CAROLINA		
Fort Wayne	512	8,141	Conway	504	7,357
Richmond	538	9,057	TEXAS		
KANSAS			Houston	674	10,949
Wichita	546	9,538	Fort Worth	626	8,370
Kansas City	898	12,389	Austin	607	11,194
KENTUCKY			San Antonio	567	8,375
Louisville	505	7,398	UTAH		
LOUISIANA			Salt Lake City	852	8,993
New Orleans	575	8,571	VERMONT		
MAINE			Burlington	460	8,976
Portland	684	10,448	VIRGINIA		
MASSACHUSETTS			Arlington	645	9,869
Boston	997	14,903	WEST VIRGINIA		
Baintree	782	13,817	Charleston	530	7,871
MICHIGAN			WISCONSIN		
Grand Rapids	708	11,157	Milwaukee	686	9,030
Detroit	713	11,268			
Lansing	574	8,846			

*
Data based on average monthly commercial consumption of 10,000 kwh at 40 kw demand and industrial consumption of 200,000 kwh at 500 kw demand. From data gathered by the Energy Information Administration for the Bureau of Labor Statistics.

20

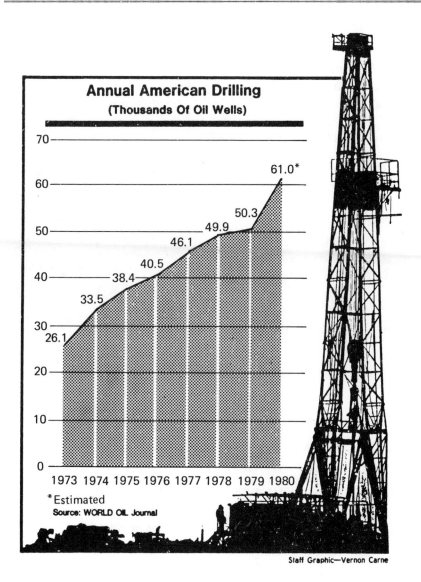

Annual American Drilling
(Thousands Of Oil Wells)

70
61.0*
60
50.3
50
49.9
46.1
40.5
40
38.4
33.5
30
26.1
20
10
0

1973 1974 1975 1976 1977 1978 1979 1980

*Estimated
Source: WORLD OIL Journal

Staff Graphic—Vernon Carne

Figure 1-11. Annual American Drilling
(Thousands of Oil Wells)

Industrial Energy Consumption

Figure 1-12 indicates that the industrial segment used approximately 39% of the total consumption, or 30.6 quads, in 1980 (1 quad = 10^{15} Btu).

For the industrial sector, 16% of the energy was used for feedstock, 39% for process steam, 26% for direct heat, and 10% for electrical drives. (Refer to Figure 1-13.)

Figure 1-12.
U.S. Energy Consumption

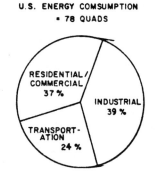

INDUSTRIAL ENERGY USE

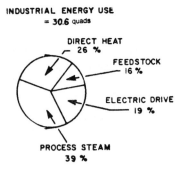

Figure 1-13.
Industrial Energy
Consumption

CODES, STANDARDS AND LEGISLATION

There are many local state and federal codes, standards and legislation which affect energy use. A review of the highlights in energy legislation activities is useful in order to give a historical perspective.

State Codes

More than three quarters of the states have adopted ASHRAE Standard 90-80 as a basis for their energy efficiency standard for new building design. The ASHRAE standard 90-80 is essentially "prescriptive" in nature. For example, the energy engineer using this standard would compute the average conductive value for the building walls and compare it against the value in the standard. If the computed value is above the recommendation, the amount of glass or building construction materials would need to be changed to meet the standard.

Another approach that California is using is the Building Energy Performance Standard (BEPS). The BEPS standard requires that the energy engineer limit Btu's/square foot to a performance value. Thus it is a performance rather than a prescriptive standard. In calculating the Btu's/sq ft each kilowatt of electricity must be multiplied by a Resource Utilization Factor (RUF) which takes into account the inefficiency of transmission of electricity.

The BEPS approach allows the energy engineer to choose any combination of ECOs which will have the effect of meeting the performance standard.

In order to verify the results, a computer energy analysis model is usually required.

Federal Legislation

The following is a summary of legislation which the energy engineer should be aware of:

23

Economic Recovery Tax Act (ERTA)—1981. As a result of the Economic Recovery Tax Act of 1981 there is a brand new, generally faster method of writing off the cost of tangible property used in business or held for the production of income. It's called the "Accelerated Cost Recovery System" or ACRS. The new system is generally applicable to eligible property (called "recovery property") placed in service on or after January 1, 1981. So it may apply to depreciable property that you've already purchased.

Recovery property is divided into four classes: 3-year, 5-year, 10-year, and 15-year property. For example:

- 3-year: Cars, light duty trucks and certain other short-lived personal property.

- 5-year: Most machinery and equipment.

- 15-year: Buildings.

For each class there is a standard set of recovery deductions (i.e., depreciation with a new name) to be taken over a fixed recovery period.

Energy Security Act—1980. The main purpose of the Act is to create a synthetic fuel industry by authorizing $20 billion to be used by the newly created Synthetic Fuel Corporation.

The Crude Oil Windfall Profits Act—1980. The main purpose of the Act is to tax oil companies $227 billion from their profits defined as windfall. In addition, the Act allows a 10% tax credit through 1985 for biomass equipment to be used to convert waste into fuel.

National Energy Conservation Policy Act (NECPA)—1978. This Act is probably the most important legislation passed.

The National Energy Conservation Policy Act of 1978, as it relates to energy audits provides for:

Utility Conservation Program for Residences—A program requiring utilities to offer energy audits to their residential customers that would identify appropriate energy conservation and solar energy measures and estimate their likely costs and savings. Utilities also will be required to offer to arrange for the installation and financing of any such measures.

Weatherization Grants for Low Income Families—Extension through 1980 of the DOE weatherization grants program for insulating lower income homes at an authorized level of $200 million in FY 1979 and 1980.

Solar Energy Loan Program—A $100 million program administered by HUD which will provide support for loans of up to $8,000 to homeowners and builders for the purchase and installation of solar heating and cooling equipment in residential units.

Energy Conservation Loan Programs—A $5 billion program of federally-supported home improvement loans for energy conservation measures; $3 billion for support of reduced interest loans up to $2,500 for elderly or moderate income families, and $2 billion for general standby financing assistance.

Grant Program for Schools and Hospitals—Grants of $900 million over the next three years to improve the energy efficiency of schools and hospitals.

Energy Audits for Public Buildings—A two-year, $65 million program for energy audits in local public buildings and public care institutions.

Appliance Efficiency Standards—Energy efficiency standards for major home appliances, such as refrigerators and air-conditioning units.

Civil Penalties Relating to Automobile Fuel Efficiency—Authority for the Secretary of Transportation to increase the civil penalties on auto manufacturers from $5 to $10 per car for each 1/10 of a mile a manufacturer's average fleet mileage fails to meet the EPCA automobile fleet average fuel economy standards.

Other Provisions—Other provisions in the Act include the following:

- Grants and standards for energy conservation in Federally-assisted housing.

- Federally-insured loans for conservation improvements in multifamily housing.
- $100 million for a Solar Demonstration program in Federal buildings.
- Conservation requirements for Federal buildings.
- $98 million for solar photovoltaic systems in Federal facilities.
- Industrial recycling targets and reporting requirements.
- Energy efficiency labeling of industrial equipment.
- A study of the energy efficiency of off-road and recreational vehicles.
- An assessment of the conservation potential of bicycles.

Solar is considered a renewable resource and is covered by most of the programs as a measure for reducing energy consumption.

In the Residential Conservation Service (RCS) Program of NEA the following opportunities are encouraged:

Conservation and Renewable Resource Measures and Practices— The RCS Program applies to existing single- to four-family dwelling units with a heating and/or cooling system. The proposed rule divides the energy conservation measures into two categories: (1) energy conservation measures which conserve energy but do not use a renewable energy source, and (2) renewable resource measures which make use of solar and wind energy. The conservation measures covered by the RCS Program, according to NECPA, include:

- caulking and weatherstripping of doors and windows
- furnace efficiency modifications
- clock thermostats
- ceiling, attic, wall and floor insulation
- water heater insulation
- storm windows and doors

26

- heat-absorbing or heat-reflective glazed windows and door materials
- load management devices

This list of measures was expanded to include measures which are eligible for residential energy tax credit provided in the Energy Tax Act of 1978. The additional measures are:

- duct insulation
- pipe insulation
- thermal windows

Replacement gas burners are eligible for tax credits but are not included in the RCS Program because they are usually installed to replace oil burners. The RCS Program does not include measures which require changing from one fossil fuel to another, or in switching from fossil fuel to electricity.

The renewable resource measures covered by the RCS Program include:

- solar domestic hot water systems
- active solar space heating systems
- combined active solar space heating and solar domestic hot water system
- passive solar space heating and cooling systems
- wind energy devices
- replacement solar swimming pool heaters

The RCS is seen as a major program to increase consumer awareness and purchase of renewable resource measures.

The Energy Tax Act of 1978 provides for:

1. Business Energy Tax Credits:
   - Business tax credits for industrial investment in alternative energy property (such as boilers for coal, nonboiler burners for alternate fuels, heat conservation equipment and recycling equipment).

27

- Denial of tax benefits for new oil and gas-fired boilers.
- Denial of investment tax credit and accelerated depreciation for new gas and oil boilers.

2. An additional 10% investment tax credit (nonrefundable except for solar equipment) is provided for investment in:

- Alternative Energy Property: This applies to boilers and other combustors which use coal or an alternative fuel, equipment to produce alternative fuels, pollution control equipment, equipment for handling and storage of alternate fuels, and geothermal equipment. The credit is not available to utilities.

- Solar or Wind Energy Property: A refundable credit for investments in equipment to use renewable energy to generate electricity or to heat or cool, or provide hot water. This credit is not available to utilities.

- Specially Defined Energy Property: This applies to equipment to improve the heat efficiency of existing industrial processes, including heat exchangers and recuperators.

Public Utility Regulatory Policies Act (PURPA)—1978. The main purpose of PURPA is to provide equitable rates to elective customers, optimize efficient use of facilities and resources by elective utilities and conserve energy supplies by elective utilities. PURPA also required the Federal Energy Regulatory Commission (FERC) to establish rules favoring industrial cogeneration facilities and requiring utilities to buy or sell power from qualified cogenerators at just and reasonable rates.

Power Plant and Industrial Fuel Use Act (FUA)—1978. The main purpose of the Act is to encourage use of coal and to minimize natural gas and oil usage by having power plants and industrial users switch back to coal. Permanent exemptions are made for industrial cogeneration facilities.

Natural Gas Policy Act (NGPA)—1978. NGPA establishes a series of maximum prices for various categories of natural gas for intrastate and interstate gas markets and lifts price controls on new gas by January 1, 1985. The Act also provides for natural gas curtailment for industrial use during emergencies.

Energy Conservation and Production Act (ECPA)—1976. This supplemental state energy conservation program provided funds for states to supplement energy conservation programs and provide for the following types of audits:

CLASS A AUDIT requires an on-site visit by a qualified energy auditor.

CLASS B AUDIT is accomplished by a questionnaire which the building owner completes and is evaluated by the State Energy Officer.

CLASS C AUDIT is accomplished by the building owner with the help of a "do-it-yourself" workbook.

Energy Policy and Conservation Act (EPCA)—1975. The main purpose of the Act was to establish a voluntary industrial energy conservation program and establish efficiency standards for automobiles and appliances.

ENERGY PROJECTIONS
TO YEAR 2000

In 1981, the DOE Office of Policy, Planning and Analysis published "Energy Projections to the Year 2000" in a supplement to the National Energy Policy Plan.

The cost to U.S. refiners for imported crude oil is expected to range from $50 to $95 per barrel (in 1981 dollars) by the year 2000.

From an actual world oil price of $37 per barrel in 1980, the projected ranges are $37–$50 by 1985, $41–$68 by 1990 and $40–$95 by 2000. The report, required by Title VIII of the DOE Organization Act (Public Law 95-91), emphasizes that the projections are based on policies and programs in effect as of June 1981 and should not be viewed as a statement of Administration energy goals. The Department will update the projections as circumstances change.

The report projects U.S. economic growth at from 2.3 to 3.8% to 1985, 2.1 to 3.6% to 1990 and 2.0 to 3.0% to the year 2000.

Energy consumption is to increase around 25% or less overall by 2000. However, a drop is forecast in oil consumption while gas, coal, renewables and electricity all increase. Coal consumption is expected to about double.

REFERENCES

1. *Monthly Energy Review,* U.S. Department of Energy, Energy Information Administration, March 1981.
2. "Energy Projections to the Year 2000," U.S. Department of Energy, 1981 (available from NTIS (No. DE–81–027525; $14).
3. "Higher Prices Fuel a Boom in Drilling," *The New York Times,* August 15, 1980.

2

Energy Economic Analysis

To justify the energy investment cost, a knowledge of life-cycle costing is required.

The life-cycle cost analysis evaluates the total owning and operating cost. It takes into account the "time value" of money and can incorporate fuel cost escalation into the economic model. This approach is also used to evaluate competitive projects. In other words, the life-cycle cost analysis considers the cost over the life of the system rather than just the first cost.

THE TIME VALUE OF MONEY CONCEPT

To compare energy utilization alternatives, it is necessary to convert all cash flow for each measure to an equivalent base. The life-cycle cost analysis takes into account the "time value" of money, thus a dollar in hand today is more valuable than one received at some time in the future. This is why a time value must be placed on all cash flows into and out of the company.

DEVELOPING CASH FLOW MODELS

The cash flow model assumes that cash flows occur at discrete points in time as lump sums and that interest is computed and payable at discrete points in time.

31

To develop a cash flow model which illustrates the effect of "compounding" of interest payments, the cash flow model is developed as follows:

End of Year 1: $\quad P + i(P) = (1 + i)\,P$

Year 2: $\quad (1 + i)P + (1 + i)Pi = (1 + i)P\,[(1 + i)]$
$$= (1 + i)^2\,P$$

Year 3: $\quad (1 + i)^3\,P$

Year n: $\quad (1 + i)^n\,P$ or $S = (1 + i)^n\,P$

Where $\quad P$ = present sum

$\qquad\quad i$ = interest rate earned at the end of each interest period

$\qquad\quad n$ = number of interest periods

$\qquad\quad S$ = future value

$(1 + i)^n$ is referred to as the "Single Payment Compound Amount" factor and is tabulated for various values of i and n in Tables 16-1 through 16-8.

The cash flow model can also be used to find the present value of a future sum S.

$$P = \left(\frac{1}{(1 + i)^n}\right) S$$

Cash flow models can be developed for a variety of other types of cash flow as illustrated in Figure 2-1.

To develop the Cash Flow Model for the "Uniform Series Compound Amount" factor, the following cash flow diagram is drawn.

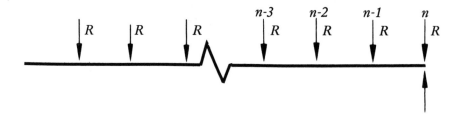

Where R is a Uniform Series of year end payments and S is the future sum of R payments for n interest periods.

The R dollars deposited at the end of the nth period earn no interest and, therefore, contribute R dollars to the fund. The R dollars deposited at the end of the $(n - 1)$ period earn interest for 1 year and will, therefore, contribute $R(1 + i)$ dollars to the fund. The R dollars deposited at the end of the $(n - 2)$ period earn interest for 2 years and will, therefore, contribute $R(1 + i)^2$. These years of earned interest in the contributions will continue to increase in this manner, and the R deposited at the end of the first period will have earned interest for $(n - 1)$ periods. The total in the fund S is, thus, equal to $R + R(1 + i) + R(1 + i)^2 + R(1 + i)^3 + R(1 + i)^4 + \ldots + R(1 + i)^{n-2} + R(1 + i)^{n-1}$. Factoring out R,

(1) $\quad S = R[1 + (1 + i) + (1 + i)^2 + \ldots + (1 + i)^{n-2} + (1 + i)^{n-1}]$

Multiplying both sides of this equation by $(1 + i)$;

(2) $\quad (1 + i)S = R[(1 + i) + (1 + i)^2 + (1 + i)^3 + \ldots + (1 + i)^{n-1} + (1 + i)^n]$

Subtracting equation (1) from (2):

$$(1 + i)S - S = R[(1 + i) + (1 + i)^2 + (1 + i)^3 \ldots$$
$$+ (1 + i)^{n-1} + (1 + i)^n] - R[1 + (1 + i)$$
$$+ (1 + i)^2 + \ldots + (1 + i)^{n-2} + (1 + i)^{n-1}]$$

$$iS = R[(1 + i)^n - 1]$$

$$S = R\left[\frac{(1 + i)^n - 1}{i}\right]$$

Interest factors are seldom calculated. They can be determined from computer programs, and interest tables included in Chapter 16, Appendix. Each factor is defined when the number of periods (n) and interest rate (i) are specified. In the case of the Gradient Present Worth factor the escalation rate must also be stated.

The three most commonly used methods in life-cycle costing are the annual cost, present worth and rate-of-return analysis.

33

Single Payment Compound Amount—SPCA

The SPCA factor is the future value of one dollar in "n" periods at interest of "i" percent.

$$S = P \times (SPCA)_i^n \qquad \text{Formula (2-1)}$$

$$SPCA = (1 + i)^n$$

Single Payment Present Worth—SPPW

The SPPW factor is the present worth of one dollar, "n" periods from now at interest of "i" percent.

$$P = S \times (SPPW)_i^n \qquad \text{Formula (2-2)}$$

$$SPPW = \frac{1}{(1 + i)^n}$$

Uniform Series Compound Amount—USCA

The USCA factor is the future value of a uniform series of one dollar deposits.

$$S = R \times (USCA)_i^n \qquad \text{Formula (2-3)}$$

$$USCA = \frac{(1 + i)^n - 1}{i}$$

Sinking Fund Payment—SFP

The SFP factor is the uniform series of deposits whose future value is one dollar.

$$R = S \times (SFP)_i^n \qquad \text{Formula (2-4)}$$

$$SFP = \frac{i}{(1 + i)^n - 1}$$

Uniform Series Present Worth—(USPW)

The USPW factor is the present value of uniform series of one dollar deposits.

$$P = R \times (USPW)_i^n \qquad \text{Formula (2-5)}$$

34

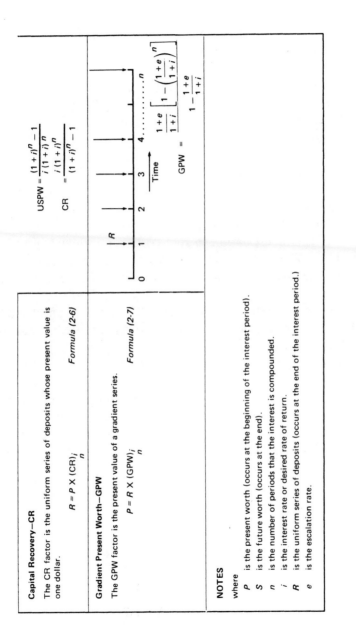

Capital Recovery—CR

The CR factor is the uniform series of deposits whose present value is one dollar.

$$R = P \times (CR)^i_n \qquad \textit{Formula (2-6)}$$

Gradient Present Worth—GPW

The GPW factor is the present value of a gradient series.

$$P = R \times (GPW)^i_n \qquad \textit{Formula (2-7)}$$

$$USPW = \frac{(1+i)^n - 1}{i(1+i)^n}$$

$$CR = \frac{i(1+i)^n}{(1+i)^n - 1}$$

$$GPW = \frac{\frac{1+e}{1+i}\left[1 - \left(\frac{1+e}{1+i}\right)^n\right]}{1 - \frac{1+e}{1+i}}$$

NOTES

where

P is the present worth (occurs at the beginning of the interest period).

S is the future worth (occurs at the end).

n is the number of periods that the interest is compounded.

i is the interest rate or desired rate of return.

R is the uniform series of deposits (occurs at the end of the interest period.)

e is the escalation rate.

Figure 2-1. Interest Factors

35

In the present worth method a minimum rate of return (i) is stipulated. All future expenditures are converted to present values using the interest factors. The alternative with lowest effective first cost is the most desirable.

A similar procedure is implemented in the annual cost method. The difference is that the first cost is converted to an annual expenditure. The alternative with lowest effective annual cost is the most desirable.

In the rate-of-return method, a trial-and-error procedure is usually required. Interpolation from the interest tables can determine what rate of return (i) will give an interest factor which will make the overall cash flow balance. The rate-of-return analysis gives a good indication of the overall ranking of independent alternates.

The effect of escalation in fuel costs can influence greatly the final decision. When an annual cost grows at a steady rate it may be treated as a gradient and the Gradient Present Worth factor can be used.

Special appreciation is given to Rudolph R. Yaneck and Dr. Robert Brown for use of their specially designed interest and escalation tables used in this text.

When life-cycle costing is used to compare several alternatives the differences between costs are important. For example, if one alternate forces additional maintenance or an operating expense to occur, then these factors as well as energy costs need to be included. Remember, what was previously spent for the item to be replaced is irrelevant. The only factor to be considered is whether the new cost can be justified based on projected savings over its useful life.

PAYBACK ANALYSIS

The simple payback analysis is sometimes used instead of the methods previously outlined. The simple payback is defined as initial investment divided by annual savings after taxes. The simple payback method does not take into account the effect of interest or escalation rate.

Since the payback period is relatively simple to calculate and due to the fact managers wish to recover their investment as rapidly as possible the payback method is frequently used.

It should be used in conjunction with other decision-making tools. When used by itself as the principal criterion it may result in choosing less profitable investments which yield high initial returns for short periods as compared with more profitable investments which provide profits over longer periods of time.

Example Problem 2-1

An electrical energy audit indicates electrical motor consumption is 4×10^6 KWH per year. By upgrading the motor spares with high efficiency motors a 10% savings can be realized. The additional cost for these motors is estimated at $80,000. Assuming an 8¢ per KWH energy charge and 20-year life, is the expenditure justified based on a minimum rate of return of 20% before taxes? Solve the problem using the present worth, annual cost, and rate-of-return methods.

Analysis

Present Worth Method

	Alternate 1 Present Method	Alternate 2 Use High Efficiency Motor Spares
(1) First Cost *(P)*	—	$80,000
Annual Cost *(R)*	$4 \times 10^{12} \times .08$	.9 X $320,000
	= $320,000	= $288,000
USPW (Table 16-4)	4.87	4.87
(2) *R* X USPW =	$1,558.400	$1,402,560
Present Worth	$1,558,400	$1,482,560
(1) + (2)		Choose Alternate with Lowest First Cost

37

Annual Cost Method

	Alternate 1	Alternate 2
(1) First Cost *(P)*	—	$80,000
Annual Cost *(R)*	$320,000	$288,000
CR (Table 16-4)	.2	.2
(2) *P* X CR	—	$16,000
Annual Cost	$320,000	$304,000
(1) + (2)		Choose Alternate with Lowest First Cost

Rate of Return Method

$$P = R\,(\text{USPW}) = (\$320,000 - \$288,000) \times \text{USPW}$$

$$\text{USPW} = \frac{80,000}{32,000} = 2.5$$

What value of i will make USPW = 2.5? i = 40% (Table 16-7).

Example Problem 2-2

Show the effect of 10% escalation on the rate-of-return analysis given the

Energy equipment investment	= $20,000
After tax savings	= $ 2,600
Equipment life *(n)*	= 15 years

Analysis

Without escalation

$$\text{CR} = \frac{R}{P} = \frac{2,600}{20,000} = .13$$

From Table 16-1, the rate of return is 10%.
With 10% escalation assumed:

$$\text{GPW} = \frac{P}{G} = \frac{20,000}{2,600} = 7.69$$

From Table 16-11, the rate of return is 21%.

Thus we see that taking into account a modest escalation rate can dramatically affect the justification of the project.

DEPRECIATION, TAXES, AND THE TAX CREDIT

Depreciation

Depreciation affects the "accounting procedure" for determining profits and losses and the income tax of a company. In other words, for tax purposes the expenditure for an asset such as a pump or motor can not be fully expensed in its first year. The original investment must be charged off for tax purposes over the useful life of the asset. A company usually wishes to expense an item as quickly as possible.

The Internal Revenue Service allows several methods for determining the annual depreciation rate.

Straight-Line Depreciation: The simplest method is referred to as a straight-line depreciation and is defined as:

$$D = \frac{P - L}{n} \qquad \text{Formula (2-8)}$$

Where:

D is the annual depreciation rate

L is the value of equipment at the end of its useful life, commonly referred to as salvage value

n is the life of the equipment which is determined by Internal Revenue Service Guidelines

P is the initial expenditure.

Sum-of-Years Digits: Another method is referred to as the sum-of-years digits. In this method the depreciation rate is determined by finding the sum of digits using the following formula:

$$N = n\frac{(n + 1)}{2} \qquad \text{Formula(2-9)}$$

Where n is the life of equipment.

Each year's depreciation rate is determined as follows:

First year
$$D = \frac{n}{N}(P - L)$$
Formula (2-10)

Second year
$$D = \frac{n-1}{N}(P - L)$$
Formula (2-11)

n year
$$D = \frac{1}{N}(P - L)$$
Formula (2-12)

Declining-Balance Depreciation: The declining-balance method allows for larger depreciation charges in the early years which is sometimes referred to as fast write-off.

The rate is calculated by taking a constant percentage of the declining undepreciated balance. The most common method used to calculate the declining balance is to predetermine the depreciation rate. Under certain circumstances a rate equal to 200% of the straight-line depreciation rate may be used. Under other circumstances the rate is limited to 1½ or ¼ times as great as straight-line depreciation. In this method the salvage value or undepreciated book value is established once the depreciation rate is preestablished.

To calculate the undepreciated book value Formula 2-13 is used:

$$D = 1 - \left(\frac{L}{P}\right)^{1/N}$$
Formula (2-13)

Where
    D is the annual depreciation rate
    L is the salvage value
    P is the first cost.

Example Problem

Calculate the depreciation rate using the straight-line, sum-of-years digit, and declining-balance methods.
    Salvage value is 0
    $n = 5$ years

$P = 150,000$

For declining balance use a 200% rate.

Straight-Line Method

$$D = \frac{P - L}{n} = \frac{150,000}{5} = \$30,000 \text{ per year}$$

Sum-of-Years Digits

$$N = \frac{n(n + 1)}{2} = \frac{5(6)}{2} = 15$$

$$D_1 = \frac{n}{N}(P) = \frac{5}{15}(150,000) = 50,000$$

	N	P
	1 =	$50,000
	2 =	40,000
	3 =	30,000
	4 =	20,000
	5 =	10,000

Declining-Balance Method

$D = 2 \times 20\% = 40\%$ (Straight-Line Depreciation Rate = 20%)

Year	Undepreciated Balance at Beginning of Year	Depreciation Charge
1	150,000	60,000
2	90,000	36,000
3	54,000	21,600
4	32,400	12,960
5	19,440	7,776
	TOTAL	138,336

Undepreciated Book Value (150,000 − 138,336) = $11,664

41

Economic Recovery Tax Act—1981

The Economic Recovery Tax Act allows for an accelerated depreciation over a shorter life. Thus energy investments will become more attractive. Details of the Act are described in Chapter 1.

Tax Considerations

Tax-deductible expenses such as maintenance, energy, operating costs, insurance and property taxes reduce the income subject to taxes.

For the after-tax life-cycle cost analysis and payback analysis the actual incurred annual savings is given as follows:

$$AS = (1-I)\,E + ID \qquad\qquad Formula\ (2\text{-}14)$$

Where:

AS = yearly annual after-tax savings (excluding effect of tax credit)

E = yearly annual energy savings (difference between original expenses and expenses after modification)

D = annual depreciation rate

I = income tax bracket

Formula 2-14 takes into account that the yearly annual energy savings is partially offset by additional taxes which must be paid due to reduced operating expenses. On the other hand, the depreciation allowance reduces taxes directly.

Tax Credit

A tax credit encourages capital investment. Essentially the tax credit lowers the income tax paid by the tax credit to an upper limit.

In addition to the investment tax credit, the Business Energy Tax Credit as a result of the National Energy Plan, can also be taken. The Business Energy Tax Credit applies to industrial investment in alternative energy property such as boilers for coal, heat conservation, and recycling equipment. The tax credit substantially increases the investment merit of the investment since it lowers the *bottom* line on the tax form. Since tax laws are in constant flux, check to determine the extent tax credits are in effect at the time of the analysis.

After-Tax Analysis

To compute a rate of return which accounts for taxes, depreciation, escalation, and tax credits a cash-flow analysis is usually required. This method analyzes all transactions including first and operating costs. To determine the after-tax rate of return a trial and error or computer analysis is required.

The Present Worth factors tables in Chapter 16, Appendix, can be used for this analysis. All money is converted to the present assuming an interest rate. The summation of all present dollars should equal zero when the correct interest rate is selected, as illustrated in Figure 2-2.

This analysis can be made assuming a fuel escalation rate by using the Gradient Present Worth interest of the Present Worth Factor.

Example Problem 2-4

Comment on the after-tax rate of return for the installation of a heat-recovery system with and without tax credit given the following:

- First Cost $100,000
- Year Savings   40,000
- Straight-line depreciation life and equipment life of 5 years
- Income tax bracket 46%

Analysis

$$D = 100,000/5 = 20,000$$
$$AS = (1-I)E + ID = .54(40,000) + .46(20,000)$$
$$= 21,600 + 9,200 = 30,800$$

Year	1 Investment	2 Tax Credit	3 After Tax Savings (AS)	4 Single Payment Present Worth Factor	(2 + 3) X 4 Present Worth
0	−P				−P
1		+TC	AS_1	$SPPW_1$	$+P_1$
2			AS_2	$SPPW_2$	P_2
3			AS_3	$SPPW_3$	P_3
4			AS_4	$SPPW_4$	P_4
Total					ΣP

$$AS = (1-I)E + ID$$
Trial & Error Solution:
Correct i when $\Sigma P = 0$

Figure 2-2. Cash Flow Rate of Return Analysis

Without Tax Credit
First Trial i = 20%

Investment	After Tax Savings	SPPW 20%	PW
0−100,000			−100,000
1	30,800	.833	25,656
2	30,800	.694	21,375
3	30,800	.578	17,802
4	30,800	.482	14,845
5	30,800	.401	12,350
			$\Sigma-$ 7,972

Since summation is negative a higher present worth factor is required. Next try is 15%.

Investment	After Tax Savings	SPPW 15%	PW
0—100,000			−100,000
1	30,800	.869	+ 26,765
2	30,800	.756	+ 23,284
3	30,800	.657	+ 20,235
4	30,800	.571	+ 17,586
5	30,800	.497	+ 15,307
			+ 3,177

Since rate of return is bracketed, linear interpolation will be used.

$$\frac{3177 + 7971}{-5} = \frac{3177-0}{15-i\%}$$

$$i = \frac{3177}{2229.6} + 15 = 16.4\%$$

With Tax Credit

Tax Credit = 10% (Investment) + 10% (Energy) = 20%

Investment	Tax Credit	After Tax Savings	SPPW 20%	PW
0—100,000				−100,000
1	20,000	30,800	.833	45,656
2		30,800	.694	21,375
3		30,800	.578	17,802
4		30,800	.482	14,845
5		30,800	.401	12,350
				+ 12,028

Next try 25%.

$$PW$$

$$-100,000$$
$$44,640$$
$$19,712$$
$$15,769$$
$$\underline{10,093}$$
$$-\ \ 9,786$$

$$\frac{12,028 + 9,786}{-5} = \frac{12,028}{20-i}$$

$$i = 22.75\%$$

THE IMPACT OF FUEL INFLATION
ON THE LIFE CYCLE ANALYSIS

The rate of return on investment becomes more attractive when life cycle costs are taken into account. Tables 16-9 through 16-12 can be used to show the impact of fuel inflation on the decision making process.

Example Problem 2-5

Develop a set of curves that indicate the capital that can be invested to give a rate of return of 15% after taxes for each $1000 saved for the following conditions.

1. The effect of escalation is not considered.
2. A 5% fuel escalation is considered.
3. A 10% fuel escalation is considered.
4. A 14% fuel escalation is considered.
5. A 20% fuel escalation is considered.

Calculate for 5-, 10-, 15-, 20-year life.

Assume straight-line depreciation over useful life, 48% income tax bracket, and no tax credit.

Answer

$$AS = (1-I)E + ID$$

$$I = 0.48, \; E = \$1000$$

$$AS = 520 + \frac{0.48P}{N}$$

Thus, the after-tax savings *(AS)* are comprised of two components. The first component is a uniform series of $520 escalating at *e* percent/year. The second component is a uniform series of $0.48P/N$.

Each component is treated individually and converted to present day values using the GPW factor and the USPW factor, respectively. The sum of these two present worth factors must equal *P*. In the case of no escalation, the formula is:

$$P = 520 \; \text{USPW} + \frac{0.48P}{N} \; \text{USPW.}$$

In the case of escalation:

$$P = 520 \; \text{GPW} + \frac{0.48P}{N} \; \text{USPW.}$$

Since there is only one unknown, the formulas can be readily solved. The results are indicated below.

	N = 5 $P	N = 10 $P	N = 15 $P	N = 20 $P
e = 0	2,570.79	3,434	3,753.62	3,829.03
e = 10%	3,362.35	5,401.84	6,872.98	7,922.35
e = 14%	3,735.87	6,523.26	8,986.38	11,176.47
e = 20%	4,364.30	8,710.68	$13,728.27	$19,698.82

Figure 2-3 illustrates the effects of escalation. This figure can be used as a quick way to determine after-tax economics of energy utilization expenditures.

Example Problem 2-6

It is desired to have an after-tax savings of 15%. Comment on the investment that can be justified if it is assumed that the fuel rate escalation should not be considered and the annual energy savings is $2000 with an equipment economic life of 12 years.

Comment on the above, assuming a 14% fuel escalation.

Answer

From Figure 2-3, for each $1000 energy savings, an investment of $3600 is justified or $7200 for a $2000 savings when no fuel increase is accounted for.

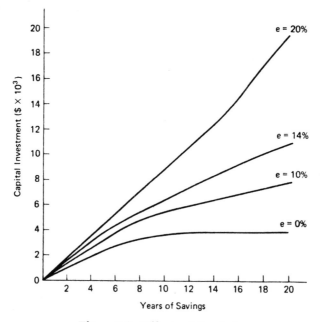

Figure 2-3. Effects of Escalation

***** EER ENERGY PRICE FORECASTS *****

REGION UNITED STATES

JANUARY 1983
UPDATES EVERY SIX MONTHS

	INDUSTRIAL ELECTRICITY			INDUSTRIAL DISTILLATE FUEL			INDUSTRIAL NATURAL GAS		
	CURRENT $ PER MM BTU	CURRENT $ PER KWH	ESCALATION RATE*	CURRENT $ PER MM BTU	CURRENT $ PER GAL	ESCALATION RATE*	CURRENT $ PER MM BTU	CURRENT $ PER MCF	ESCALATION RATE*
1980	10.30	0.04	BASE	6.67	0.93	BASE	2.76	2.92	BASE
1981	11.81	0.04		7.80	1.08	6.6	3.24	3.31	13.0
1982	13.07	0.04		7.59	1.05	6.5	3.78	3.86	17.7
1983	14.01	0.05	13.4	8.04	1.12	8.4	4.62	4.72	15.6
1984	15.89	0.05	13.4	8.57	1.19	8.3	5.22	5.33	14.5
1985	17.93	0.06	13.1	9.12	1.26	8.5	6.40	6.53	13.8
1986	19.37	0.07	11.4	10.23	1.42	9.4	7.12	7.27	13.1
1987	20.92	0.07	10.0	11.48	1.59	9.9	7.93	8.09	
1988	22.60	0.08	10.0	12.88	1.79	9.9	8.82	9.01	
1989	24.41	0.08	9.7	14.45	2.00	10.3	9.82	10.03	
1990	26.37	0.09	9.5	16.21	2.25	10.5	10.93	11.16	
1991	28.05	0.10	9.1	17.89	2.48	10.5	12.08	12.34	12.8
1992	29.74	0.10	8.8	19.73	2.74	10.5	13.36	13.64	12.5
1993	31.74	0.11	8.5	21.78	3.02	10.5	14.77	15.09	12.2
1994	33.76	0.11	8.3	24.01	3.33	10.5	16.33	16.67	12.2
1995	35.91	0.12	8.2	26.51	3.68	10.5	18.05	18.43	12.0
1996	37.87	0.13	7.9	28.27	3.92	10.2	19.25	19.66	11.6
1997	39.93	0.14	7.8	30.15	4.18	9.7	20.50	20.96	11.2
1998	42.11	0.14	7.6	32.16	4.46	9.5	21.90	22.85	11.0
1999	44.40	0.15	7.5	34.30	4.76	9.3	23.36	23.85	10.7
2000	46.82	0.16	7.4	36.58	5.07		24.91	25.43	10.4

* AVERAGE ANNUAL ESCALATION RATE OF PRICES, PERCENT FROM BASE YEAR TO CURRENT YEAR

COPYRIGHT 1983

Source: *EER Energy Price Forecasts, PO Box 14227, Atlanta, GA 30324*

Figure 2-4. Typical Fuel Price Projections

With a 14% fuel escalation rate on investment of $6000 justified for each $1000 energy savings, thus $12,000 can be justified for $2000 savings. Thus, a 66% higher expenditure is economically justifiable and will yield the same after-tax rate of return of 15% when a fuel escalation of 14% is considered.

In order to estimate future fuel costs a computer simulation, forecasting service or historical data may be used. One such forecast is illustrated in Figure 2-4. It should be noted that even though larger price projections are made in the short term the average long-term projections are in the order of 7 to 10% depending on the fuel type. The average escalation number from this figure is then used as illustrated in the previous examples.

3

Energy Auditing
and Accounting

TYPES OF ENERGY AUDITS

The simplest definition for an energy audit is as follows: An energy audit serves the purpose of identifying where a building or plant facility uses energy and identifies energy conservation opportunities.

There is a direct relationship to the cost of the audit (amount of data collected and analyzed) and the number of energy conservation opportunities to be found. Thus, a first distinction is made between cost of the audit which determines the type of audit to be performed.

The second distinction is made between the type of facility. For example, a building audit may emphasize the building envelope, lighting, heating, and ventilation requirements. On the other hand, an audit of an industrial plant emphasizes the process requirements.

Most energy audits fall into three categories or types, namely, Walk-Through, Mini-Audit, or Maxi-Audit.

Walk-Through—This type of audit is the least costly and identifies preliminary energy savings. A visual inspection of the facility is made to determine maintenance and operation energy saving opportunities plus collection of information to determine the need for a more detailed analysis.

51

Mini-Audit—This type of audit requires tests and measurements to quantify energy uses and losses and determine the economics for changes.

Maxi-Audit—This type of audit goes one step further than the mini-audit. It contains an evaluation of how much energy is used for each function such as lighting, process, etc. It also requires a model analysis, such as a computer simulation, to determine energy use patterns and predictions on a year-round basis, taking into account such variables as weather data.

As noted in the audit definition, there are two essential parts, namely, data acquisition and data analysis.

Data Acqusition

This phase requires the accumulation of utility bills, establishing a baseline to provide historical documentation and a survey of the facility.

All energy flows should be accounted for; thus all "energy in" should equal "energy out." This is referred to as an energy balance.

All energy costs should be determined for each fuel type. The energy survey is essential. Instrumentation commonly used in conducting a survey is discussed at the conclusion of the chapter.

Data Analysis

As a result of knowing how energy is used, a complete list of "Energy Conservation Opportunities" (ECO's) will be generated. The life-cycle costing techniques presented in Chapter 2 will be used to determine which alternative should be given priority.

A very important phase of the overall program is to continuously monitor the facility even after the ECO's have been implemented. Documentation of the cost avoidance or savings is essential to the audit.

Remember in order to have a continuous on-going program, *individuals must be made accountable* for energy use. As part of the audit recommendations should be made as to where to add "root" or submetering.

ENERGY USE PROFILES

The energy audit process for a building emphasizes building envelope, heating and ventilation, air-conditioning, plus lighting functions. For an industrial facility the energy audit approach includes process consideration. Figures 3-1 through 3-3 illustrate how energy is used for a typical industrial plant. It is important to account for total consumption, cost, and how energy is used for each commodity such as steam, water, air and natural gas. This procedure is required to develop the appropriate energy conservation strategy.

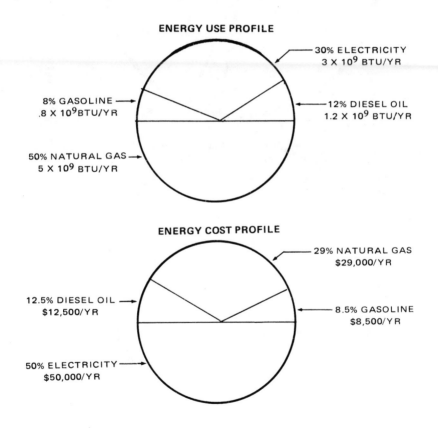

Figure 3-1. Energy Use and Cost Profile

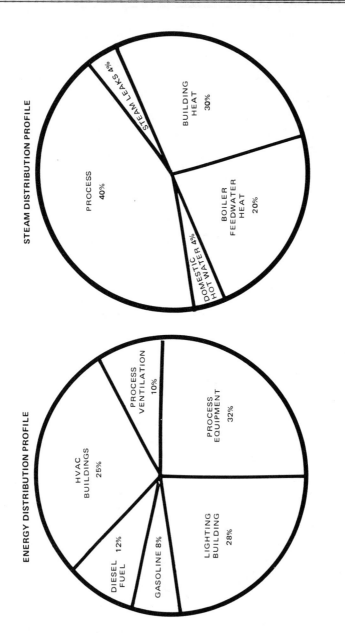

Figure 3-2. Energy Profile by Function

The top portion of Figure 3-1 illustrates how much energy is used by fuel type and its relative percentage. The pie chart below shows how much is spent for each fuel type. Using a pie-chart presentation or nodal flow diagram can be very helpful in visualizing how energy is being used.

Figure 3-2 on the other hand shows how much of the energy is used for each function such as lighting, process, and building heating and ventilation. Pie charts similar to the right-hand side of the figure should be made for each category such as air, steam, electricity, water and natural gas.

Figure 3-3 illustrates an alternate representation for the steam distribution profile.

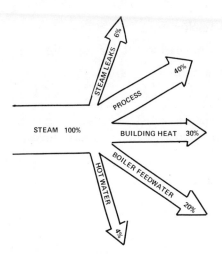

Figure 3-3. Steam Distribution Nodal Diagram

Several audits are required to construct the energy use profiles, such as:

Envelope Audit—This audit surveys the building envelope for losses or gains due to leaks, building construction, doors, glass, lack of insulation, etc.

Functional Audit—This audit determines the amount of energy required for a particular function and identifies energy conservation opportunities. Functional audits include:

- Heating, ventilation and air-conditioning
- Building
- Lighting
- Domestic hot water
- Air distribution

Process Audit— This audit determines the amount of energy required for each process function and identifies energy conservation opportunities. Process functional audits include:

- Process machinery
- Heating, ventilation and air-conditioning process
- Heat treatment
- Furnaces

Transportation Audit—This audit determines the amount of energy required for forklift trucks, cars, vehicles, trucks, etc.

Utility Audit—This audit analyzes the monthly, daily or yearly energy usage for each utility.

ENERGY USERS

Energy use profiles for several end-users are summarized in Tables 3-1 through 3-11.

Table 3-1. Energy Use in Apartment Buildings

	Range (%)	*Norms (%)*
Environmental Control	50 to 80	70
Lighting and Wall Receptacles	10 to 20	15
Hot Water	2 to 5	3
Special Functions		
Laundry, Swimming Pool, Restaurants,		
Parking, Elevators, Security Lighting	5 to 20	10

Source: Federal Energy Commission
Data Estimates, December 24, 1977

Table 3-2. Energy Use in Bakeries

Housekeeping Energy	*Percent*
Space Heating	21.5
Air Conditioning	1.6
Lighting	1.4
Domestic Hot Water	1.8
TOTAL	26.3
Process Energy	*Percent*
Baking Ovens	49.0
Pan Washing	10.6
Mixers	4.1
Freezers	3.3
Cooking	2.0
Fryers	1.8
Proof Boxes	1.8
Other Processes	1.1
TOTAL	73.7

Data are for a 27,000-square-foot bakery in
Washington, D.C.

Table 3-3. Energy Use in Die Casting Plants

Housekeeping Energy	Percent
Space Heating	24
Air Conditioning	2
Lighting	2
Domestic Hot Water	2
TOTAL	30

Process Energy	Percent
Melting Hearth	30
Quiet Pool	20
Molding Machines	10
Air Compressors	5
Other Processes	5
TOTAL	70

Source: Federal Energy Commission
Data Estimates, January 11, 1978

Table 3-4. Energy Use in Hospital Buildings

	Range (%)	Norms (%)
Environmental Control	40 to 65	58
Lighting and Wall Receptacles ·	10 to 20	15
Laundry	8 to 15	12
Food Service, Kitchen Operations	5 to 10	7
Medical Equipment, Sterilization, Incinerator, Parking, Elevators, Security Lighting	5 to 15	8

Source: Federal Energy Commission
Data Estimates, November 3, 1977

Table 3-5. Energy Use in Hotels and Motels

	Range (%)	Norms (%)
Space Heating	45 to 70	60
Lighting	5 to 15	11
Air Conditioning	3 to 15	10
Refrigeration	0 to 10	4
Special Functions	5 to 20	15
Laundry, Kitchen, Restaurant, Swimming Pool, Garage, Security Lighting, Hot Water		

Source: Federal Energy Commission
Data Estimates, December 8, 1977

Table 3-6. Energy Use in Retail Stores

	Range (%)	Norms (%)
HVAC	20 to 50	30
Lighting	40 to 75	60
Special Functions	5 to 20	10
Elevators, General Power, Parking Security Lighting, Hot Water		

Source: Federal Energy Commission
Data Estimates, December 14, 1977

Table 3-7. Energy Use in Restaurants

	Table Restaurant Norms (%)	Fast Food Restaurant Norms (%)
HVAC	32	36
Lighting	8	26
Special Functions		
Food Preparation	45	27
Food Storage	2	6
Sanitation	12	1
Other	1	4

Source: Federal Energy Commission
Data Estimates, December 8, 1977

Table 3-8. Energy Use in Schools

	Range (%)	Norms (%)
Environmental Control	45 to 80	65
Lighting and Wall Receptacles	10 to 20	15
Food Service	5 to 10	7
Hot Water	2 to 5	3
Special Functions	0 to 20	10

Source: Federal Energy Commission
Data Estimates, November 3, 1977

Table 3-9. Energy Use in Transportation Terminals

	Range (%)	Norms (%)
Space Heating	50 to 75	60
Lighting	5 to 25	15
Air Conditioning	5 to 25	15
Special Functions	3 to 20	10
Elevators, General Power, Parking, Security Lighting, Hot Water		

Source: Federal Energy Commission
Data Estimates, December 10, 1977

Table 3-10. Energy Use in Warehouses and Storage Facilities

(Vehicles Not Included)	Range (%)	Norms (%)*
Space Heating	45 to 80	67
Air Conditioning	3 to 10	6
Lighting	4 to 12	7
Refrigeration	0 to 40	12
Special Functions	5 to 15	8
Elevators, General Power, Parking, Security Lighting, Hot Water		

* Norms for a warehouse or storage facility are strongly dependent on the products and their specific requirements for temperature and humidity control.

Source: Federal Energy Commission
Data Estimates, December 21, 1977

Table 3-11. Comparative Energy Use by System

		Heating & Ventilation	Cooling & Ventilation	Lighting	Power & Process	Domestic Hot Water
Schools	A	4	3	1	5	—
	B	1	4	2	5	3
	C	1	4	2	5	3
Colleges	A	5	2	1	4	3
	B	1	3	2	5	4
	C	1	5	2	4	3
Office Bldg.	A	3	1	2	4	5
	B	1	3	2	4	5
	C	1	3	2	4	5
Commercial Stores	A	3	1	2	4	5
	B	2	3	1	4	5
	C	1	3	2	4	5
Religious Bldg.	A	3	2	1	4	5
	B	1	3	2	4	5
	C	1	3	2	4	5
Hospitals	A	4	1	2	5	3
	B	1	3	4	5	2
	C	1	5	3	4	2

Climatic Zone A: Fewer than 2500 degree days
Climate Zone B: 2500—5500 degree days
Climate Zone C: 5500—9500 degree days

Source: Guidelines For Saving Energy In Existing Buildings ECM—1

Note: Numbers indicate energy consumption relative to each other
   (1) greatest consumption
   (5) least consumption

62

THE ENERGY SURVEY

As part of the data acquisition phase, a detailed survey should be conducted. The various types of instrumentation commonly used in the survey are discussed in this section.

Infrared Equipment

Some companies may have the wrong impression that infrared equipment can meet most of their instrumentation needs.

The primary use of infrared equipment in an energy utilization program is to detect building or equipment losses. Thus it is just one of the many options available.

Several energy managers find infrared in use in their plant prior to the energy utilization program. Infrared equipment, in many instances, was purchased by the electrical department and used to detect electrical hot spots.

Infrared energy is an invisible part of the electromagnetic spectrum. It exists naturally and can be measured by remote heat-sensing equipment. Within the last four years lightweight portable infrared systems became available to help determine energy losses. Differences in the infrared emissions from the surface of objects cause color variations to appear on the scanner. The hotter the object, the more infrared radiated. With the aid of an isotherm circuit the intensity of these radiation levels can be accurately measured and quantified. In essence the infrared scanning device is a diagnostic tool which can be used to determine building heat losses. Equipment costs range from $400 to $50,000.

An overview energy scan of the plant can be made through an aerial survey using infrared equipment. Several companies offer aerial scan services starting at $1500. Aerial scans can determine underground stream pipe leaks, hot gas discharges, leaks, etc.

Since IR detection and measurement equipment have gained increased importance in the energy audit process, a summary of the fundamentals are reviewed in this section.

The visible portion of the spectrum runs from .4 to .75 micrometers (μm). The infrared or thermal radiation begins at this point and extends to approximately 1000 μm. Objects such as people, plants, or buildings will emit radiation with wavelengths around 10 μm. (See Figure 3-4.)

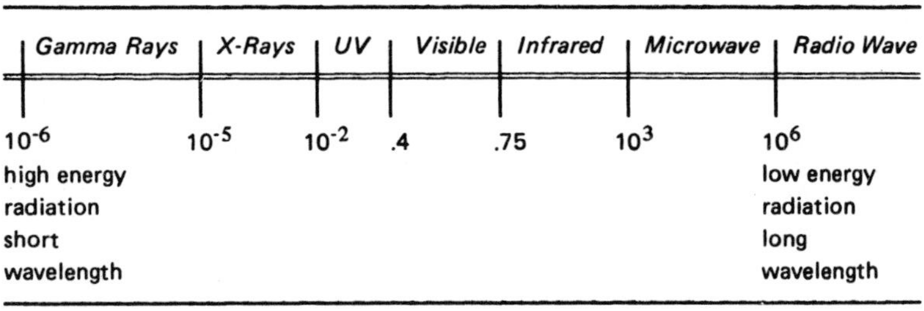

Gamma Rays	X-Rays	UV	Visible	Infrared	Microwave	Radio Wave

10^{-6} 10^{-5} 10^{-2} .4   .75 10^{3} 10^{6}

high energy radiation short wavelength   low energy radiation long wavelength

Figure 3-4. Electromagnetic Spectrum

Infrared instruments are required to detect and measure the thermal radiation. To calibrate the instrument a special "black body" radiator is used. A black body radiator absorbs all the radiation that impinges on it and has an absorbing efficiency or emissivity of 1.

The accuracy of temperature measurements by infrared instruments depends on the three processes which are responsible for an object acting like a black body. These processes—absorbed, reflected, and transmitted radiation—are responsible for the total radiation reaching an infrared scanner.

The real temperature of the object is dependent only upon its emitted radiation.

Corrections to apparent temperatures are made by knowing the emissivity of an object at a specified temperature.

The heart of the infrared instrument is the infrared detector. The detector absorbs infrared energy and converts it into electrical voltage or current. The two principal types of detectors are the thermal and photo type. The thermal detector generally requires a given period of time to develop an image on photographic film. The photo detectors are more sensitive and have a higher response time. Television-like displays on a cathode ray tube permit studies of dynamic thermal events on moving objects in real time.

There are various ways of displaying signals produced by infrared detectors. One way is by use of an isotherm contour. The lightest areas of the picture represent the warmest areas of the subject and the darkest areas represent the coolest portions. These instruments can show thermal variations of less than 0.1°C and can cover a range of −30°C to over 2000°C.

The isotherm can be calibrated by means of a black body radiator so that a specific temperature is known. The scanner can then be moved and the temperatures of the various parts of the subject can be made.

MEASURING ELECTRICAL
SYSTEM PERFORMANCE

The ammeter, voltmeter, wattmeter, power factor meter, and footcandle meter are usually required to do an electrical survey. These instruments are described below.

Ammeter and Voltmeter

To measure electrical currents, ammeters are used. For most audits, alternating currents are measured. Ammeters used in audits are portable and are designed to be easily attached and removed.

There are many brands and styles of snap-on ammeters commonly available that can read up to 1000 amperes continuously. This range can be extended to 4000 amperes continuously for some models with an accessory step-down current transformer.

The snap-on ammeters can be either indicating or recording with a printout. After attachment, the recording ammeter can keep recording current variations for as long as a full month on one roll of recording paper. This allows studying current variations in a conductor for extended periods without constant operator attention.

The ammeter supplies a direct measurement of electrical current which is one of the parameters needed to calculate electrical energy. The second parameter required to calculate energy is voltage, and it is measured by a voltmeter.

Several types of electrical meters can read the voltage or current. A voltmeter measures the difference in electrical potential between two points in an electrical circuit.

In series with the probes are the galvanometer and a fixed resistance (which determine the voltage scale). The current through this fixed resistance circuit is then proportional to the voltage and the galvanometer deflects in proportion to the voltage.

The voltage drops measured in many instances are fairly constant and need only be performed once. If there are appreciable fluctuations, additional readings or the use of a recording voltmeter may be indicated.

Most voltages measured in practice are under 600 volts and there are many portable voltmeter/ammeter clamp-ons available for this and lower ranges.

Wattmeter and Power Factor Meter

The portable wattmeter can be used to indicate by direct reading electrical energy in watts. It can also be calculated by measuring voltage, current and the angle between them (power factor angle).

The basic wattmeter consists of three voltage probes and a snap-on current coil which feeds the wattmeter movement.

The typical operating limits are 300 kilowatts, 650 volts, and 600 amperes. It can be used on both one- and three-phase circuits.

The portable power factor meter is primarily a three-phase instrument. One of its three voltage probes is attached to each conductor phase and a snap-on jaw is placed about one of the phases. By disconnecting the wattmeter circuitry, it will directly read the power factor of the circuit to which it is attached.

It can measure power factor over a range of 1.0 leading to 1.0 lagging with "ampacities" up to 1500 amperes at 600 volts. This

66

range covers the large bulk of the applications found in light industry and commerce.

The power factor is a basic parameter whose value must be known to calculate electric energy usage. Diagnostically it is a useful instrument to determine the sources of poor power factor in a facility.

Portable digital KWH and KW demand units are now available.

Digital read-outs of energy usage in both KWH and KW demand or in dollars and cents, including instantaneous usage, accumulated usage, projected usage for a particular billing period, alarms when over-target levels are desired for usage, and control-outputs for load-shedding and cycling are possible.

Continuous displays or intermittent alternating displays are available at the touch of a button of any information needed such as the cost of operating a production machine for one shift, one hour or one week.

Footcandle Meter

Footcandle meters measure illumination in units of footcandles through light-sensitive barrier layer of cells contained within them. They are usually pocket size and portable and are meant to be used as field instruments to survey levels of illumination. Footcandle meters differ from conventional photographic lightmeters in that they are color and cosine corrected.

TEMPERATURE MEASUREMENTS

To maximize system performance, knowledge of the temperature of a fluid, surface, etc. is essential. Several types of temperature devices are described in this section.

Thermometer

There are many types of thermometers that can be used in an energy audit. The choice of what to use is usually dictated by cost, durability, and application.

For air-conditioning, ventilation and hot-water service applications (temperature ranges 50°F to 250°F) a multipurpose portable battery-operated thermometer is used. Three separate probes are usually provided to measure liquid, air or surface temperatures.

For boiler and oven stacks (1000°F) a dial thermometer is used. Thermocouples are used for measurements above 1000°F.

Surface Pyrometer

Surface pyrometers are instruments which measure the temperature of surfaces. They are somewhat more complex than other temperature instruments because their probe must make intimate contact with the surface being measured.

Surface pyrometers are of immense help in assessing heat losses through walls and also for testing steam traps.

They may be divided into two classes: low-temperature (up to 250°F) and high-temperature (up to 600°F to 700°F). The low-temperature unit is usually part of the multipurpose thermometer kit. The high-temperature unit is more specialized, but needed for evaluating fired units and general steam service.

There are also noncontact surface pyrometers which measure infrared radiation from surfaces in terms of temperature. These are suitable for general work and also for measuring surfaces which are visually but not physically accessible.

A more specialized instrument is the optical pyrometer. This is for high-temperature work (above 1500°F) because it measures the temperature of bodies which are incandescent because of their temperature.

Psychrometer

A psychrometer is an instrument which measures relative humidity based on the relation of the dry-bulb temperature and the wet-bulb temperature.

Relative humidity is of prime importance in HVAC and drying operations. Recording psychrometers are also available. Above 200°F humidity studies constitute a specialized field of endeavor.

Portable Electronic Thermometer

The portable electronic thermometer is an adaptable temperature measurement tool. The battery-powered basic instrument, when housed in a carrying case, is suitable for laboratory or industrial use.

A pocket-size digital, battery-operated thermometer is especially convenient for spot checks or where a number of rapid readings of process temperatures need to be taken.

Thermocouple Probe

No matter what sort of indicating instrument is employed, the thermocouple used should be carefully selected to match the application and properly positioned if a representative temperature is to be measured. The same care is needed for all sensing devices—thermocouple, bimetals, resistance elements, fluid expansion, and vapour pressure bulbs.

Suction Pyrometer

Errors arise if a normal sheathed thermocouple is used to measure gas temperatures, especially high ones. The suction pyrometer overcomes these by shielding the thermocouple from wall radiation and drawing gases over it at high velocity to ensure good convective heat transfer. The thermocouple thus produces a reading which approaches the true temperature at the sampling point rather than a temperature between that of the walls and the gases.

MEASURING COMBUSTION SYSTEMS

To maximize combustion efficiency it is necessary to know the composition of the flue gas. By obtaining a good air-fuel ratio substantial energy will be saved.

Combustion Tester

Combustion testing consists of determining the concentrations of the products of combustion in a stack gas. The products of combustion usually considered are carbon dioxide and carbon monoxide. Oxygen is tested to assure proper excess air levels.

The definitive test for these constituents is an Orsat apparatus. This test consists of taking a measured volume of stack gas and measuring successive volumes after intimate contact with selective absorbing solutions. The reduction in volume after each absorption is the measure of each constituent.

The Orsat has a number of disadvantages. The main ones are that it requires considerable time to set up and use and its operator must have a good degree of dexterity and be in constant practice.

Instead of an Orsat, there are portable and easy to use absorbing instruments which can easily determine the concentrations of the constituents of interest on an individual basis. Setup and operating times are minimal and just about anyone can learn to use them.

The typical range of concentrations are CO_2, 0–20%; O_2, 0–21%; and CO, 0–0.5%. The CO_2 or O_2 content, along with knowledge of flue gas temperature and fuel type, allows the flue gas loss to be determined off standard charts.

Boiler Test Kit

The boiler test kit contains the following:

CO_2 Gas analyzer
O_2 Gas analyzer
     Inclined monometer
CO   Gas analyzer.

The purpose of the components of the kit is to help evaluate fireside boiler operation. Good combustion usually means high carbon dioxide (CO_2), low oxygen (O_2), and little or no trace of carbon monoxide (CO).

Gas Analyzers

The gas analyzers are usually of the Fyrite type. The Fyrite type differs from the Orsat apparatus in that it is more limited in application and less accurate. The chief advantages of the Fyrite are that it is simple and easy to use and is inexpensive. This device is many times used in an energy audit. Three readings using the Fyrite analyzer should be made and the results averaged.

Draft Gauge

The draft gauge is used to measure pressure. It can be the pocket type, or the inclined monometer type.

Smoke Tester

To measure combustion completeness the smoke detector is used. Smoke is unburned carbon, which wastes fuel, causes air pollution, and fouls heat-exchanger surfaces. To use the instrument, a measured volume of flue gas is drawn through filter paper with the probe. The smoke spot is compared visually with a standard scale and a measure of smoke density is determined.

Combustion Analyzer

The combustion electronic analyzer permits fast, close adjustments. The unit contains digital displays. A standard sampler assembly with probe allows for stack measurements through a single stack or breaching hole.

MEASURING HEATING, VENTILATION AND AIR-CONDITIONING (HVAC) SYSTEM PERFORMANCE

Air Velocity Measurement

The following suggests the preference, suitability, and approximate costs of particular equipment.

- *Smoke pellets*—limited use but very low cost. Considered to be useful if engineering staff has experience in handling.
- *Anemometer* (deflecting vane)—good indication of air movement with acceptable order of accuracy. Considered useful (approximately $50).
- *Anemometer* (revolving vane)—good indicator of air movement with acceptable accuracy. However, easily subject to damage. Considered useful (approximately $100).
- *Pitot tube*—a standard air measurement device with good levels of accuracy. Considered essential. Can be purchased in various lengths—12" about $20, 48" about $35. Must be used with a monometer. These vary considerably in cost, but could be on the order of $20 to $60.
- *Impact tube*—usually packaged air flow meter kits, complete with various jets for testing ducts, grills, open areas, etc. These units are convenient to use and of sufficient accuracy. The costs vary around $150 to $300, and therefore this order of cost could only be justified for a large system.
- *Heated thermocouple*—these units are sensitive and accurate but costly. A typical cost would be about $500 and can only be justified for regular use in a large plant.
- *Hot wire anemometer*—not recommended. Too costly and too complex.

Temperature Measurement

The temperature devices most commonly used are as follows.
- *Glass thermometers*—considered to be the most useful to temperature measuring instruments—accurate and convenient but fragile. Cost runs from $5 each for 12" long mercury in glass. Engineers should have a selection of various ranges.
- *Resistance thermometers*—considered to be very useful for A/C testing. Accuracy is good and they are reliable and convenient to use. Suitable units can be purchased from $150 up, some with a selection of several temperature ranges.

72

- *Thermocouples*—similar to resistance thermocouple, but do not require battery power source. Chrome-Alum or iron types are the most useful and have satisfactory accuracy and repeatability. Costs start from $50 and go up.
- *Bimetallic thermometers*—considered unsuitable.
- *Pressure bulb thermometers*—more suitable for permanent installation. Accurate and reasonable in cost—$40 up.
- *Optical pyrometers*—only suitable for furnace settings and therefore limited in use. Cost from $300 up.
- *Radiation pyrometers*—limited in use for A/C work and costs from $500 up.
- *Indicating crayons*—limited in use and not considered suitable for A/C testing—costs around $2/crayon.
- *Thermographs*—use for recording room or space temperature and gives a chart indicating variations over a 12- or 168-hour period. Reasonably accurate. Low cost at around $30 to $60. (Spring-wound drive.)

Pressure Measurement (Absolute and Differential)

Common devices used for measuring pressure in HVAC applications (accuracy, range, application, and limitations are discussed in relation to HVAC work) are as follows.
- *Absolute pressure manometer*—not really suited to HVAC test work.
- *Diaphragm*—not really suited to HVAC test work.
- *Barometer (Hg manometer)*—not really suited to HVAC test work.
- *Micromanometer*—not usually portable, but suitable for fixed measurement of pressure differentials across filter, coils, etc. Cost around $30 and up.
- *Draft gauges*—can be protable and used for either direct pressure or pressure differential. From $30 up.
- *Manometers*—can be portable. Used for direct pressure reading and with pitot tubes for air flows. Very useful. Costs from $20 up.

73

- *Swing Vane gauges*—can be portable. Usually used for air flow. Costs about $30.
- *Bourdon tube gauges*—very useful for measuring all forms of system fluid pressures from 5 psi up. Costs vary greatly, from $10 up. Special types for refrigeration plants.

Humidity Measurement

The data given below indicate the type of instruments available for humidity measurement. The following indicates equipment suitable for HVAC applications.

- *Psychrometers*—basically these are wet and dry bulb thermometers. They can be fixed on a portable stand or mounted in a frame with a handle for revolving in air. Costs are low ($10 to $30) and are convenient to use.
- *Dewpoint hygrometers*—not considered suitable for HVAC test work.
- *Dimensional change*—device usually consists of a "hair," which changes in length proportionally with humidity changes. Not usually portable, fragile, and only suitable for limited temperature and humidity ranges.
- *Electrical conductivity*—can be compact and portable but of a higher cost (from $200 up). Very convenient to use.
- *Electrolytic*—as above. But for very low temperature ranges. Therefore unsuitable for HVAC test work.
- *Gravemeter*—not suitable.

IDENTIFYING STEAM AND UTILITY COSTS

Steam and utility costs are significant for most plants. It is important to quantify usage, fuel costs as a function of production. Figure 3-5 illustrates a typical Steam and Utility Cost Report. This report enables the plant manager to evaluate the total Btu of fuel consumed, the total fuel cost, and the total steam generation cost as a function of production. This report is issued monthly.

STEAM AND UTILITY COST REPORT

Preparatory Production [＿＿＿＿＿] Lbs C & F   Month [＿＿＿＿＿] 19 [＿]

Tire Production ＿＿＿＿＿＿＿＿＿＿＿   Operating Days [＿＿]

A. STEAM   Quantities

1. Actual Steam Generated   [＿＿＿＿＿ 000] lbs
   a) Press and Temp ＿＿ psi ＿＿°F ＿＿ 000 lbs
   b) Press and Temp ＿＿ psi ＿＿°F ＿＿ 000 lbs
   c) Press and Temp ＿＿ psi ＿＿°F ＿＿ 000 lbs

2. Factor of Evaporation
   (ASME Standard)   a) ＿＿＿＿
                         b) ＿＿＿＿
                         c) ＿＿＿＿

3. Equivalent Steam (No. 1 x No. 2) [＿＿＿＿ 000] lbs

4. Blowdown   [＿＿＿＿＿] lbs
   Heat Loss in BTU from Blowdown ＿＿＿＿＿ BTU
   BTU/Lb-Gal-Cu Ft

	Total $	Total	
5. Coal	[＿＿]	[＿＿]	[＿＿] tons
6. Oil	[＿＿]	[＿＿]	[＿＿] gals
7. Gas	[＿＿]	[＿＿]	[＿＿] MCF
8. Misc	[＿＿]	[＿＿]	[＿＿] Units

9. Total BTU of Fuel Consumed [＿＿＿＿ 000,000] BTU

10. Boiler Efficiency (970.3 x No. 3) + No. 4 = [＿＿] %
    ─────────────────
          No. 9

11. Total Fuel Cost (No. 5 + No. 6 + No. 7 + No. 8) $ [＿＿＿＿]

12. Operating Labor (include wage benefits) $ [＿＿＿＿]

13. Operating Supplies (water, chemicals, etc.) $ ＿＿＿＿

14. Electricity for Power Plant Auxiliaries ＿＿＿ KWH $ ＿＿＿＿

15. Maintenance Charges $ ＿＿＿＿

16. Other Miscellaneous Charges $ ＿＿＿＿

17. Total Operating Cost (No. 11 thru No. 16) $ [＿＿＿＿]

18. Total Steam Generation Cost (No. 17 + fixed Cost) $ [＿＿＿＿]

B. OPERATING CONDITIONS
   AND DATA
   IN BOILER PLANT

Quantity of
Feedwater Makeup [＿＿ 000] (Gal.)

Water Treating Chemicals

a) Phosphate   ＿＿＿＿＿＿ lbs
b) Salt   ＿＿＿＿＿＿ lbs
c) ＿＿＿＿＿   ＿＿＿＿＿＿ lbs
d) ＿＿＿＿＿   ＿＿＿＿＿＿ lbs
e) ＿＿＿＿＿   ＿＿＿＿＿＿ lbs
f) ＿＿＿＿＿   ＿＿＿＿＿＿ lbs
g) ＿＿＿＿＿   ＿＿＿＿＿＿ lbs

Ave. Feedwater Temp. ＿＿＿＿ °F

Max Steam Demand ＿＿＿＿ lbs/hr

Equivalent Steam Consumption
Per 100 lbs C&F ＿＿＿＿＿＿ lbs

Units ＿＿＿＿＿＿ Fuel ＿＿＿＿＿＿

Unit Cost
$/1000 Lbs
Equiv. Steam

Cost

$ ＿＿＿＿

$ ＿＿＿＿

$ ＿＿＿＿

$ ＿＿＿＿

Figure 3-5. Steam and Utility Cost Report

Since each plant has the same report, plant to plant comparisons are made and the effectiveness of the energy use is measured.

CALCULATING THE ENERGY
CONTENT OF THE PROCESS

Knowing the energy content of the plant's process is the first step in understanding how to reduce its cost. Using energy more efficiently reduces the product cost, thus increasing profits. In order to account for the process energy content, all energy that enters and leaves a plant during a given period must be measured. Figure 3-6 illustrates Energy Content of a Process Report. The report applies to any manufacturing operation, whether it be a pulp mill, steel mill, or assembly line. This report enables one to quickly identify energy inefficient operations.

Attention can then be focused on which equipment should be replaced and what maintenance programs should be initiated. This report also focuses attention on the choice of raw materials. By using Btu's per unit of production, measureable goals can be set. This report will also identify opportunities where energy usage can be reduced. The energy content of raw materials can be estimated by using the heating values indicated in Table 3-12.

Example Problem 3-1

Comment on energy content by modifying process No. 1 as follows:

Monthly Usage Rate

	Process No. 1	*Modified Process No. 1*
Ethane	30,000 lb	50,000 lb
Steam	250 X 10^6 lb	200 X 10^6 lb
Electricity	0.5 X 10^6 Kwh	0.8 X 10^6 Kwh
Natural gas	350 X 10^6 ft³	300 X 10^6 ft³

Assuming a Btu content of steam of 1077 Btu/lb, compute the net energy content per process.

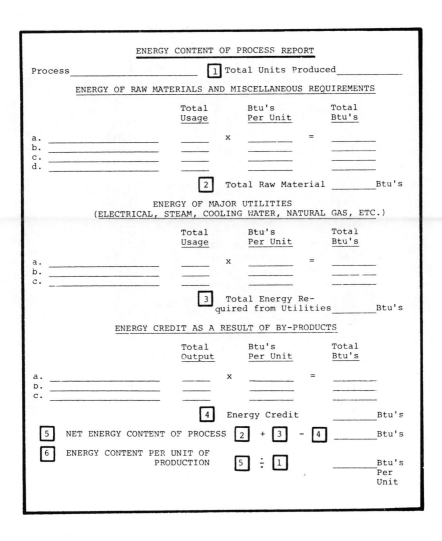

Figure 3-6. Energy Content of Process Report

Table 3-12. Heat of Combustion for Raw Materials

	Formula	Gross Heat of Combustion Btu/lb
Raw Material		
Carbon	C	14,093
Hydrogen	H_2	61,095
Carbon monoxide	CO	4,347
Paraffin Series		
Methane	CH_4	23,875
Ethane	C_2H_4	22,323
Propane	C_3H_8	21,669
n-Butane	C_4H_{10}	21,321
Isobutane	C_4H_{10}	21,271
n-Pentane	C_5H_{12}	21,095
Isopentane	C_5H_{12}	21,047
Neopentane	C_5H_{12}	20,978
n-Hexane	C_6H_{14}	20,966
Olefin Series		
Ethylene	C_2H_4	21,636
Propylene	C_3H_6	21,048
n-Butene	C_4H_8	20,854
Isobutene	C_4H_8	20,737
n-Pentene	C_5H_{10}	20,720
Aromatic Series		
Benzene	C_6H_6	18,184
Toluene	C_7H_8	18,501
Xylene	C_8H_{10}	18,651
Miscellaneous Gases		
Acetylene	C_2H_2	21,502
Naphthalene	$C_{10}H_8$	17,303
Methyl alcohol	CH_3OH	10,258
Ethyl alcohol	C_2H_5OH	13,161
Ammonia	NH_3	9,667

Source: NBS Handbook 115.

78

Answer

Process No. 1
Energy of Raw Materials

	Total Usage	Btu's Per Unit	Total Btu's
Ethane	30,000 lb	From Table 3-12 22,323 Btu/lb	0.6 X 10^9

Energy of Major Utilities

Steam	250 X 10^6 lb	1077 Btu/lb	269.2 X 10^9
Electricity	0.5 X 10^6 Kwh	From Table 1-2 10,000 Btu/Kwh	5 X 10^9
Natural gas	350 X 10^6 ft^3	From Table 1-1 1000 Btu/ft^3	350 X 10^9
	Net energy content for Process No. 1		624 X 10^9

Modified Process No. 1
Energy of Raw Materials

Ethane	50,000 lb	From Table 3-12 22,323 Btu/lb	1.1 X 10^9

Energy of Major Utilities

Steam	200 X 10^6 lb	1077 Btu/lb	215.4 X 10^9
Electricity	0.8 X 10^6 lb	From Table 1-2 10,000 Btu/Kwh	8 X 10^9
Natural gas	300 X 10^6 ft^3	From Table 1-1 1000 Btu/ft^3	300 X 10^9
	Net energy content for Process No. 1		524 X 10^9

Modifying Process No. 1 saves 100 X 10^9 Btu's per month.

4

Electrical System Optimization

THE POWER TRIANGLE

The total power requirement of a load is made up of two components: namely, the resistive part and the reactive part. The resistive portion of a load can not be added directly to the reactive component since it is essentially 90 degrees out of phase with the other. The pure resistive power is known as the watt, while the reactive power is referred to as the reactive volt amperes. To compute the total volt ampere load it is necessary to analyze the power triangle indicated below:

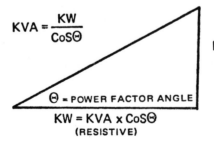

$$KVA = \frac{KW}{Cos\Theta}$$

$$KVAR = KVA\ Sin\Theta$$
(REACTIVE)

Θ = POWER FACTOR ANGLE

$$KW = KVA \times Cos\Theta$$
(RESISTIVE)

$$K = 1000$$
$$W = \text{Watts}$$
$$VA = \text{Volt Amperes}$$
$$VAR = \text{Volt Amperes Reactive}$$
$$\Theta = \text{Angle Between KVA and KW}$$
$$Co S\Theta = \text{Power Factor}$$
$$\text{Tan } \Theta = \frac{KVAR}{KW}$$

The windings of transformers and motors are usually connected in a wye or delta configuration. The relationships for line and phase voltages and currents are illustrated by Figure 4-1.

For a balanced 3-phase load

$$\text{Power} = \underbrace{\sqrt{3} \; V_L \; I_L}_{} \; Co S\Theta \qquad \qquad \textit{Formula (4-1)}$$

$$\text{Watts} = \qquad \begin{array}{cc} \text{Volt} & \text{Power} \\ \text{Amperes} & \text{Factor} \end{array}$$

For a balanced 1-phase load

$$P = V_L \; I_L \; Co S\Theta \qquad \qquad \textit{Formula (4-2)}$$

The primary windings of 13.8 Kv — 480 Volt unit substations are usually delta-connected with the secondary wye-connected.

MOTOR HORSEPOWER

The standard power rating of a motor is referred to as a horsepower. In order to relate the motor horsepower to a kilowatt (KW) multiply the horsepower by .746 (Conversion Factor) and divide by the motor efficiency.

$$KVA = \frac{HP \times .746}{\eta \times P.F.} \qquad \qquad \textit{Formula (4-3)}$$

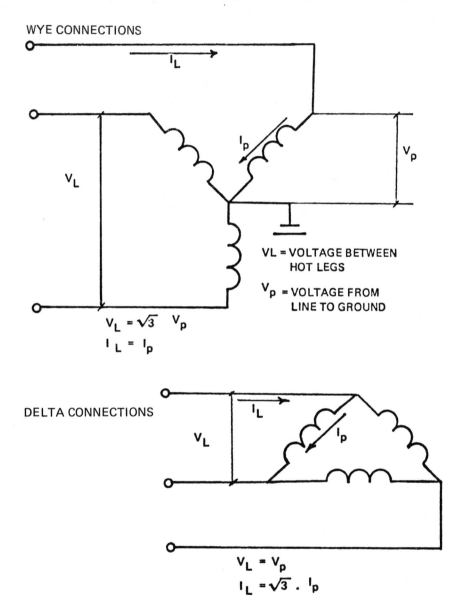

WYE CONNECTIONS

I_L

I_p

V_L

V_p

VL = VOLTAGE BETWEEN
HOT LEGS

V_p = VOLTAGE FROM
LINE TO GROUND

$$V_L = \sqrt{3}\ V_p$$
$$I_L = I_p$$

DELTA CONNECTIONS

V_L

I_L

I_p

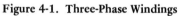

$$V_L = V_p$$
$$I_L = \sqrt{3} \cdot I_p$$

Figure 4-1. Three-Phase Windings

82

HP = Motor Horsepower
η = Efficiency of Motor
P.F. = Power Factor of Motor

Motor efficiencies and power factors vary with load. Typical values are shown in Table 4-1. Values are based on totally enclosed fan-cooled motors (TEFC) running at 1800 RPM "T" frame.

Table 4-1.

HP RANGE	3-30	40-100
η% at		
½ Load	83.3	89.2
¾ Load	85.8	90.7
Full Load	86.2	90.9
P.F. at		
½ Load	70.1	79.2
¾ Load	79.2	85.4
Full Load	83.5	87.4

POWER FLOW CONCEPT

Power flowing is analogous to water flowing in a pipe. To supply several small water users, a large pipe services the plant at a high pressure. Several branches from the main pipe service various loads. Pressure reducing stations lower the main pressure to meet the requirements of each user. Similarly, a large feeder at a high voltage services a plant. Through switchgear breakers, the main feeder is distributed into smaller feeders. The switchgear breakers serve as a protector for each of the smaller feeders. Transformers are used to lower the voltage to the nominal value needed by the user.

ELECTRICAL EQUIPMENT

Electrical equipment commonly specified is as follows:

- *Switchgear-Breakers*—used to distribute power.
- *Unit Substation*—used to step down voltage. Consists of a high voltage disconnect switch, transformer and low-voltage breakers. Typical 480-volt transformer sizes are 300 KVA, 500 KVA, 750 KVA, 1000 KVA, 2500 KVA and 3000 KVA.
- *Motor Control Center (M.C.C.)*—a structure which houses starters and circuit breakers or fuses for motor control. It consists of the following:
  (1) Thermal overload relays which guard against motor overloads;
  (2) Fuse disconnect switches or breakers which protect the cable and motor and can be used as a disconnecting means;
  (3) Contactors (relays) whose contacts are capable of opening and closing the power source to the motor.

MOTORS

- *Squirrel Cage Induction Motors* are commonly used. These motors require three power loads. For two-speed applications several different types of motors are available. Depending on the process requirements such as constant horsepower or constant torque, the windings of the motor are connected differently. The theory of two-speed operation is based on Formula 4-4.

$$\text{Frequency} = \frac{\text{No. of poles} \times \text{speed}}{120} \qquad \textit{Formula (4-4)}$$

Thus, if the frequency is fixed, the effective number of motor poles should be changed to change the speed. This can be accomplished by the manner in which the windings are connected. Two-speed motors require six power leads.

- *D. C. Motors* are used where speed control is essential. The speed of a D.C. Motor is changed by varying the field voltage through

84

a rheostat. A D.C. Motor requires two power wires to the armature and two smaller cables for the field.

- *Synchronous Motors* are used when constant speed operation is essential. Synchronous motors are sometimes cheaper in the large horsepower categories when slow speed operation is required. Synchronous motors also are considered for power factor correction. A .8 P.F. synchronous motor will supply corrective KVARs to the system. A synchronous motor requires A.C. for power and D.C. for the field. Since many synchronous motors are self-excited, only the power cables are required to the motor.

IMPORTANCE OF POWER FACTOR

Transformer size is based on KVA. The closer Θ equals $0°$ or power factor approaches unity, the smaller the KVA. Many times utility companies have a power factor clause in their contract with the customer. The statement usually causes the customer to pay an additional power rate if the power factor of the plant deviates substantially from unity. The utility company wishes to maximize the efficiency of their transformers and associated equipment.

POWER FACTOR CORRECTION

One problem facing the energy engineer is to estimate the power factor of a new plant and to install equipment such as capacitor banks or synchronous motors so that the overall power factor will meet the utility company's objectives.

Capacitor banks lower the total reactive KVAR by the value of the capacitors installed.

A second problem is to retrofit an existing plant such that the overall power factor desired is obtained.

Example Problem 4-1

A total motor horsepower load of 854 is made up of motors ranging from 40–100 horsepower. Calculate the connected KVA.

85

It is desired to operate the plant at a power factor of .95. What approximate capacitor bank is required?

Answer

$$KVA = \frac{HP \times .746}{Motor\ Eff. \times Motor\ P.F.}$$

From Table 4-1 at full load

$$Eff. = .909\ and\ P.F. = .87$$

$$KVA = \frac{854 \times .746}{.909 \times .87} = 806$$

The plant is operating at a power factor of .87. The power factor of .87 corresponds to an angle of 29°.

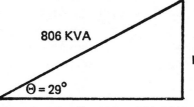

806 KVA

$\Theta = 29°$

KW = 806 CoSΘ = 806 x .87 = 701

KVAR = 806 Sin 29°
= 806 x .48 = 386

KVAR = 386

A power factor of .95 is required.

$$CoS\Theta = .95$$
$$\Theta = 18°$$

The KVAR of 386 needs to be reduced by adding capacitors.

Remember KW does not change with different power factors, but KVA does.

Thus, the desired power triangle would look as follows:

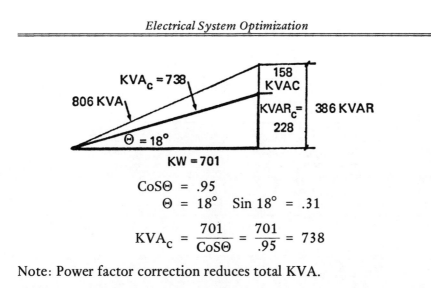

$$CoS\Theta = .95$$
$$\Theta = 18° \quad Sin\ 18° = .31$$

$$KVA_c = \frac{701}{CoS\Theta} = \frac{701}{.95} = 738$$

Note: Power factor correction reduces total KVA.

$$KVAR_c = 738\ Sin\ 18° = 738 \times .31 = 228$$

$$Capacitance\ Bank = 386-228$$
$$158\ KVAC$$

Example Problem 4-2

The client wishes to know the expected power factor for a new plant. The lighting load is 40 KW. The plant is comprised of two identical modules (2 motors for each equipment number listed). Remember that KVAs at different power factors can not be added directly.)

Motor List — Module 1

Motor No.	Description	HP	Voltage	Phase
AG-1	Agitator Motor	60	460	3
CF-3	Centrifuge Motor	100	460	3
FP-4	Feed Pump Motor	30	460	3
TP-5	Transfer Pump Motor	10	460	3
CTP-6	Cooling Tower	25	460	3
	Feed Pump Motor			*(Continued)*

Motor List — Module 1 (Continued)

Motor No.	Description	HP	Voltage	Phase
CT-9	Cooling Tower Motor	20	460	3
HF-10	H&V Supply Fan Motor	40	460	3
HF-11	H&V Exhaust Fan Motor	20	460	3
BC-13	Brine Compressor Motor	50	460	3
C-16	Conveyor Motor	20	460	3
H-17	Hoist Motor	5	460	3

Answer

Based on the motor list, the plant power factor is estimated as follows:

Module 1

Lighting KW_3 = 40 Total

Motors 3-30	Motors 40-100
30	60
10	100
25	40
20	$\underline{50}$
20	250
20	
$\underline{5}$	
130	

At Full Load:

$$P.F. = 83.5 \qquad\qquad P.F. = 87.4$$
$$\eta = 86.2 \qquad\qquad \eta = 90.9$$

$$KVA_1 = \frac{130 \times .746}{.83 \times .86} = 135 \qquad KVA_2 = \frac{250 \times .746}{.90 \times .87} = 238$$

$$KW_1 = KVA\ Co S\Theta \qquad\qquad KW_2 = KVA_2\ Co S\Theta$$
$$= KVA\ .83 \qquad\qquad\qquad = KVA\ .87$$
$$= 112 \qquad\qquad\qquad\qquad = 207$$

$$KVAR_1 = KVA_1\ Sin\Theta \qquad\qquad KVAR_2 = KVA_2\ Sin\Theta$$

$$\Theta = 33° \qquad\qquad\qquad\qquad \Theta = 29°$$

$$KVAR_1 = KVA_1 \times .54 \qquad\qquad KVAR_2 = KVA_2 \times .48$$
$$= 135 \times .54 \qquad\qquad\qquad KVAR_2 = 115$$
$$= 73$$

$$KW_{total} = KW_1 + KW_2 + KW_1 + KW_2 + KW_3 = 678\ KW$$
$$\qquad\qquad \text{Module 1} \qquad\qquad \text{Module 2}$$

$$KVAR_{total} = KVAR_1 + KVAR_2 + KVAR_1 + KVAR_2$$
$$\qquad\qquad\qquad \text{Module 1} \qquad\qquad \text{Module 2}$$
$$= 73 + 115 + 73 + 115 = 376$$

$$KVA_{total} = \sqrt{(678)^2 \times (376)^2} = 774\ KVA$$

$$Co S\Theta = \frac{KW_{total}}{KVA_{total}} = \frac{678}{774} = .87$$

WHERE TO LOCATE CAPACITORS

As indicated, the primary purpose of capacitors is to reduce the power consumption. Additional benefits are derived by capacitor location. Figure 4-2 indicates typical capacitor locations. Maximum benefit of capacitors is derived by locating them as close as possible to the load. At this location, its kilovars are confined to the smallest possible segment, decreasing the load current. This, in turn, will reduce power losses of the system substantially. Power losses are proportional to the square of the current. When power losses are reduced, voltage at the motor increases; thus, motor performance also increases.

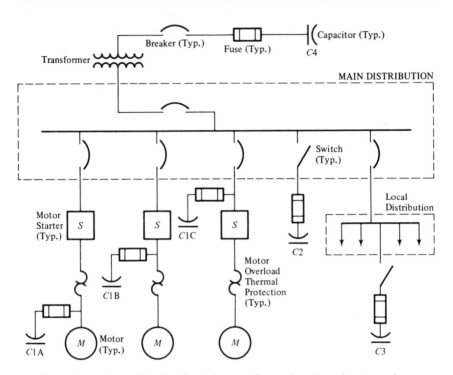

Figure 4-2. Power Distribution Diagram Illustrating Capacitor Locations.

Locations $C1A$, $C1B$ and $C1C$ of Figure 4-2 indicate three different arrangements at the load. Note that in all three locations, extra switches are not required, since the capacitor is either switched with the motor starter or the breaker before the starter. Case $C1A$ is the recommended for new installation, since the maximum benefit is derived and the size of the motor thermal protector is reduced. In Case $C1B$, as in Case $C1A$, the capacitor is energized only when the motor is in operation. Case $C1B$ is recommended in cases where the installation is existing and the thermal protector does not need to be re-sized. In position $C1C$, the capacitor is permanently connected to the circuit, but does not require a separate switch, since it can be disconnected by the breaker before the starter.

90

It should be noted that the rating of the capacitor should *not* be greater than the no-load magnetizing KVAR of the motor. If this condition exists, damaging overvoltage or transient torques can occur. This is why most motor manufacturers specify maximum capacitor ratings to be applied to specific motors.

The next preference for capacitor locations as illustrated by Figure 4-2 is at locations $C2$ and $C3$. In these locations, a breaker or switch will be required. Location $C4$ requires a high voltage breaker. The advantage of locating capacitors at power centers or feeders is that they can be grouped together. When several motors are running intermittently, the capacitors are permitted to be on line all the time, reducing the total power regardless of load.

ENERGY EFFICIENT MOTORS

Energy efficient motors are now available. These motors are approximately 30% more expensive than their standard counterpart. Based on the energy cost it can be determined if the added investment is justified. With the emphasis on energy conservation, new lines of energy efficient motors are being introduced. Figures 4-3 and 4-4 illustrate a typical comparison between energy efficient and standard motors.

LIGHTING BASICS

By understanding the basics of lighting design, several ways to improve the efficiency of lighting systems will become apparent.

There are two common lighting methods used: One is called the "Lumen" method, while the other is the "Point by Point" method. The Lumen method assumes an equal footcandle level throughout the area. This method is used frequently by lighting designers since it is simplest; however, it wastes energy, since it is the light "at the task" which must be maintained and not the light in the surrounding areas. The "Point by Point" method calculates the lighting requirements for the task in question.

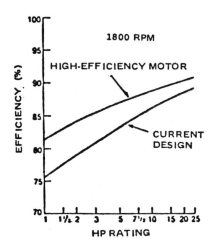

Figure 4-3. Efficiency vs Horsepower Rating (Dripproof Motors)

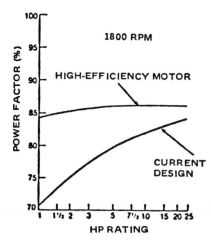

Figure 4-4. Power Factor vs Horsepower Rating (Dripproof Motors)

Lumen Method

A footcandle is the illuminance on a surface of one square foot in area having a uniformly distributed flux of one lumen. From this definition, the "Lumen Method" is developed and illustrated by Formula 4-5.

$$N = \frac{F_1 \times A}{Lu \times L_1 \times L_2 \times Cu} \qquad \textit{Formula (4-5)}$$

Where

N is the number of lamps required.

F_1 is the required footcandle level at the task. A footcandle is a measure of illumination; one standard candle power measured one foot away.

A is the area of the room in square feet.

Lu is the Lumen output per lamp. A Lumen is a measure of lamp intensity: its value is found in the manufacturer's catalogue.

Cu is the coefficient of utilization. It represents the ratio of the Lumens reaching the working plane to the total Lumens generated by the lamp. The coefficient of utilization makes allowances for light absorbed or reflected by walls, ceilings, and the fixture itself. Its values are found in the manufacturer's catalogue.

L_1 is the lamp depreciation factor. It takes into account that the lamp Lumen depreciates with time. Its value is found in the manufacturer's catalogue.

L_2 is the luminaire (fixture) dirt depreciation factor. It takes into account the effect of dirt on a luminaire, and varies with type of luminaire and the atmosphere in which it is operated.

The lumen method formula illustrates several ways lighting efficiency can be improved. First, the type of lamp which directly affects the lumen output should be analyzed. Figure 4-5 illustrates lumen outputs of various types of lamp.

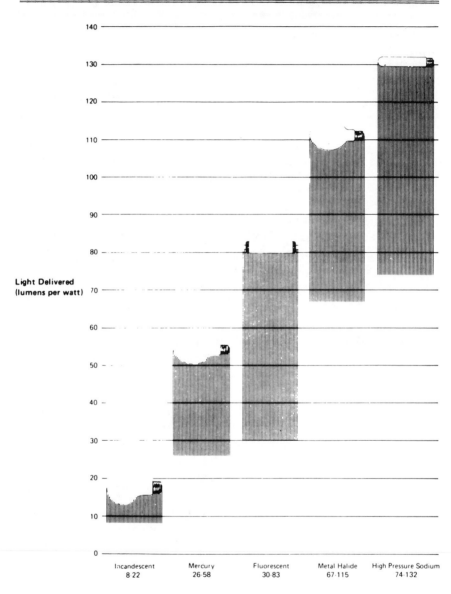

Figure 4-5. Efficiency of Various Light Sources

Many manufacturers are producing highly efficient lamp substitutes for existing lighting systems. For example, for a premium of less than 15%, highly efficient 40-watt fluorescent lamps can replace existing lamps. The end result is a 3–5% reduction in illumination with a 12% reduction in energy usage. Thus there is usually a trade-off in increased first cost coupled with a reduction in illumination and energy usage.

Lamp characteristics of several widely used types are described in this section.

Efficient Use of Incandescents

Even though incandescent lighting is not the most energy-efficient per se, efficiencies can be achieved by choosing the right type and wattage of bulb.

An important rule of thumb in determining efficiencies of incandescents is that efficiency increases as the wattage increases. For example, one 100-watt incandescent bulb has an output of 18 lumens per watt, while a 60-watt bulb produces only 14 lumens per watt. Thus, the substitution of one 100-watt bulb (1800 lumens) for two 60-watt bulbs (1680 lumens) produces more light and uses less electricity. This type of substitution saves energy, reduces maintenance, frees circuitry, and therefore should be made wherever possible.

A factor that has a significant impact on the efficiency of incandescents is the *life span* of the bulb, which is measured in hours. Not only do incandescents have the shortest life span of all available lamps, but near the end of the bulb lifetime the light output has depreciated by 20% of its original output. Light output is reduced because, as the coiled tungsten filament in incandescent bulbs emits light, molecules of the metal burn off. These molecules become deposited on the surface of the bulb, and slowly cause the bulb to darken. As it darkens, the bulb consumes the same amount of energy as it did when new, yet it produces less light. The bulb eventually burns out when the filament ruptures. Energy and, in the long run, money can be saved by replacing darkened bulbs before they burn out. *Long-life* bulbs (which last from 2500 to 3500 hours) are the least efficient incandescents of all, because light output is sacrificed in favor of long life.

A tinted bulb has a lower light output than a standard incandescent bulb of the same wattage. This is because the coating on the bulb inhibits the transmission of light. And the higher prices of tinted bulbs further illustrate that energy efficiency is often less expensive from the start.

Efficient Types of Incandescents for Limited Use

Attempts to increase the efficiency of incandescent lighting while maintaining good color rendition have led to the manufacture of a number of energy-saving incandescent lamps for limited residential use.

Tungsten Halogen—These lamps vary from the standard incandescent by the addition of halogen gases to the bulb. Halogen gases keep the glass bulb from darkening by preventing the filament from evaporating, and thereby increase lifetime up to four times that of a standard bulb. The lumen-per-watt rating is approximately the same for both types of incandescents, but tungsten halogen lamps average 94% efficiency throughout their extended lifetime, offering significant energy and operating cost savings. However, tungsten halogen lamps require special fixtures, and during operation, the surface of the bulb reaches very high temperatures, so they are not commonly used in the home.

Reflector or R-Lamps—Reflector lamps (R-lamps) are incandescents with an interior coating of aluminum that directs the light to the front of the bulb. Certain incandescent light fixtures, such as recessed or directional fixtures, trap light inside. Reflector lamps project a cone of light out of the fixture and into the room, so that more light is delivered where it is needed. In these fixtures, a 50-watt reflector bulb will provide better lighting and use less energy when substituted for a 100-watt standard incandescent bulb.

Reflector lamps are an appropriate choice for task lighting, because they directly illuminate a work area, and for accent lighting. Reflector lamps are available in 25, 30, 50, 75, and 150 watts. While they have a lower initial efficiency (lumens per watt) than regular incandescents, they direct light more effectively, so that more light is actually delivered than with regular incandescents. See Figure 4-6.

PAR Lamps— Parabolic aluminized reflector (PAR) lamps are reflector lamps with a lens of heavy, durable glass, which makes them an appropriate choice for outdoor flood and spot lighting. They are available in 75, 150, and 250 watts. They have longer lifetimes with less depreciation than standard incandescents.

*ER Lamps—*Ellipsoidal reflector (ER) lamps are ideally suited for recessed fixtures, because the beam of light produced is focused two inches ahead of the lamp to reduce the amount of light trapped in the fixture. In a directional fixture, a 75-watt ellipsoidal reflector lamp delivers more light than a 150-watt R-lamp. See Figure 4-6.

Fluorescent Lighting

Unlike incandescent bulbs, fluorescent lamps do not depend on the buildup of heat for light; rather, they convert energy to light by using an electric charge to "excite" gaseous atoms within the fluorescent tube. The charge is sparked in the ballast and flows through cathodes in either end of the tube. The resulting gaseous discharge causes the phosphor coating on the inside of the tube to "fluoresce," and emit strong visible light. Because the buildup of heat is not requisite to the creation of the light, the energy wasted as heat is significantly less than is wasted by incandescent lighting.

Efficient Use of Fluorescents

Energy savings are to be found simply in the use of fluorescents rather than in the choice of lamps or wattages, because the efficiency rating and lifespan of fluorescents remain consistently high over a

Standard Incandescent

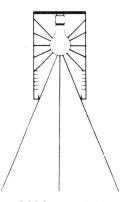

A high percentage
of light output
is trapped in fixture

R-Lamp

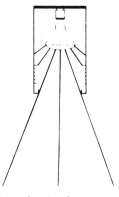

An aluminum
coating directs light
out of the fixture

ER Lamp

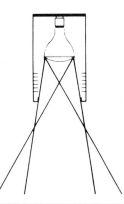

The beam is focused 2 inches
ahead of the lamp, so that very
little light is trapped in the fixture

Figure 4-6. Comparison of Incandescent Lamps

range of wattages. However, the efficiency of a fluorescent lamp will increase as the length of the tube increases. Therefore, wherever practical, large fixtures should be used to save energy.

The ballast consumes a small but constant amount of energy, even when a tube has been removed. Disconnecting the ballast or unplugging a fixture not in use is a good way to conserve electricity. As one ballast is often shared by two or more lamps where fluorescent lights are arranged in strips, removing one lamp will cause the others activated by the same ballast to go out. An electrician can make the simple adjustments required to prevent this.

While turning off incandescent lamps not in use is a commonly recognized way to achieve energy savings, misunderstandings abound as to the efficiencies of turning fluorescent lamps on and off. Like incandescents, fluorescent lamps should be turned off when not in use, even if only for a few minutes. No energy is required to turn a light off, and the initial charge required to turn a fluorescent back on does not use a significant amount of energy unless the switch is flipped back and forth in rapid succession. Fluorescent lamp life is rated according to the number of hours of operation per start, and while it was once true that the greater the number of hours operated per start, the longer the lamp life, recent technology has increased lamp life ratings to an extent that makes the number of starts far less important than they were 10 or 20 years ago. As a rule, if a space is to remain unoccupied for more than 15 minutes, fluorescent lamps should be turned off.

HID Lighting

Because outdoor security and safety lights burn for long hours—sometimes from dusk to dawn—the potential for energy savings is great. This is especially true for outdoor home lighting, which commonly uses incandescent floodlights instead of more energy-efficient High Intensity Discharge (HID) lamps.

The greater efficiency of HID lamps can be seen from the lumen-per-watt information given in Figure 4-5. Considering the higher wattages needed for outdoor security lamps and the long hours they burn, the higher initial investment required for energy-efficient HID lamps makes sense.

Three types of HID lamps are on the market: mercury, metal halide, and high-pressure sodium. All three are more energy-efficient than the standard incandescent bulbs, and are commonly used in business and industry for lighting large areas, such as parking lots, arenas, and lobbies. Each type requires a ballast designed specifically for it.

Mercury—Of the three, mercury lamps are the most commonly used outdoor lighting source. They have the lowest installation cost and a very long life. They are available in 40, 50, 75, 100, and 250 watts, and while comparable in size to incandescent lamps of the same wattage, produce twice as much light. Clear mercury lamps have poor color rendition—they accentuate blue tones—but color-corrected deluxe cool white or deluxe warm white lamps give objects a more familiar appearance. For long-burning outdoor safety lighting, where efficiency is generally more important than color rendition, mercury lamps are a good choice.

Metal Halide—Metal halide lamps are more efficient and have a better color rendition than mercury lamps. They are widely used for general commercial interior and exterior lighting, but because the lowest wattage available is 175 watts, they provide greater levels of illumination than ordinarily required by a homeowner.

High-Pressure Soldim—These are the most energy-efficient light source currently on the market. Homeowners generally would use them only for outdoor lighting, however, because the lowest wattage is 70 watts, and its high lumen output is generally too bright for interior home use. In addition, the color produced by these lamps is golden-white, which grays the color of red and blue objects. However, they are excellent sources for lighting large outdoor areas.

Apart from color rendition, HID lamps pose one potential drawback—a start-up delay of from 1 to 7 minutes from the time they are switched on until they fully illuminate. However, the continued refinement of HID lamps, particularly the metal halide and

100

the high pressure sodium lamps, is expected to make them practical alternatives.

Other Efficiency Criteria

Other lighting efficiency criteria are illustrated by Formula (4-5).

Footcandle Level—The footcandle level required is that at the task. Footcandle levels can be lowered to one-third of the levels for surrounding areas such as aisles. (A minimum 20-footcandle level should be maintained.)

The placement of the lamp is also important. If the luminaire can be lowered or placed at a better location the lamp wattage may be reduced.

Coefficient of Utilization (Cu)—The color of the walls, ceiling, and floors, the type of luminaire, and the characteristics of the room determine the *Cu*. This value is determined based on manufacturer's literature. The *Cu* can be improved by analyzing components such as lighter colored walls and more efficient luminaires for the space.

Lamp Depreciation Factor and *Dirt Depreciation Factor*—These two factors are involved in the maintenance program. Choosing a luminaire which resists dirt build-up, group relamping and cleaning the luminaire, will keep the system in optimum performance. Taking these factors into account can reduce the number of lamps initially required.

The light loss factor *(LLF)* takes into account that the lamp lumen depreciates with time (L_1), that the lumen output depreciates due to dirt build-up (L_2), and that lamps burn out (L_3). Formula 4-6 illustrates the relationship of these factors.

$$LLF = L_1 \times L_2 \times L_3 \qquad \text{Formula (4-6)}$$

To reduce the number of lamps required which in turn reduces energy consumption, it is necessary to increase the overall light loss factor. This is accomplished in several ways. One is to choose the

101

luminaire which minimizes dust build-up. The second is to improve the maintenance program to replace lamps prior to burn-out. Thus if it is known that a group relamping program will be used at a given percentage of rated life, the appropriate lumen depreciation factor can be found from manufacturer's data. It may be decided to use a shorter relamping period in order to increase (L_1) even further. If a group relamping program is used (L_3) is assumed to be unity.

Figure 4-7 illustrates the effect of dirt build-up on (L_2) for a dustproof luminaire. Every luminaire has a tendency for dirt build-up. Manufacturer's data should be consulted when estimating (L_2) for the luminaire in question.

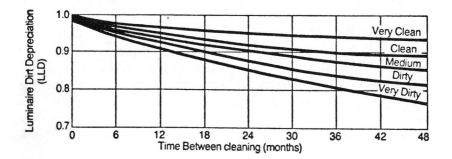

Figure 4-7. Effect of Dirt Build-Up on Dustproof Luminaires for Various Atmospheric Conditions

In addition to these items consideration should be given to the following:

Operation—Significant energy savings can be achieved simply by using light sources efficiently. The following checklist suggests ways to increase lighting efficiency.

- Lower wattages, eliminate unnecessary light, and rely on daylight whenever possible.
- Replace two bulbs with one having a comparable number of lumens.

102

- Place fixtures on separate switches so they can be operated independently of each other.
- Convert decorative outdoor gas lamps to standard incandescent or mercury lamps. If the lamp cannot be converted, limit its use but maintain safety and security by using other sources to light the area.

Timing Devices, Photocells, and Dimmers—The practice of inadvertently leaving lights burning when they are not needed can be eliminated by installing small timing devices or photocells to indoor and outdoor lighting. Timing devices can be set to automatically turn lights off and on at predetermined times, for example, on at dusk and off at dawn.

Photocells offer an automatic way to turn lights on and off in direct response to the amount of natural light available. Photocells are small photoelectric cells, sometimes less than an inch in diameter, that are sensitive to sunlight. When light strikes a photocell, it is converted to electrical energy. The brighter the light, the stronger the resulting electrical charge. Photocells are designed so that when a certain amount of natural daylight strikes them, the electrical charge created by the light triggers a mechanism that switches lights off; when daylight wanes, on stormy days, or when clouds temporarily decrease the amount of natural light available, the photocell switches lights on again. The reliability and low initial cost of photocells makes them an excellent tool for conserving energy.

Nearly all types of light sources can be dimmed by controlling the amount of power applied to them. Dimmer controls permit the occupant to adjust the level of light in a room over the full range, from off to the highest illumination level, in response to the varying light levels required at different times of the day, in different rooms of a building, and by different activities in a room. Dimmers can achieve energy savings if they are used regularly.

Two types of energy-saving dimmer controls are currently on the market: solid-state and variable-automatic transformers. Both save energy by reducing the amount of electricity delivered to the light. In addition, because bulbs are operating on reduced voltage, the life of the bulb is extended. Dimmer fixtures replace standard light switches, and are inexpensive and easy to install. New models on the market can be attached to table lamps as well.

Dimmers for fluorescent lamps must be used with a special dimmer ballast, which replaces the standard ballast. Because the use of fluorescent lamps in itself provides energy savings, the use of dimmers with fluorescent lamps is not as widespread as with incandescent bulbs. Currently, dimming equipment is available for only 30-watt and 40-watt rapid-start fluorescent lamps.

The precursor of photocells and dimmers is a dimming mechanism still sold occasionally, called a rheostat. Because it works by transforming into heat that portion of the electrical energy not used for light, it does not save energy and thus is not recommended for use.

As an alternative to full-range dimmers, there are "hi-lo" switches on the market that provide two settings for overhead lamps. While they do not provide the flexibility of a full-range dimmer, they do achieve energy savings, and are comparable to the three-way bulbs used in table lamps.

When used regularly, the costs of these lighting controls are more than offset by the electricity saved from the lower levels of illumination they make possible.

ELECTRICAL RATE TARIFF

The basic electrical rate charges contain the following elements:

Billing Demand—The maximum kilowatt requirement over a 15-, 30-, or 60-minute interval.

Load Factor—The ratio of the average load over a designated period to the peak demand load occurring in that period.

Power Factor—The ratio of resistive power to apparent power. Traditionally electrical rate tariffs have a decreasing kilowatt hour (KWH) charge with usage. This practice is likely to gradually phase out. New tariffs are containing the following elements:

Time of Day—Discounts are allowed for electrical usage during off-peak hours.

Ratchet Rate—The billing demand is based on 80–90% of peak demand for any one month. The billing demand will remain at that ratchet for 12 months even though the actual demand for the succeeding months may be less.

Energy Management Systems

One of the biggest energy cost-saving potentials is to reduce peak demands through Energy Management Systems (EMS). The simplest method to reduce peak loads is to manually schedule activities so that big power users do not operate at the same time. This is sometimes possible during initial plant start-up where one system can be operated when another is down. The second method relies upon automatic controls which shut off nonessential users during peak periods. Nonessential users such as heating, ventilation and airconditioning equipment can be automatically controlled through packaged equipment such as Energy Management Systems.

EMS—Load Demand Controller

The load-demand controller is basically a comparator. A comparison is made between the actual rate of energy usage to a predetermined ideal rate of energy usage during the demand interval. As the actual usage rate approaches the ideal usage rate, the controller determines if the present demand will be exceeded. If the determination is positive, the controller will begin to shed loads based upon a predetermined priority. The control action usually occurs during the last few minutes of the demand interval. The loads are automatically restored when the new demand interval is started. Figure 4-8 illustrates a typical demand chart before and after the installation of a demand controller.

There are several types of demand controller on the market.

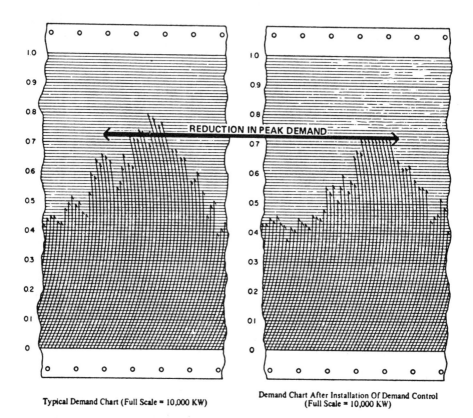

Typical Demand Chart (Full Scale = 10,000 KW)

Demand Chart After Installation Of Demand Control
(Full Scale = 10,000 KW)

**Figure 4-8. Typical Demand Chart Compared
After Installation of a Demand Control**

Careful consideration should be given to the type of controller specified. Cost and proper application are the primary criteria for choosing
the demand controller for the plant. Several commonly used types
are:

- Packaged solid-state controllers
- Time-interval reset type
- Sliding-window type
- Computer

106

The time-interval reset type controller needs two input signals as illustrated in Figure 4-9. The inputs represent kilowatt-hour energy usage and an interval signal pulse. Before purchasing this unit, it should be verified that the utility company will permit use of the synchronous pulse with the demand controller. Several utilities do not permit this signal to be used; thus another type of controller may be required.

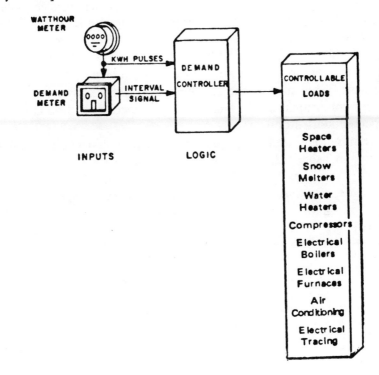

Figure 4-9. Time-Interval Reset Type Controller

The sliding-window type does not require a synchronous pulse. In the sliding-window type power consumption is monitored continu-

107

ously. The width of the sliding window can be set from 5 to 60 minutes in 5-minute increments. The energy consumed during this interval is proportional to demand. Loads are shed and restored based on predetermined upper and lower thresholds. Predicted usage is calculated based on the consumption during the most recent quarter window width. A scheduling clock can be built into these units allowing for turning loads on and off during specific times of the day.

Computers used for load shedding can also serve several other functions. Computer systems presently offered can be used in conjunction with security and fire-alarm protection. These units also allow for better reporting of energy consumption through audit reports. One feature of computer load shedding is the floating target adjustment. This allows the system to automatically respond to changing conditions in the facility such as weekends, holidays, and unexpected work stoppages.

Terminology used in specifying an Energy Management System is defined in Table 4-2.

Table 4-2. Glossary of Terms for Energy Management Systems

Algorithm:
A set of well-defined rules or procedures for solving a problem or providing an output from a specific set of inputs.

Analog-to-Digital Converter:
A circuit or device whose input is information in analog form and whose output is the same information in digital form.

Architecture:
The general organization and structure of hardware and software.

ASCII:
American Standard Code for Information Interchange. An 8-bit coded character set to be used for the general interchange of data among information processing systems, communications systems, process control systems, and associated equipment.

Automatic Temperature Control (ATC):
A local loop network of pneumatic or electric/electronic devices which are interconnected to control temperature.

BASIC:
An acronym for Beginners All-Purpose Symbolic Instruction Code, a high-level, English-like programming language used for general applications.

Baud:
A unit of signalling speed equal to the number of discrete conditions, or signal events, per second.

Bit:
An acronym for binary digit. The smallest unit of information which can be represented. A bit may be in one of two

Table 4-2. *(Continued)*

Bit: *(Continued)*
states, represented by the binary digits 0 and 1.

Bootstrap:
A technique or device designed to bring a computer into a desired state by means of its own action.

Buffer:
A temporary data storage device used to compensate for a difference in data flow rate or event times, when transmitting data from one device to another.

Bus:
A circuit path (or parallel paths) over which data or instructions are transferred to all points in the computer system. Computers have several separate busses: the data, address, and control busses are those of greatest importance.

Byte:
A group of eight bits.

Central Memory:
Core or semiconductor memory which communicates directly with a CPU.

Central Processing Unit (CPU):
The portion of a computer that performs the interpretation and execution of instructions. It does *not* include memory or I/O.

Character:
One of a set of elementary symbols which normally include both alpha and numeric codes plus punctuation marks and any other symbol which may be read, stored or written.

Clock:
A device or a part of a device that generates all the timing pulses for the coordination of a digital system. System clocks usually generate two or more

clock phases. Each phase is a separate, square wave pulse train output.

Command Line Mnemonic (CLM):
A computer language consisting of a set of fixed, simplified English commands designed to assist operators unfamiliar with computer technology in operating the equipment.

Command Line Mnemonic Interpreter (CLMI):
Software used to implement the CLM language.

Control Point Adjustment (CPA):
The procedure of changing the operating point of a local loop controller from a remote location.

Control Sequence:
Equipment operating order established upon a correlated set of data environment conditions.

Control Strategy:
A procedure for controlling the operation of heating, ventilating and air conditioning (HVAC) equipment in an energy efficient manner.

Crowbar:
An electronic circuit which can rapidly sense an overvoltage condition and provide a solid-state low impedance path to eliminate this transient condition.

Data Environment (DE):
The sensors and control devices connected to a controller from the equipment and systems sampled or controlled.

Data Transmission Media (DTM):
Transmission equipment including cables and interface modules (excluding MODEMs) permitting transmission of digital and analog information.

109

Table 4-2. *(Continued)*

Deck:

In HVAC terminology, the air discharge of the hot or cold coil in a duct serving a conditioned space.

Demand:

The term used to describe the maximum rate of use of electrical energy averaged over a specific interval of time and usually expressed in kilowatts.

Demultiplexer:

A device used to separate two or more signals previously combined by compatible multiplexer for transmission over a single circuit.

Diagnostic Program:

Machine-executable instructions used to detect and isolate component malfunctions.

Direct Digital Control (DDC):

Sensing and control of processes directly with digital control electronics.

Digital-to-Analog (D/A) Converter:

A hardware device which converts a digital signal into a voltage or current proportional to the digital input.

Direct Memory Access (DMA):

Provision for transfer of data directly between central memory and an external device.

Disk Storage:

A bulk storage, random access device for storing digital information. Usually constructed of a thin rotating circular plate having a magnetizable coating, a read/write head and associated control equipment.

Distributed Processing System:

A system of multiple processors each performing its own task, yet working together as a complete system under the supervision of a central computer, to perform multiple associated tasks.

Download:

The transfer of digital data or programs from a host computer to another data processing system such as from central computer to microcomputer.

Executive Software:

The main system program designed to establish priorities and to process and control other programs.

Facility Engineer:

Person in charge of maintaining and operating the physical plant. In the Navy it is the Public Works Officer.

Fall-Back Mode:

The preselected operating mode of a controller or the operating sequence of each local control loop when the controller to which it is connected ceases to function.

Firmware:

An instruction set resident in ROM or PROM for accomplishing a special program or procedure.

FORTRAN:

An acronym for FORmula TRANSlation. A high-level, English-like programming language used for technical applications.

Hardware:

Equipment such as a CPU, memory, peripherals, sensors, and relays.

Initialize:

To set counters, switches, and addresses to zero or other starting values at the beginning of or at prescribed points in a computer program.

Input/Output (I/O) Devices:

Digital hardware that transmit or receive data.

Interactive:

Functions performed by a process where the machine prompts or other-

Table 4-2. *(Continued)*

Interactive: *(Continued)*

wise assists an operator to program the device while it continues to perform all other tasks as scheduled.

Interpreter:

A language translator which converts individual source statements into machine instructions by translating and executing each statement as it is encountered.

Interrupt:

An external or internal signal requesting that current operations be suspended to perform more important tasks.

Large Scale Integration (LSI):

The technology of manufacturing integrated circuits capable of performing complex functions. Devices of this class contain 100 or more logic gates.

Line Conditioning:

Electronic modification of the characteristic response of a line to meet certain standards. The characteristics include frequency response, signal levels, noise suppression, impedance, and time delay.

Line Driver:

A hardware element which enables signals to be directly transmitted over circuits to other devices some distance away.

Local Loop Control:

The controls for any system or subsystem which will continue to function when the EMCS microprocessor controller is nonoperative.

Machine Language:

The binary code corresponding to the instruction set recognized the CPU.

Memory:

Any device that can store logic 1 and logic 0 bits in such a manner that a single bit or group of bits can be accessed and retrieved.

Memory Address:

A binary number that specifies the precise memory location of a stored word.

Microcomputer:

A computer system based on a microprocessor and containing all the memory and interface hardware necessary to perform calculations and specified transformations.

Microprocessor:

A central processing unit fabricated as one integrated circuit.

MODEM:

An acronym for MOdulator/DEModulater. A hardware device used for changing digital information to and from an analog form to allow transmission over voice grade circuits.

Multi-Tasking:

The procedure allowing a computer to perform a number of programs simultaneously under the management of the operating system.

Non-Volatile Memory:

Memory which retains information in the absence of applied power (i.e.; magnetic core, ROM, and PROM).

Object Code:

A term used to describe the machine language version of a program.

Operating System:

A complex software system which manages the computer and its components and allows for human interaction.

Optical Isolation:

Electrical isolation of a portion of an electronic circuit by using optical

111

Table 4-2. *(Concluded)*

Optical Isolation: *(Continued)*
semiconductors and modulated light to carry the signal.

Point:
A single connected monitor or control device (i.e., relay, temperature sensor).

Program:
A sequence of instructions causing the computer to perform a specified function.

Protocol:
A formal set of conventions governing the format and relative timing of message exchange between two terminals.

Random Access Memory (RAM):
Volatile semiconductor data storage device in which data may be stored or retrieved. Access time is effectively independent of data location.

ROM, PROM, EPROM, EEPROM:
Read-Only-Memory, Programmable ROM, Erasable PROM, Electronically Erasable PROM. All are nonvolatile semiconductor memory.

Real Time:
A situation in which a computer moni-
tors, evaluates, reaches decisions, and effects controls within the response time of the fastest phenomenon.

Register:
A digital device capable of retaining information.

Resistance Temperature Detector (RTD):
A temperature sensor based on a linear relationship between resistance and temperature.

Software:
A term used to describe all programs whether in machine, assembly, or high-level language.

Throughput:
The total capability of equipment to process or transmit data during a specified time period.

Volatile Memory:
A semiconductor device in which the stored digital data is lost when power is removed.

Zone:
An area composed of a building, a portion of a building, or a group of buildings.

5

Waste Heat Recovery

INTRODUCTION

Waste heat is heat which is generated in a process but then "dumped" to the environment even though it could still be reused for some useful and economic purpose.

The essential quality of heat is not the amount but rather its "value."

The strategy of how to recover this heat depends in part on the temperature of the waste heat gases and the economics involved.

This chapter will present the various methods involved in traditionally recovering waste heat.

Portions of material used in Chapters 5 and 6 are based upon the *Waste Heat Management Guidebooks* published by the U.S. Department of Commerce/National Bureau of Standards. The authors express appreciation to Kenneth G. Kreider; Michael B. McNeil; W. M. Rohrer, Jr.; R. Ruegg; B. Leidy and W. Owens who have contributed extensively to this publication.

113

SOURCES OF WASTE HEAT

Sources of waste energy can be divided according to temperature into three temperature ranges. The high temperature range refers to temperatures above 1200F. The medium temperature range is between 450F and 1200F, and the low temperature range is below 450F.

High and medium temperature waste heat can be used to produce process steam. If one has high temperature waste heat, instead of producing steam directly, one should consider the possibility of using the high temperature energy to do useful work before the waste heat is extracted. Both gas and steam turbines are useful and fully developed heat engines.

In the low temperature range, waste energy which would be otherwise useless can sometimes be made useful by application of mechanical work through a device called the heat pump.

HIGH TEMPERATURE HEAT RECOVERY

The combustion of hydrocarbon fuels produces product gases in the high temperature range. The maximum theoretical temperature possible in atmospheric combustors is somewhat under 3500F, while measured flame temperatures in practical combustors are just under 3000F. Secondary air or some other dilutant is often admitted to the combustor to lower the temperature of the products to the required process temperature, for example to protect equipment, thus lowering the practical waste heat temperature.

Table 5-1 gives temperatures of waste gases from industrial process equipment in the high temperature range. All of these result from direct fuel fired processes.

MEDIUM TEMPERATURE HEAT RECOVERY

Table 5-2 gives the temperatures of waste gases from process equipment in the medium temperature range. Most of the waste heat

114

Table 5-1

Type of Device	Temperature F
Nickel refining furnace	2500–3000
Aluminum refining furnace	1200–1400
Zinc refining furnace	1400–2000
Copper refining furnace	1400–1500
Steel heating furnaces	1700–1900
Copper reverberatory furnace	1650–2000
Open hearth furnace	1200–1300
Cement kiln (Dry process)	1150–1350
Glass melting furnace	1800–2800
Hydrogen plants	1200–1800
Solid waste incinerators	1200–1800
Fume incinerators	1200–2600

in this temperature range comes from the exhausts of directly fired process units. Medium temperature waste heat is still hot enough to allow consideration of the extraction of mechanical work from the waste heat, by a steam or gas turbine. Gas turbines can be economically utilized in some cases at inlet pressures in the range of 15 to 30 lb/in^2 g. Steam can be generated at almost any desired pressure and steam turbines used when economical.

Table 5-2

Type of Device	Temperature F
Steam boiler exhausts	450–900
Gas turbine exhausts	700–1000
Reciprocating engine exhausts	600–1100
Reciprocating engine exhausts (turbocharged)	450–700
Heat treating furnaces	800–1200
Drying and baking ovens	450–1100
Catalytic crackers	800–1200
Annealing furnace cooling systems	800–1200

LOW TEMPERATURE HEAT RECOVERY

Table 5-3 lists some heat sources in the low temperature range. In this range it is usually not practicable to extract work from the source, though steam production may not be completely excluded if there is a need for low pressure steam. Low temperature waste heat may be useful in a supplementary way for preheating purposes. Taking a common example, it is possible to use economically the energy from an air conditioning condenser operating at around 90F to heat the domestic water supply. Since the hot water must be heated to about 160F, obviously the air conditioner waste heat is not hot enough. However, since the cold water enters the domestic water system at about 50F, energy interchange can take place raising the water to something less than 90F. Depending upon the relative air conditioning lead and hot water requirements, any excess condenser heat

Table 5-3

Source	Temperature F
Process steam condensate	130–190
Cooling water from:	
Furnace doors	90–130
Bearings	90–190
Welding machines	90–190
Injection molding machines	90–190
Annealing furnaces	150–450
Forming dies	80–190
Air compressors	80–120
Pumps	80–190
Internal combustion engines	150–250
Air conditioning and	
refrigeration condensers	90–110
Liquid still condensers	90–190
Drying, baking and curing ovens	200–450
Hot processed liquids	90–450
Hot processed solids	200–450

can be rejected and the additional energy required by the hot water provided by the usual electrical or fired heater.

WASTE HEAT RECOVERY APPLICATIONS

To use waste heat from sources such as those above, one often wishes to transfer the heat in one fluid stream to another (e.g., from flue gas to feedwater or combustion air). The device which accomplishes the transfer is called a heat exchanger. In the discussion immediately below is a listing of common uses for waste heat energy and in some cases, the name of the heat exchanger that would normally be applied in each particular case.

The equipment that is used to recover waste heat can range from something as simple as a pipe or duct to something as complex as a waste heat boiler.

Some applications of waste heat are as follows:

• Medium to high temperature exhaust gases can be used to preheat the combustion air for:

Boilers using air-preheaters

Furnaces using recuperators

Ovens using recuperators

Gas turbines using regenerators

• Low to medium temperature exhaust gases can be used to preheat boiler feedwater or boiler makeup water using *economizers,* which are simply gas-to-liquid water heating devices.

• Exhaust gases and cooling water from condensers can be used to preheat liquid and/or solid feedstocks in industrial processes. Finned tubes and tube-in-shell *heat exchangers* are used.

• Exhaust gases can be used to generate steam in *waste heat boilers* to produce electrical power, mechanical power, process steam, and any combination of above.

• Waste heat may be transferred to liquid or gaseous process units directly through pipes and ducts or indirectly through a secondary fluid such as steam or oil.

• Waste heat may be transferred to an intermediate fluid by heat exchangers or waste heat boilers, or it may be used by circulating the hot exit gas through pipes or ducts. Waste heat can be used to operate an absorption cooling unit for air conditioning or refrigeration.

117

THE WASTE HEAT RECOVERY SURVEY

In order to identify source of waste heat a survey is usually made. Figure 5-1 illustrates a survey form which can be used for the Waste Heat Audit. It is important to record flow and temperature of waste gases.

Composition data is required for heat recovery and system design calculations. Be sure to note contaminants since this factor could limit the type of heat recovery equipment to apply. Contaminants can foul or plug heat exchangers.

Operation schedule affects the economics and type of equipment to be specified. For example, an incinerator that is only used one shift per day may require a different method of recovering discharges than if it were used three shifts a day. A heat exchanger used for waste heat recovery in this service would soon deteriorate due to metal fatigue. A different type of heat recovery incinerator utilizing heat storage materials such as rock or ceramic would be more suitable.

WASTE HEAT RECOVERY CALCULATIONS

From the heat balance (Chapter 6), the heat recovered from the source is determined by Formula 5-1.

$$q = m \, c_p \, \Delta T \qquad \qquad \textit{Formula (5-1)}$$

where: q = heat recovered, Btu/hr
     m = mass flow rate lbs/hr
     c_p = specific heat of fluid, Btu/lb°F
     ΔT = temperature change of gas or liquid during heat recovery °F

If the flow is air, then Formula 5-1 can be expressed as

$$q = 1.08 \, \text{CFM} \, \Delta T \qquad \qquad \textit{Formula (5-2)}$$

where: CFM = volume flow rate in standard CFM

118

SURVEY FORM FOR INDUSTRIAL PROCESS UNITS

NAME OF PROCESS UNIT _____ INVENTORY NUMBER _____

LOCATION OF PROCESS UNIT, PLANT NAME _____ BUILDING _____

MANUFACTURER _____ MODEL _____ SERIAL NUMBER _____

| | | FIRING RATE | HHV | TEMPERATURE OF | | | FLUE GAS COMPOSITION % VOLUME | | | | |
	NAME			COMB. AIR	FUEL	STACK	CO_2	O_2	CO	CH	N_2
PRIMARY FUEL											
FIRST ALTERNAT.											
SECOND ALTERNAT.											

	FLOW PATH 1	FLOW PATH 2	FLOW PATH 3	FLOW PATH 4
FLUID COMPOSITION				
FLOW RATE				
INLET TEMPERATURE				
OUTLET TEMPERATURE				
DESCRIPTION				

ANNUAL HOURS OPERATION _____ ANNUAL CAPACITY FACTOR, % _____

ANNUAL FUEL CONSUMPTION: PRIMARY FUEL _____; FIRST ALTERN. _____ SEC. ALTERN. _____

PRESENT FUEL COST: PRIMARY FUEL _____; FIRST ALTERN. _____ SEC. ALTERN. _____

ANNUAL ELECTRICAL ENERGY CONSUMPTION, KWHR. _____

PRESENT ELECTRICAL ENERGY RATE _____

CONTAMINANTS _____

Figure 5-1. Waste Heat Survey

119

If the flow is water, then Formula 5-1 can be expressed as

$$q = 500 \text{ GPM } \Delta T \qquad \qquad \textit{Formula (5-3)}$$

where: GPM = volume flow rate in gallons per minute

Example Problem 5-1

A waste heat audit survey indicates 10,000 lb/hr of water at 190° is discharged to the sewer. How much heat can be saved by utilizing this fluid as makeup to the boiler instead of the 70°F feed-water supply? Fuel cost is $6 per million Btu, boiler efficiency .8, and hours of operation 4000.

Analysis

$q = mc_p \ \Delta T = 10{,}000 \times (190{-}70) = 1.2 \times 10^6$ Btu/hr
Savings $= 1.2 \times 10^6 \times 4000 \times \$6/10^6/.8 = \$36{,}000$

Heat Transfer by Convection

Convection is the transfer of heat between a fluid, gas, or liquid. Formula 5-4 is indicative of the basic form of convective heat transfer. U_0, in this case, represents the convection film conductance, Btu/ft^2 · hr · °F.

Heat transferred for heat exchanger applications is predominantly a combination of conduction and convection expressed as:

$$q = U_0 \, A\Delta T_m \qquad \qquad \textit{Formula (5-4)}$$

where: q = rate of heat flow by convection, Btu/hr
$\quad \ \ U_0$ = is the overall heat transfer coefficient Btu/ft^2 · hr · °F
$\quad \ \ A$ = is the area of the tubes in square feet
$\quad \ \ \Delta T_m$ = is the logarithmic mean temperature difference and represents the situation where the temperature of two fluids change as they transverse the surface.

$$\Delta T_m = \frac{\Delta T_1 - \Delta T_2}{\text{Log}_e \ [\Delta T_1/\Delta T_2]} \qquad \qquad \textit{Formula (5-5)}$$

120

A. COUNTERFLOW

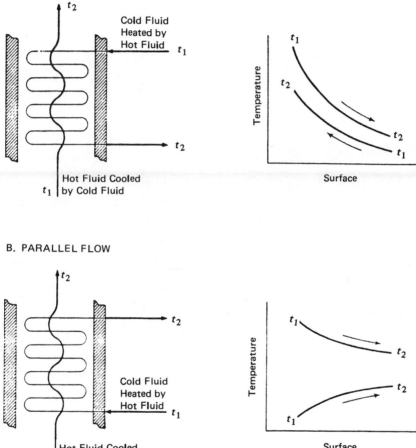

B. PARALLEL FLOW

Figure 5-2. Temperature Relationships for Heat Exchangers

121

To understand the different logarithmic mean temperature relationships, Figure 5-2 should be used. Referring to Figure 5-2, the ΔT_m for the counterflow heat exchanger is:

$$\Delta T_m = \frac{(t_1 - t_2') - (t_2 - t_1')}{\text{Log}_e\ [t_1 - t_2'/t_2 - t_1']} \qquad Formula\ (5\text{-}6)$$

The ΔT_m for the parallel flow heat exchanger is:

$$\Delta T_m = \frac{(t_1 - t_1') - (t_2 - t_2')}{\text{Log}_e\ [t_1 - t_1'/t_2 - t_2']} \qquad Formula\ (5\text{-}7)$$

HEAT TRANSFER BY RADIATION

Radiation is the transfer of heat energy by electromagnetic means between two materials whose surfaces "see" each other. The governing equation is known as the Stefan-Boltzmann equation, and is written:

$$q = \sigma F_e F_a A\ (T^4_{aba_1} - T^4_{aba_2}) \qquad Formula\ (5\text{-}8)$$

Again q and A are as defined for conduction and convection and T_{aba_1} and T_{aba_2} are the absolute temperatures of the two surfaces involved. The factor F_e is a function of the condition of the radiation surfaces and in some cases the areas of the surfaces. The factor F_a is the configuration factor and is a function of the areas and their positions. Both F_e and F_a are dimensionless. The Stefan-Boltzmann constant, σ is equal to 0.1714 But/h·ft^2·R^4. Although radiation is usually associated with solid surfaces, certain gases can emit and absorb radiation. These include the so-called nonpolar molecular gases such as H_2O, CO_2, CO, SO_2, NH_3 and the hydrocarbons. Some of these gases are present in every combustion process.

It is convenient in certain calculations to express the heat transferred by radiation in the form of

$$q = h_r A\ (T_1 - T_2) \qquad Formula\ (5\text{-}9)$$

where the coefficient of radiation, b_r, is defined as

$$b_r = \frac{\sigma F_e F_a A (T^4{}_{aba_1} - T^4{}_{aba_2})}{(T_1 - T_2)}$$ *Formula (5-10)*

Note that b_r is still dependent on the factors F_e and F_a, and also on the absolute temperatures to the fourth power. The main advantage is that in Formula 5-10 the rate of heat flow by radiation is a function of the temperature difference and can be combined with the coefficient of convection to determine the total heat flow to or from a surface.

WASTE HEAT RECOVERY EQUIPMENT

Industrial heat exchangers have many pseudonyms. They are sometimes called recuperators, regenerators, waste heat steam generators, condensers, heat wheels, temperature and moisture exchangers, etc. Whatever name they may have, they all perform one basic function: the transfer of heat.

Heat exchangers are characterized as single or multipass gas to gas, liquid to gas, liquid to liquid, evaporator, condenser, parallel flow, counterflow, or crossflow. The terms single or multipass refer to the heating or cooling media passing over the heat transfer surface once or a number of times. Multipass flow involves the use of internal baffles. The next three terms refer to the two fluids between which heat is transferred in the heat exchanger, and imply that no phase changes occur in those fluids. Here the term "fluid" is used in the most general sense. Thus, we can say that these terms apply to non-evaporator and noncondensing heat exchangers. The term evaporator applies to a heat exchanger in which heat is transferred to an evaporating (boiling) liquid, while a condenser is a heat exchanger in which heat is removed from a condensing vapor. A parallel flow heat exchanger is one in which both fluids flow in approximately the same direction whereas in counterflow the two fluids move in opposite directions. When the two fluids move at right angles to each other, the heat exchanger is considered to be of the crossflow type.

123

The principal methods of reclaiming waste heat in industrial plants make use of heat exchangers. The heat exchanger is a system which separates the stream containing waste heat and the medium which is to absorb it, but allows the flow of heat across the separation boundaries. The reasons for separating the two streams may be any of the following:

(1) A pressure difference may exist between the two streams of fluid. The rigid boundaries of the heat exchanger can be designed to withstand the pressure difference.

(2) In many, if not most, cases the one stream would contaminate the other, if they were permitted to mix. The heat exchanger prevents mixing.

(3) Heat exchangers permit the use of an intermediate fluid better suited than either of the principal exchange media for transporting waste heat through long distances. The secondary fluid is often steam, but another substance may be selected for special properties.

(4) Certain types of heat exchangers, specifically the heat wheel, are capable of transferring liquids as well as heat. Vapors being cooled in the gases are condensed in the wheel and later re-evaporated into the gas being heated. This can result in improved humidity and/or process control, abatement of atmospheric air pollution, and conservation of valuable resources.

The various names or designations applied to heat exchangers are partly an attempt to describe their function and partly the result of tradition within certain industries. For example, a recuperator is a heat exchanger which recovers waste heat from the exhaust gases of a furnace to heat the incoming air for combustion. This is the name used in both the steel and the glass making industries. The heat exchanger performing the same function in the steam generator of an electric power plant is termed an air preheater, and in the case of a gas turbine plant, a regenerator.

However, in the glass and steel industries the word regenerator refers to two chambers of brick checkerwork which alternately absorb heat from the exhaust gases and then give up part of that heat to the incoming air. The flows of flue gas and of air are periodically reversed by valves so that one chamber of the regenerator is being heated by the products of combustion while the other is being cooled by the incoming air. Regenerators are often more expensive to buy and more expensive to maintain than are recuperators, and their application is primarily in glass melt tanks and in open hearth steel furnaces.

It must be pointed out, however, that although their functions are similar, the three heat exchangers mentioned above may be structurally quite different as well as different in their principal modes of heat transfer. A more complete description of the various industrial heat exchangers follows later in this chapter and details of their differences will be clarified.

The specification of an industrial heat exchanger must include the heat exchange capacity, the temperatures of the fluids, the allowable pressure drop in each fluid path, and the properties and volumetric flow of the fluids entering the exchanger. These specifications will determine construction parameters and thus the cost of the heat exchanger. The final design will be a compromise between pressure drop, heat exchanger effectiveness, and cost. Decisions leading to that final design will balance out the cost of maintenance and operation of the overall system against the fixed costs in such a way as to minimize the total. Advice on selection and design of heat exchangers is available from vendors.

The essential parameters which should be known in order to make an optimum choice of waste heat recovery devices are:

- Temperature of waste heat fluid
- Flow rate of waste heat fluid
- Chemical composition of waste heat fluid
- Minimum allowable temperature of waste heat fluid
- Temperature of heated fluid
- Chemical composition of heated fluid
- Maximum allowable temperature of heated fluid

- Control temperature, if control required

In the rest of this chapter, some common types of waste heat recovery devices are discussed in some detail.

GAS TO GAS HEAT EXCHANGERS

Recuperators

The simplest configuration for a heat exchanger is the metallic radiation recuperator which consists of two concentric lengths of metal tubing as shown in Figure 5-3.

The inner tube carries the hot exhaust gases while the external annulus carries the combustion air from the atmosphere to the air inlets of the furnace burners. The hot gases are cooled by the incoming combustion air which now carries additional energy into the combustion chamber. This is energy which does not have to be supplied by the fuel; consequently, less fuel is burned for a given furnace loading. The saving in fuel also means a decrease in combustion air and therefore stack losses are decreased not only by lowering the stack gas temperatures, but also by discharging smaller quantities of exhaust gas. This particular recuperator gets its name from the fact that a substantial portion of the heat transfer from the hot gases to the surface of the inner tube take place by radiative heat transfer. The cold air in the annulus, however, is almost transparent to infrared radiation so that only convection heat transfer takes place to the incoming air. As shown in the diagram, the two gas flows are usually parallel, although the configuration would be simpler and the heat transfer more efficient if the flows were opposed in direction (or counterflow). The reason for the use of parallel flow is that recuperators frequently serve the additional function of cooling the duct carrying away the exhaust gases, and consequently extending its service life.

The inner tube is often fabricated from high temperature materials such as stainless steels of high nickel content. The large temperature differential at the inlet causes differential expansion,

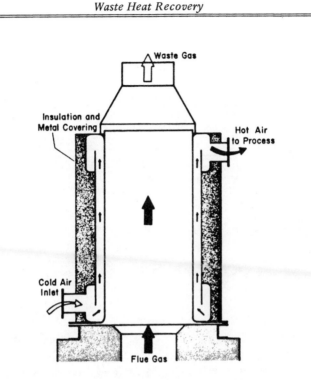

Figure 5-3. Diagram of Metallic Radiation Recuperator

since the outer shell is usually of a different and less expensive material. The mechanical design must take this effect into account. More elaborate designs of radiation recuperators incorporate two sections; the bottom operating in parallel flow and the upper section using the more efficient counterflow arrangement. Because of the large axial expansions experienced and the stress conditions at the bottom of the recuperator, the unit is often supported at the top by a free standing support frame with an expansion joint between the furnace and recuperator.

A second common configuration for recuperators is called the tube type or convective recuperator. As seen in the schematic diagram of Figure 5-4, the hot gases are carried through a number of

127

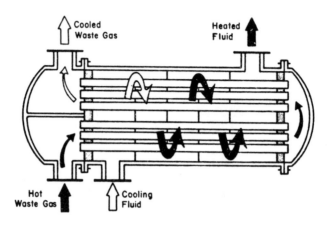

Figure 5-4. Diagram of Convective-Type Recuperator

parallel small diameter tubes, while the incoming air to be heated enters a shell surrounding the tubes and passes over the hot tubes one or more times in a direction normal to their axes.

If the tubes are baffled to allow the gas to pass over them twice, the heat exchanger is termed a two-pass recuperator; if two baffles are used, a three-pass recuperator, etc. Although baffling increases both the cost of the exchanger and the pressure drop in the combustion air path, it increases the effectiveness of heat exchange. Shell- and tube-type recuperators are generally more compact and have a higher effectiveness than radiation recuperators, because of the larger heat transfer area made possible through the use of multiple tubes and multiple passes of the gases.

The principal limitation on the heat recovery of metal recuperators is the reduced life of the liner at inlet temperatures exceeding 2000F. At this temperature, it is necessary to use the less efficient arrangement of parallel flows of exhaust gas and coolant in order to maintain sufficient cooling of the inner shell. In addition, when furnace combustion air flow is dropped back because of reduced load, the heat transfer rate from hot waste gases to preheat combustion air becomes excessive, causing rapid surface deterioration. Then, it is usually necessary to provide an ambient air by-pass to cool the exhaust gases.

In order to overcome the temperature limitations of metal recuperators, ceramic tube recuperators have been developed, whose materials allow operation on the gas side to 2800F and on the preheated air side to 2200F on an experimental basis, and to 1500F on a more or less practical basis. Early ceramic recuperators were built of tile and joined with furnace cement, and thermal cycling caused cracking of joints and rapid deterioration of the tubes. Later developments introduced various kinds of short silicon carbide tubes which can be joined by flexible seals located in the air headers. This kind of patented design illustrated in Figure 5-5 maintains the seals at comparatively low temperatures and has reduced the seal leakage rates to a few percent.

Earlier designs had experienced leakage rates from 8 to 60 percent. The new designs are reported to last two years with air preheat temperatures as high as 1300F, with much lower leakage rates.

An alternative arrangement for the convective type recuperator, in which the cold combustion air is heated in a bank of parallel vertical tubes which extend into the flue gas stream, is shown schematically in Figure 5-6. The advantage claimed for this arrangement is the ease of replacing individual tubes, which can be done during full capacity furnace operation. This minimizes the cost, the inconvenience, and possible furnace damage due to a shutdown forced by recuperator failure.

For maximum effectiveness of heat transfer, combinations of radiation type and convective type recuperators are used, with the convective type always following the high temperature radiation recuperator. A schematic diagram of this arrangement is seen in Figure 5-7.

Although the use of recuperators conserves fuel in industrial furnaces, and although their original cost is relatively modest, the purchase of the unit is often just the beginning of a somewhat more

129

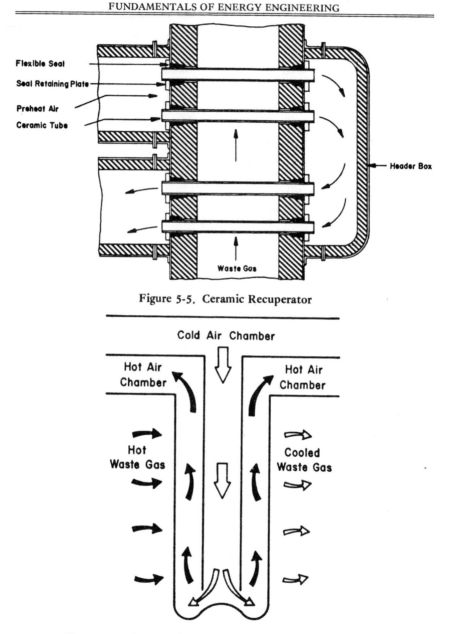

Figure 5-5. Ceramic Recuperator

Figure 5-6. Diagram of Vertical Tube-Within-Tube Recuperator

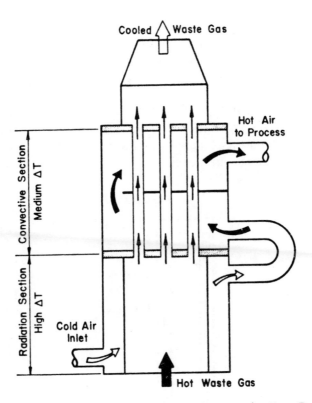

Figure 5-7. Diagram of Combined Radiation and Convective Type Recuperator

extensive capital improvement program. The use of a recuperator, which raises the temperature of the incoming combustion air, may require purchase of high temperature burners, larger diameter air lines with flexible fittings to allow for expansion, cold air lines for cooling the burners, modified combustion controls to maintain the required air/fuel ratio despite variable recuperator heating, stack dampers, cold air bleeds, controls to protect the recuperator during blower failure or power failures, and larger fans to overcome the additional pressure drop in the recuperator. It is vitally important to

131

protect the recuperator against damage due to excessive temperatures, since the cost of rebuilding a damaged recuperator may be as high as 90 percent of the initial cost of manufacture and the drop in efficiency of a damaged recuperator may easily increase fuel costs by 10 to 15 percent.

Figure 5-8 shows a schematic diagram of one radiant tube burner fitted with a radiation recuperator. With such a short stack, it is necessary to use two annuli for the incoming air to achieve reasonable heat exchange efficiencies.

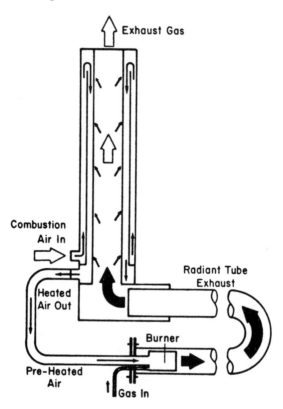

Figure 5-8. Diagram of a Small Radiation-Type Recuperator Fitted to a Radiant Tube Burner

132

Recuperators are used for recovering heat from exhaust gases to heat other gases in the medium to high temperature range. Some typical applications are in soaking ovens, annealing ovens, melting furnaces, afterburners and gas incinerators, radiant-tube burners, reheat furnaces, and other gas to gas waste heat recovery applications in the medium to high temperature range.

Heat Wheels

A rotary regenerator (also called an air preheater or a heat wheel) is finding increasing applications in low to medium temperature waste heat recovery. Figure 5-9 is a sketch illustrating the application of a heat wheel. It is a sizable porous disk, fabricated from some material having a fairly high heat capacity, which rotates between two side-by-side ducts; one a cold gas duct, the other a hot gas duct. The axis of the disk is located parallel to, and on the partition between the two ducts. As the disk slowly rotates, sensible heat (and in some cases, moisture containing latent heat) is transferred to the disk by the hot air and as the disk rotates, from the disk to the cold air. The overall efficiency of sensible heat transfer for this kind of regenerator can be as high as 85 percent. Heat wheels have been built as large as 70 feet in diameter with air capacities up to 40,000 ft^3/min. Multiple units can be used in parallel. This may help to prevent a mismatch between capacity requirements and the limited number of sizes available in packaged units. In very large installations such as those required for preheating combustion air in fixed station electrical generating stations, the units are custom designed.

The limitation on temperature range for the heat wheel is primarily due to mechanical difficulties introduced by uneven expansion of the rotating wheel when the temperature differences mean large differential expansion, causing excessive deformations of the wheel and thus difficulties in maintaining adequate air seals between duct and wheel.

Heat wheels are available in four types. The first consists of a metal frame packed with a core of knitted mesh stainless steel or

133

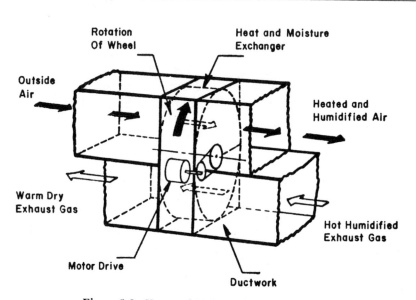

Figure 5-9. Heat and Moisture Recovery Using a
Heat Wheel Type Regenerator

aluminum wire, resembling that found in the common metallic kitchen pot scraper; the second, called a laminar wheel, is fabricated from corrugated metal and is composed of many parallel flow passages; the third variety is also a laminar wheel but is constructed from a ceramic matrix of honeycomb configuration. This type is used for higher temperature applications with a present-day limit of about 1600F. The fourth variety is of laminar construction but the flow passages are coated with a hygroscopic material so that latent heat may be recovered. The packing material of the hygroscopic wheel may be any of a number of materials. The hygroscopic material is often termed a dessicant.

Most industrial stack gases contain water vapor, since water vapor is a product of the combustion of all hydrocarbon fuels and since water is introduced into many industrial processes, and part of the process water evaporates as it is exposed to the hot gas stream. Each pound of water requires approximately 1000 Btu for its evaporation at atmospheric pressure, thus each pound of water vapor leaving in the exit stream will carry 1000 Btu of energy with it. This latent heat may be a substantial fraction of the sensible energy in the exit gas stream. A hygroscopic material is one such as lithium chloride (LiCl) which readily absorbs water vapor. Lithium chloride is a solid which absorbs water to form a hydrate, $LiCl \cdot H_2O$, in which one molecule of lithium chloride combines with one molecule of water. Thus, the ratio of water to lithium chloride in $LiCl \cdot H_2O$ is 3/7 by weight. In a hygroscopic heat wheel, the hot gas stream gives up part of its water vapor to the coating; the cool gases which enter the wheel to be heated are drier than those in the inlet duct and part of the absorbed water is given up to the incoming gas stream. The latent heat of the water adds directly to the total quantity of recovered waste heat. The efficiency of recovery of water vapor can be as high as 50 percent.

Since the pores of heat wheels carry a small amount of gas from the exhaust to the intake duct, cross contamination can result. If this contamination is undesirable, the carryover of exhaust gas can be partially eliminated by the addition of a purge section where a small amount of clean air is blown through the wheel and then exhausted to the atmosphere, thereby clearing the passages of exhaust gas. Figure 5-10 illustrates the features of an installation using a purge section. Note that additional seals are required to separate the purge ducts. Common practice is to use about six air changes of clean air for purging. This limits gas contamination to as little as 0.04 percent and particle contamination to less than 0.2 percent in laminar wheels, and cross contamination to less than 1 percent in packed wheels. If inlet gas temperature is to be held constant, regardless of heating loads and exhaust gas temperatures, then the heat wheel must be driven at variable speed. This requires a variable speed drive and a speed control system using an inlet air temperature sensor as the control element. This feature, however, adds considerably to the cost and complexity of the system. When operating with outside air in periods of high humidity and sub-zero temperatures, heat

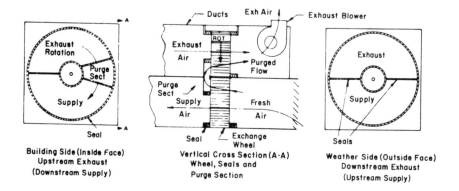

Figure 5-10. Heat Wheel Equipped with Purge Section to
Clear Contaminants from the Heat Transfer Surface

wheels may require preheat systems to prevent frost formation. When handling gases which contain water-soluble, greasy or adhesive contaminants or large concentrations of process dust, air filters may be required in the exhaust system upstream from the heat wheel.

One application of heat wheels is in space heating situations where unusually large quantities of ventilation air are required for health or safety reasons. As many as 20 or 30 air changes per hour may be required to remove toxic gases or to prevent the accumulation of explosive mixtures. Comfort heating for that quantity of ventilation air is frequently expensive enough to make the use of heat wheels economical. In the summer season the heat wheel can be used to cool the incoming air from the cold exhaust air, reducing the air conditioning load by as much as 50 percent. It should be pointed out that in many circumstances where large ventilating requirements are mandatory, a better solution than the installation of heat wheels may be the use of local ventilation systems to reduce the hazards and/or the use of infrared comfort heating at principal work areas.

Heat wheels are finding increasing use for process heat recovery in low and moderate temperature environments. Typical applications would be curing or drying ovens and air preheaters in all sizes for industrial and utility boilers.

136

Air Preheaters

Passive gas to gas regenerators, sometimes called air preheaters, are available for applications which cannot tolerate any cross contamination. They are constructed of alternate channels (see Figure 5-11) which put the flows of the heating and the heated gases in close contact with each other, separated only by a thin wall of conductive metal. They occupy more volume and are more expensive to construct than are heat wheels, since a much greater heat transfer surface area is required for the same efficiency. An advantage, besides the absence of cross-contamination, is the decreased mechanical complexity since no drive mechanism is required. However, it becomes more difficult to achieve temperature control with the passive regeneration and, if this is a requirement, some of the advantages of its basic simplicity are lost.

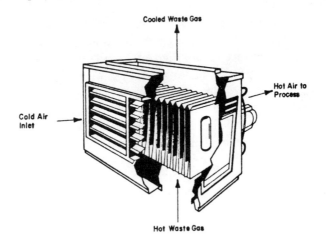

Figure 5-11. A Passive Gas to Gas Regenerator

Gas-to-gas regenerators are used for recovering heat from exhaust gases to heat other gases in the low to medium temperature range. A list of typical applications follows:

- Heat and moisture recovery from building heating and ventilation systems
- Heat and moisture recovery from moist rooms and swimming pools
- Reduction of building air conditioner loads
- Recovery of heat and water from wet industrial processes
- Heat recovery from steam boiler exhaust gases
- Heat recovery from gas and vapor incinerators
- Heat recovery from baking, drying, and curing ovens
- Heat recovery from gas turbine exhausts
- Heat recovery from other gas-to-gas applications in the low through high temperature range.

Heat-Pipe Exchangers

The heat pipe is a heat transfer element that has only recently become commercial, but it shows promise as an industrial waste heat recovery option because of its high efficiency and compact size. In use, it operates as a passive gas-to-gas finned-tube regenerator. As can be seen in Figure 5-12, the elements form a bundle of heat pipes which extend through the exhaust and inlet ducts in a pattern that resembles the structured finned coil heat exchangers. Each pipe, however, is a separate sealed element consisting of an annular wick on the inside of the full length of the tube, in which an appropriate heat transfer fluid is entrained.

Figure 5-13 shows how the heat absorbed from hot exhaust gases evaporates the entrained fluid, causing the vapor to collect in the center core. The latent heat of vaporization is carried in the vapor to the cold end of the heat pipe located in the cold gas duct. Here the vapor condenses giving up its latent heat. The condensed liquid is then carried by capillary (and/or gravity) action back to the hot end where it is recycled. The heat pipe is compact and efficient because: (1) the finned-tube bundle is inherently a good configuration for convective heat transfer in both gas ducts, and (2) the evapora-

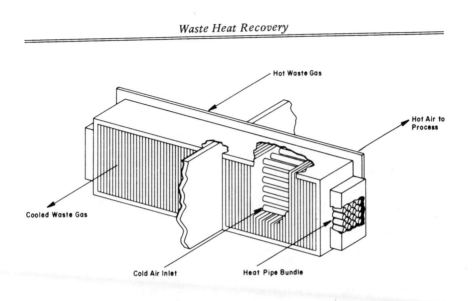

Hot Waste Gas

Hot Air to Process

Cooled Waste Gas

Cold Air Inlet

Heat Pipe Bundle

Figure 5-12. Heat Pipe Bundle Incorporated in Gas to Gas Regenerator

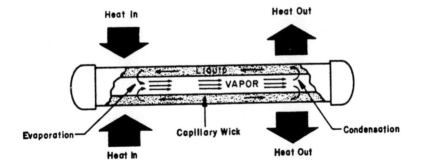

Heat In

Heat Out

LIQUID

VAPOR

Evaporation

Capillary Wick

Condensation

Heat In

Heat Out

Figure 5-13. Heat Pipe Schematic

139

tive-condensing cycle within the heat tubes is a highly efficient way of transferring the heat internally. It is also free from cross contamination. Possible applications include:

- Drying, curing and baking ovens
- Waste steam reclamation
- Air preheaters in steam boilers
- Air dryers
- Brick kilns (secondary recovery)
- Reverberatory furnaces (secondary recovery)
- Heating, ventilating and air conditioning systems

GAS OR LIQUID TO LIQUID REGENERATORS

Finned-Tube Heat Exchangers

When waste heat in exhaust gases is recovered for heating liquids for purposes such as providing domestic hot water, heating the feedwater for steam boilers, or for hot water space heating, the finned-tube heat exchanger is generally used. Round tubes are connected together in bundles to contain the heated liquid and fins are welded or otherwise attached to the outside of the tubes to provide additional surface area for removing the waste heat in the gases.

Figure 5-14 shows the usual arrangement for the finned-tube exchanger positioned in a duct and details of a typical finned-tube construction. This particular type of application is more commonly known as an economizer. The tubes are often connected all in series but can also be arranged in series-parallel bundles to control the liquid side pressure drop. The air side pressure drop is controlled by the spacing of the tubes and the number of rows of tubes within the duct.

Finned-tube exchangers are available prepackaged in modular sizes or can be made up to custom specifications very rapidly from

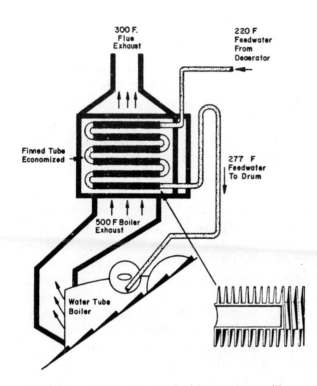

Figure 5-14. Finned-Tube Gas to Liquid Regenerator (Economizer)

standard components. Temperature control of the heated liquid is usually provided by a bypass duct arrangement which varies the flow rate of hot gases over the heat exchanger. Materials for the tubes and the fins can be selected to withstand corrosive liquids and/or corrosive exhaust gases.

Finned-tube heat exchangers are used to recover waste heat in the low to medium temperature range from exhaust gases for heating liquids. Typical applications are domestic hot water heating, heating boiler feedwater, hot water space heating, absorption-type refrigeration or air conditioning, and heating process liquids.

141

Shell and Tube Heat Exchanger

When the medium containing waste heat is a liquid or a vapor which heats another liquid, then the shell and tube heat exchanger must be used since both paths must be sealed to contain the pressures of their respective fluids. The shell contains the tube bundle, and usually internal baffles, to direct the fluid in the shell over the tubes in multiple passes. The shell is inherently weaker than the tubes so that the higher pressure fluid is circulated in the tubes while the lower pressure fluid flows through the shell. When a vapor contains the waste heat, it usually condenses, giving up its latent heat to the liquid being heated. In this application, the vapor is almost invariably contained within the shell. If the reverse is attempted, the condensation of vapors within small diameter parallel tubes causes flow instabilities. Tube and shell heat exchangers are available in a wide range of standard sizes with many combinations of materials for the tubes and shells.

Typical applications of shell and tube heat exchangers include heating liquids with the heat contained by condensates from refrigeration and air conditioning systems; condensate from process steam; coolants from furnace doors, grates, and pipe supports; coolants from engines, air compressors, bearings, and lubricants; and the condensates from distillation processes.

Waste Heat Boilers

Waste heat boilers are ordinarily water tube boilers in which the hot exhaust gases from gas turbines, incinerators, etc., pass over a number of parallel tubes containing water. The water is vaporized in the tubes and collected in a steam drum from which it is drawn off for use as heating or processing steam.

Figure 5-15 indicates one arrangement that is used, where the exhaust gases pass over the water tubes twice before they are exhausted to the air. Because the exhaust gases are usually in the medi-

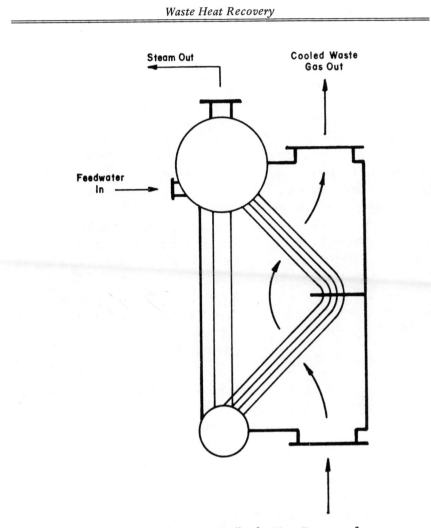

**Figure 5-15. Waste Heat Boiler for Heat Recovery from
Gas Turbines or Incinerators**

um temperature range and in order to conserve space, a more compact
boiler can be produced if the water tubes are finned in order to
increase the effective heat transfer area on the gas side. The diagram
shows a mud drum, a set of tubes over which the hot gases make a

143

double pass, and a steam drum which collects the steam generated above the water surface. The pressure at which the steam is generated and the rate of steam production depend on the temperature of the hot gases entering the boiler, the flow rate of the hot gases, and the efficiency of the boiler. The pressure of a pure vapor in the presence of its liquid is a function of the temperature of the liquid from which it is evaporated. The steam tables tabulate this relationship between saturation pressure and temperature. Should the waste heat in the exhaust gases be insufficient for generating the required amount of process steam, it is sometimes possible to add auxiliary burners which burn fuel in the waste heat boiler or to add an afterburner to the exhaust gas duct just ahead of the boiler. Waste heat boilers are built in capacities from less than a thousand to almost a million ft^3/min. of exhaust gas.

Typical applications of waste heat boilers are to recover energy from the exhausts of gas turbines, reciprocating engines, incinerators, and furnaces.

Gas and Vapor Expanders

Industrial steam and gas turbines are in an advanced state of development and readily available on a commercial basis. Recently special gas turbine designs for low pressure waste gases have become available; for example, a turbine is available for operation from the top gases of a blast furnace. In this case, as much as 20 MW of power could be generated, representing a recovery of 20 to 30 percent of the available energy of the furnace exhaust gas stream. Maximum top pressures are of the order of 40 lb/in^2 g.

Perhaps of greater applicability than the last example are steam turbines used for producing mechanical work or for driving electrical generators. After removing the necessary energy for doing work, the steam turbine exhausts partially spent steam at a lower pressure than the inlet pressure. The energy in the turbine exhaust stream can then be used for process heat in the usual ways. Steam turbines are classified as back-pressure turbines, available with allowable exit pressure operation above 400 lb/in²g, or condensing turbines which operate below atmospheric exit pressures. The steam used for driving the turbines can be generated in direct fired or waste heat boilers. A list of typical applications for gas and vapor expanders follows:

- Electrical power generation
- Compressor drives
- Pump drives
- Fan drives

Heat Pumps

In the commercial options previously discussed in this chapter, we find waste heat being transferred from a hot fluid to a fluid at a lower temperature. Heat must flow spontaneously "downhill"; that is, from a system at high temperature to one at a lower temperature. This can be expressed scientifically in a number of ways; all the variations of the statement of the second law of thermodynamics. The practical impact of these statements is that energy as it is transformed again and again and transferred from system to system, becomes less and less available for use. Eventually that energy has such low intensity (resides in a medium at such low temperature) that it is no longer available at all to perform a useful function. It has been taken as a general rule of thumb in industrial operations that fluids with temperatures less than 250F are of little value for waste heat extraction; flue gases should not be cooled below 250F (or, better, 300F to provide a safe margin), because of the risk of condensation of corrosive liquids. However, as fuel costs continue to rise, such waste heat can be used economically for space heating and other low temperature applications. It is possible to reverse the direction of spontaneous energy flow by the use of a thermodynamic system known as a heat pump.

This device consists of two heat exchangers, a compressor and an expansion device. A liquid or a mixture of liquid and vapor of a pure chemical species flows through an evaporator, where it absorbs

145

heat at low temperature and in doing so is completely vaporized. The low temperature vapor is compressed by a compressor which requires external work. The work done on the vapor raises its pressure and temperature to a level where its energy becomes available for use. The vapor flows through a condenser where it gives up its energy as it condenses to a liquid. The liquid is then expanded through a device back to the evaporator where the cycle repeats. The heat pump was developed as a space heating system where low temperature energy from the ambient air, water, or earth is raised to heating system temperatures by doing compression work with an electric motor-driven compressor. The performance of the heat pump is ordinarily described in terms of the coefficient of performance or COP, which is defined as:

$$COP = \frac{\text{Heat transferred in condenser}}{\text{Compressor work}} \qquad Formula\ (5\text{-}10)$$

which in an ideal heat pump is found as:

$$COP = \frac{T_H}{T_H - T_L} \qquad Formula\ (5\text{-}11)$$

where T_L is the temperature at which waste heat is extracted from the low temperature medium and T_H is the high temperature at which heat is given up by the pump as useful energy. The coefficient of performance expresses the economy of heat transfer.

In the past, the heat pump has not been applied generally to industrial applications. However, several manufacturers are now redeveloping their domestic heat pump systems as well as new equipment for industrial use. The best applications for the device in this new context are not yet clear, but it may well make possible the use of large quantities of low-grade waste heat with relatively small expenditures of work.

Table 5-4. Operation and Application Characteristics of Industrial Heat Exchangers

COMMERCIAL HEAT TRANSFER EQUIPMENT	Low Temperature Sub-Zero – 250°F	Intermediate Temp. 250°F – 1200°F	High Temperature 1200°F – 2000°F	Recovers Moisture	Large Temperature Differentials Permitted	Packaged Units Available	Can Be Retrofit	No Cross-Contamination	Compact Size	Gas-to-Gas Heat Exchange	Gas-to-Liquid Heat Exchanger	Liquid-to-Liquid Heat Exchanger	Corrosive Gases Permitted with Special Construction
Radiation Recuperator		●			●	1	●	●		●			●
Convection Recuperator		●	●		●	●	●	●		●			●
Metallic Heat Wheel	●	●		2		●	●	3	●	●			●
Hygroscopic Heat Wheel	●			●		●	●	3	●	●			
Ceramic Heat Wheel		●	●		●	●	●		●	●			●
Passive Regenerator	●	●			●	●	●	●		●			●
Finned-Tube Heat Exchanger	●	●			●	●	●	●	●		●		4
Tube Shell-and-Tube Exchanger	●	●			●	●	●	●	●		●	●	
Waste Heat Boilers	●	●			●	●	●				●		4
Heat Pipes	●	●			5	●	●	●	●	●			●

1. Off-the-shelf items available in small capacities only.
2. Controversial subject. Some authorities claim moisture recovery. Do not advise depending on it.
3. With a purge section added, cross-contamination can be limited to less than 1% by mass.
4. Can be constructed of corrosion-resistant materials, but consider possible extensive damage to equipment caused by leaks or tube ruptures.
5. Allowable temperatures and temperature differential limited by the phase equilibrium properties of the internal fluid.

Summary

Table 5-4 presents the collation of a number of significant attributes of the most common types of industrial heat exchangers in matrix form. This matrix allows rapid comparisons to be made in selecting competing types of heat exchangers. The characteristics given in the table for each type of heat exchanger are: allowable temperature range, ability to transfer moisture, ability to withstand large temperature differentials, availability as packaged units, suitability for retrofitting, and compactness and the allowable combinations of heat transfer fluids.

6

Utility System Optimization

BASIS OF THERMODYNAMICS

Thermodynamics deals with the relationships between heat and work. It is based on two basic laws of nature; the first and second laws of thermodynamics. The principles are used in the design of equipment such as steam engines, turbines, pumps, and refrigerators, and in practically every process involving a flow of heat or a chemical equilibrium.

First Law: The first law states that energy can neither be created nor destroyed, thus, it is referred to as the law of conservation of energy. Formula 6-1 expresses the first law for the steady state condition.

$$E_2 - E_1 = Q - W \qquad \text{Formula (6-1)}$$

where

$E_2 - E_1$ is the change in stored energy at the boundary states 1 and 2 of the system

Q is the heat added to the system

W is the work done by the system.

Figure 6-1 illustrates a thermodynamic process where mass enters and leaves the system. The potential energy (Z) and the kinetic

149

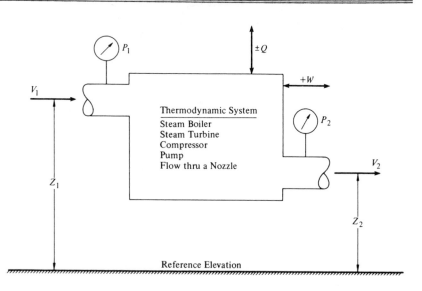

Figure 6-1. System Illustrating Conservation of Energy

energy $(V^2/64.2)$ plus the enthalpy represents the stored energy of the mass. Note, Z is the elevation above the reference point in feet, and V is the velocity of the mass in ft/sec. In the case of the steam turbine, the change in Z, V, and Q are small in comparison to the change in enthalpy. Thus, the energy equation reduces to

$$W/778 = h_1 - h_2 \qquad \textit{Formula (6-2)}$$

where

W is the work done in ft · lb/lb
h_1 is the enthalpy of the entering steam, Btu/lb
h_2 is the enthalpy of the exhaust steam, Btu/lb
And 1 Btu equals 778 ft · lb.

Second Law: The second law qualifies the first law by discussing the conversion between heat and work. All forms of energy, including work, can be converted to heat, but the converse is not generally true. The Kelvin-Planck statement of the second law of thermody-

150

namics says essentially the following: Only a portion of the heat from a heat work cycle, such as a steam power plant, can be converted to work. The remaining heat must be rejected as heat to a sink of lower temperature; to the atmosphere, for instance.

The Clausis statement, which also deals with the second law, states that heat, in the absence of some form of external assistance, can only flow from a hotter to a colder body.

THE CARNOT CYCLE

The Carnot cycle is of interest because it is used as a comparison of the efficiency of equipment performance. The Carnot cycle offers the maximum thermal efficiency attainable between any given temperatures of heat source and sink. A thermodynamic cycle is a series of processes forming a closed curve on any system of thermodynamic coordinates. The Carnot cycle is illustrated on a temperature-entropy diagram, Figure 6-2A, and on the Mollier Diagram for superheated steam, Figure 6-2B.

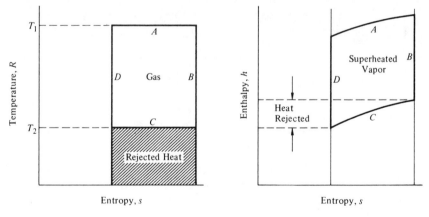

Figure 6-2A. Temperature–entropy   Figure 6-2B. Mollier diagram;
diagram; gas.   superheated vapor.

Figure 6-2. Carnot Cycles

151

The cycle consists of the following:

1. Heat addition at constant temperature, resulting in expansion work and changes in enthalpy.
2. Adiabatic isentropic expansion (change in entropy is zero) with expansion work and an equivalent decrease in enthalpy.
3. Constant temperature heat rejection to the surroundings, equal to the compression work and any changes in enthalpy.
4. Adiabatic isentropic compression returning to the starting temperature with compression work and an equivalent increase in enthalpy.

The Carnot cycle is an example of a reversible process and has no counterpart in practice. Nevertheless, this cycle illustrates the principles of thermodynamics. The thermal efficiency for the Carnot cycle is illustrated by Formula (6-3):

$$\text{Thermal efficiency} = \frac{T_1 - T_2}{T_1} \qquad \textit{Formula (6-3)}$$

where

T_1 = Absolute temperature of heat source, °R (Rankine)
T_2 = Absolute temperature of heat sink, °R

and absolute temperature is given by Formula 6-4.

Absolute temperature = 460 + temperature in Fahrenheit.

Formula (6-4)

T_2 is usually based on atmospheric temperature, which is taken as 500°R.

PROPERTIES OF STEAM PRESSURE
AND TEMPERATURE

Water boils at 212°F when it is in an open vessel under atmospheric pressure equal to 14.7 psia (pounds per square inch, absolute). Absolute pressure is the amount of pressure exerted by a system on its boundaries and is used to differentiate it from gage pressure. A pressure gage indicates the difference between the pressure of the system and atmospheric pressure.

$$\text{psia} = \text{psig} + \text{atmospheric pressure in psia} \qquad \textit{Formula (6-5)}$$

Changing the pressure of water changes the boiling temperature. Thus, water can be vaporized at 170°F, at 300°F, or any other temperature, as long as the applied pressure corresponds to that boiling point.

SOLID, LIQUID AND VAPOR STATES OF A LIQUID

Water, as well as other liquids, can exist in three states: solid, liquid, and vapor. In order to change the state from ice to water or from water to steam, heat must be added. The heat required to change a solid to liquid is called the *latent heat of fusion.* The heat required to change a liquid to a vapor is called the *latent heat of vaporization.*

In condensing steam, heat must be removed. The quantity is exactly equal to the latent heat that went into the water to change it to steam.

Heat supplied to a fluid, during the change of state to a vapor, will not cause the temperature to rise; thus, it is referred to as the latent heat of vaporization. Heat given off by a substance when it condenses from steam to a liquid is called *sensible* heat. Physical properties of water, such as the latent heat of vaporization, also change with variations in pressure.

USE OF THE STEAM TABLES

Steam properties are illustrated in Chapter 16 by Tables 16-13, 16-14 and 16-15 (ASME Steam Tables). Table 16-13 is referred to as the steam table for saturated steam. Steam properties are shown and correlated to temperature. Table 16-14 is another form of the steam table, where steam properties are shown and correlated to pressure. (Pressure must be converted to psia.) The properties of superheated steam are indicated by Table 16-15.

The term hf_g is the latent heat or enthalpy of vaporization. Thus, from Table 16-14, at atmospheric pressure the latent heat of vaporization is 970.3 Btu/lb. At 200 psia the latent heat of vaporization is 842.8 Btu/lb.

The enthalpy hf represents the amount of heat required to raise one pound of water from 32°F to a liquid state at another temperature. As an example, from Table 16-13, to raise water from 32°F to 170°F will require 137.9 Btu/lb.

Example Problem 6-1

How much heat is required to raise 100 pounds of water at 126°F to 170°F?

Answer

From Table 16-13:

At 126°F — hf = 93.9 Btu/lb
At 170°F — hf = 137.9 Btu/lb
$Q = 100(137.9 - 93.9) = 44 \times 10^2$ Btu.

USE OF THE SPECIFIC HEAT CONCEPT

Another physical property of a material is the *specific heat.* The specific heat is defined as the amount of heat in Btu required to raise one pound of a substance one degree F. For water, it can be seen from the previous examples that one Btu of heat is required to raise one lb water 1°F; thus, the specific heat of water Cp = 1. Specific heats for other materials are illustrated in Table 6-1. This leads to two equations.

$$Q = wCp\ \Delta T \hspace{3cm} Formula\ (6\text{-}6)$$

154

Table 6-1. Specific Heat of Various Substances

SUBSTANCE	SPECIFIC HEAT	SUBSTANCE	SPECIFIC HEAT
SOLIDS		LIQUIDS	
ALUMINUM.	0.230	ALCOHOL....	0.600
ASBESTOS	0.195	AMMONIA........	1.100
BRASS....	0.086	BRINE, CALCIUM (20% SOLUTION).	0.730
BRICK....	0.220	BRINE, SODIUM (20% SOLUTION)	0.810
BRONZE...	0.086	CARBON TETRACHLORIDE.......	0.200
CHALK......	0.215	CHLOROFORM...............	0.230
CONCRETE...	0.270	ETHER........	0.530
COPPER......	0.093	GASOLINE....	0.700
CORK...........	0.485	GLYCERINE.	0.576
GLASS, CROWN......	0.161	KEROSENE....	0.500
GLASS, FLINT............	0.117	MACHINE OIL.	0.400
GLASS, THERMOMETER...	0.199	MERCURY.....	0.033
GOLD................	0.030	PETROLEUM......	0.500
GRANITE......	0.192	SULPHURIC ACID..	0.336
GYPSUM.	0.259	TURPENTINE....	0.470
ICE.............	0.480	WATER.......	1.000
IRON, CAST......	0.130	WATER, SEA..	0.940
IRON, WROUGHT..	0.114		
LEAD.............	0.031	GASES	
LEATHER........	0.360	AIR......	0.240
LIMESTONE.....	0.216	AMMONIA...	0.520
MARBLE........	0.210	BROMINE......	0.056
MONEL METAL..	0.128	CARBON DIOXIDE....	0.200
PORCELAIN....	0.255	CARBON MONOXIDE.	0.243
RUBBER........	0.481	CHLOROFORM.......	0.144
SILVER..	0.055	ETHER..........	0.428
STEEL...	0.118	HYDROGEN...	3.410
TIN.....	0.045	METHANE...	0.593
WOOD.	0.330	NITROGEN.....	0.240
ZINC...	0.092	OXYGEN............	0.220
		SULPHUR DIOXIDE......	0.154
		STEAM (SUPERHEATED, 1 PSI)....	0.450

Reprinted by permission of The Trane Company.

where

Q = quantity of heat, Btu
w = weight of substance, lb
Cp = specific heat of substance, Btu per lb $^\circ$F
ΔT = temperature change of substance $^\circ$F

$$q = MCp \, \Delta T \qquad \qquad \text{\textit{Formula (6-7)}}$$

where

Q = quantity of heat, Btu/hr (Btu/hr is sometimes abbreviated as Btuh)
M = flow rate, lbs/hr
Cp = specific heat, Btu per lb $^\circ$F
ΔT = temperature change of substance$^\circ$F

155

Sample Problem 6-2

Check answer to Sample Problem 6-1 using Formula 6-6.

Answer

$$Q = WCp\ \Delta T$$
$$= 100 \times 1 \times (170 - 126) = 44 \times 10^2 \text{ Btu.}$$

The saturated vapor enthalpy h_g represents the amount of heat necessary to change water at $32°F$ to steam at a specified temperature and pressure.

Sample Problem 6-3

How much heat is required to raise 100 pounds of water at $126°F$ to 15 psig steam?

Answer

At $126°F - h_f = 93.9$ Btu/lb
15 psig corresponds to 30 psia at $250°F$
At 30 psia, $h_g = 1164.1$ Btu/lb
$Q = 100(1164.1 - 93.9) = 1070.2\ 10^2$ Btu.

USE OF THE LATENT HEAT CONCEPT

When water is raised from $32°F$ to $212°F$ steam, the total heat required is comprised of two components:

(a) The heat required to raise the temperature of water from $32°F$ to $212°F$; $h_f = 180.17$ Btu/lb from Table 16-13.
(b) The heat required to evaporate the water at $212°F$; $h_{fg} = 970.3$ Btu/lb from Table 16-13.

Thus, $h_g = h_f + h_{fg} = 180.17 + 970.3 = 1150.5$ Btu/lb, which agrees with the value of h_g found in Steam Table 16-13.

Example Problem 6-4

30 psig steam is used for a heat exchanger and returns to the system as 30 psig condensate. What amount of heat is given off to the process fluid?

Answer

30 psig = 45 psia. This corresponds to an h_{fg} of approximately 928 Btu/lb.

THE USE OF THE SPECIFIC VOLUME CONCEPT

Another property of water is the specific volume v_f of water and the specific volume v_g of steam. *Specific volume* is defined as the space occupied by one pound of a material. For example, at 50 psia, water occupies 0.01727 ft^3 per lb and steam occupies 8.515 ft^3 per lb.

The *specific weight* is simply the weight of one cubic foot of a material and is the reciprocal of the specific volume.

THE MOLLIER DIAGRAM

A visual tool for understanding and using the properties of steam is illustrated by the Mollier Diagram, Figure 6-3. The Mollier Diagram enables one to find the relationship between temperature, pressure, enthalpy, and entropy, for steam. Constant temperature and pressure curves illustrate the effect of various processes on steam.

For a constant temperature process (isothermal), the change in entropy is equal to the heat added (or subtracted) divided by the temperature at which the process is carried out. This is a simple way of explaining the physical meaning of entropy. The *change* in entropy is of interest. Referring to Table 16-13, the value of entropy at 32°F is zero. Increases in entropy are a measure of the portion of heat in a process which is unavailable for conversion to work. Entropy

MOLLIER CHART

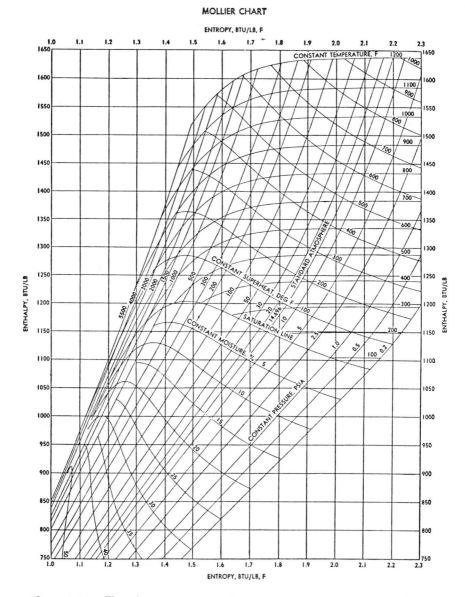

Figure 6-3. Mollier Diagram *(Courtesy Babcock & Wilcox Company and ASME)*

has a close relationship to the second law of thermodynamics, discussed later in this chapter.

Another item of interest from the Mollier Diagram is the saturation line. The saturation line indicates temperature and pressure relationships corresponding to saturated steam. Below this curve, steam contains a % moisture, as indicated by the % moisture curves. Steam at temperatures above the saturation curve is referred to as superheated. As an example, this chart indicates that at 212°F and 14.696 psia, the enthalpy h_g is 1150 Btu/lb, which agrees with Steam Table 16-14.

SUPERHEATED STEAM

If additional heat is added to raise the temperature of steam over the point at which it was evaporated, the steam is termed superheated. Thus, steam at the same temperature as boiling water is saturated steam. Steam at temperatures higher than boiling water, at the same pressure, is superheated steam.

Example Problem 6-5

Using the Mollier Diagram, find the enthalpy of steam at 14.696 psia and 300°F.

Answer

Follow the constant pressure line until it intersects the 300°F curve. The answer to the right is 1192 Btu/lb.

In addition to the Mollier Diagram, the Superheated Steam Table 16-15 is helpful.

Steam cannot be superheated in the presence of water because the heat supplied will only evaporate the water. Thus, the water will be evaporated prior to becoming superheated. Superheated steam is condensed by first cooling it down to the boiling point corresponding to its pressure. When the steam has been de-superheated, further removal of heat will result in its condensation.

159

In the generation of power, superheated steam has many uses.

HEAT BALANCE

A heat balance is an analysis of a process which shows where all the heat comes from and where it goes. This is a vital tool in assessing the profit implications of heat losses and proposed waste heat utilization projects. The heat balance for a steam boiler, process furnace, air conditioner, etc., must be derived from measurements made during actual operating periods. Chapter 3 provides information on the instrumentation available to make these measurements. The measurements that are needed to get a complete heat balance involve: energy inputs, energy losses to the environment, and energy discharges.

Energy Input

Energy enters most process equipment either as chemical energy in the form of fossil fuels, of sensible enthalpy of fluid streams, of latent heat in vapor streams, or as electrical energy.

For each input it is necessary to meter the quantity of fluid flowing or the electrical current. This means that if accurate results are to be obtained, submetering for each flow is required (unless all other equipment served by a main meter can be shut down so that the main meter can be used to measure the inlet flow to the unit). It is not necessary to continuously submeter every flow since temporary installations can provide sufficient information. In the case of furnaces and boilers that use pressure ratio combustion controls, the control flow meters can be utilized to yield the correct information. It should also be pointed out that for furnaces and boilers only the fuel need be metered. Tests of the exhaust products provide sufficient information to derive the oxidant (usually air) flow if accurate fuel flow data are available. For electrical energy inflows, the current is measured with an ammeter, or a kilowatt hour meter may be installed as a submeter. Ammeters using split core transformers are available for measuring alternating current flow without opening the line. These are particularly convenient for temporary installations.

160

In addition to measuring the flow for each inlet stream it is necessary to know the chemical composition of the stream. For air, water, and other pure substances no tests for composition are required, but for fossil fuels the composition must be determined by chemical analysis or secured from the fuel supplier. For vapors one should know the quality—this is the mass fraction of vapor present in the mixture of vapor and droplets. Measurement of quality is made with a vapor calorimeter which requires only a small sample of the vapor stream.

Other measurements that are required are the entering temperatures of the inlet stream of fluid and the voltages of the electrical energy entering (unless kilowatt-hour meters are used).

The testing routines discussed above involve a good deal of time, trouble, and expense. However, they are necessary for accurate analyses and may constitute the critical element in the engineering and economic analyses required to support decisions to expend capital on waste heat recovery equipment.

Energy Losses

Energy loss from process equipment to the ambient environment is usually by radiative and convective heat transfer. Radiant heat transfer, that is, heat transfer by light or other electromagnetic radiation, is discussed in the section of Chapter 3 dealing with infrared thermography. Convective heat transfer, which takes place by hot gas at the surface of the hot material being displaced by cooler gas, may be analyzed using Newton's law of cooling.

$$Q = UA \ (T_s - T_o) \qquad\qquad Formula \ (6\text{-}8)$$

where

Q = rate of heat loss in energy units Btu/h
U = heat transfer coefficient in Btu/h$\cdot$ft$^2 \cdot$F
A = area of surface losing heat in ft^2
T_s = surface temperature
T_o = ambient temperature

161

Although heat flux meters are available, it is usually easier to measure the quantities above and derive the heat loss from the equations. The problems encountered in using the equation involve the measurement of surface temperatures and the finding of accurate values for the heat transfer coefficient.

Unfortunately the temperature distribution over the surface of a process unit can be very nonuniform so that an estimate of the overall average is quite difficult. New infrared measurement techniques, which are discussed in Chapter 3, make the determination somewhat more accurate. The heat transfer coefficient is not only a strong function of surface and ambient temperatures but also depends on geometric considerations and surface conditions. Thus for given surface and ambient temperatures a flat vertical plate will have a different h_{cr} value than will a horizontal or inclined plate.

Energy Discharges

The composition, discharge rate and temperature of each outflow from the process unit are required in order to complete the heat balance. For a fuel-fired unit, only the composition of the exhaust products, the flue gas temperature and the fuel input rate to the unit are required to derive:

(1) air input rate
(2) exhaust gas flow rate
(3) energy discharge rate from exhaust stack

The composition of the exhaust products can be determined from an Orsat analysis, a chromatographic test, or less accurately from a determination of the volumetric fraction of oxygen or CO_2. Figure 6-4 can be used for determining the quantity of excess or deficiency of air in the combustible mixture. It is based on the fact that chemical reactions occur with fixed ratios of reactants to form given products.

For example, natural gas with the following composition:

CO_2 — 0.7% volume
O_2 — 0.0

162

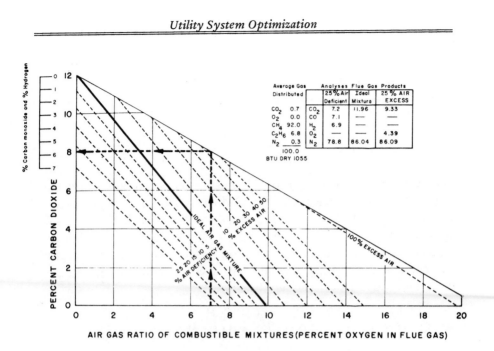

Figure 6-1. Natural Gas Combustion Chart

$$CH_4 - 92.0$$
$$C_2H_6 - 6.8$$
$$N_2 - 0.5$$
$$\overline{100.00}$$

is burned to completion with the theoretical amount of air indicated in the volume equation below:

FUEL: $0.92\ CH_4 + 0.068\ C_2H_6$
$(1\ ft^3)$ $+ 0.007\ CO_2 + 0.005\ N_2$
plus
AIR: $2.078\ O_2 + 7.813\ N_2 \rightarrow$
$(8.891\ ft^3)$
yields
DRY PRODUCTS: $1.063\ CO_2 + 7.818\ N_2$
        $+ 2.044\ H_2O$ *Formula (6-9)*

163

The equation is based upon the laws of conservation of mass and elemental chemical species. The ratio of N_2 and O_2 in the combustible mixture comes about from the approximate volumetric ratio of N_2 to O_2 in air, i.e.,

$$20.9\% \ O_2, 79.1\% \ N_2 \text{ or } \frac{79.1}{20.9} = 3.76 = \frac{\text{Volume } N_2}{\text{Volume } O_2}.$$

For gases the coefficients of the chemical equation represent relative volumes of each species reacting. Ordinarily excess air is provided to the fuel so that every fuel molecule will react with the necessary number of oxygen molecules even though the physical mixing process is imperfect. If 10 percent excess air were supplied, this mixture of reactants and products would give a chemical equation appropriately modified as given below:

FUEL: $0.92 \ CH_4 + 0.068 \ C_2H_6$
(1 ft^3) $+ 0.007 \ CO_2 + 0.005 \ N_2$
plus
AIR: $2.86 \ O_2 + 8.594 \ N_2 \rightarrow$
(10.880 ft^3)
yields
DRY PRODUCTS: $1.063 \ CO_2 + 0.208 \ O_2 + 8.599 \ N_2$
                    $+ 2.044 \ H_2O$ *Formula (6-10)*

where an additional term representing the excess oxygen appears in the products, along with a corresponding increase in the nitrogen.

As an example let us assume that an oxygen meter has indicated a reading of 7% for the products of combustion from the natural gas whose composition was given previously and that the exhaust gas temperature was measured as 700°F. Figure 6-4 is used as indicated to determine that 45% excess air is mixed with fuel. The combustion equation then becomes:

FUEL: $0.92 \ CH_4 + 0.068 \ C_2H_6$
(1 ft^3) $+ 0.007 \ CO_2 + 0.005 \ N_2$
plus

AIR: $3.013\ O_2 + 11.329\ N_2 \rightarrow$

yields

DRY PRODUCTS: $1.063\ CO_2 + 0.935\ O_2 + 1.334\ N_2$
($15.376\ ft^3$) $+ 2.044\ H_2O$

For each $1\ ft^3$ of fuel, $14.342\ ft^3$ of air is supplied and $15{,}376\ ft^3$ of exhaust products (at mixture temperature) are formed. Each cubic foot of fuel contains 1055 Btu of energy, so the fuel energy input is $250{,}000 \times 1055 = 263{,}750{,}000$ Btu/h. From Figure 6-5 we compute the exhaust gas losses at $700°F$ as:

CO_2: $250{,}000 \times 1.063 \times 17.5\ = 4{,}651{,}000$ Btu/h
H_2O: $250{,}000 \times 2.044 \times 14.0\ = 7{,}154{,}000$
O_2: $250{,}000 \times 0.935 \times 12.3\ = 2{,}875{,}000$
N_2: $250{,}000 \times 11.334 \times 11.5\ = \underline{32{,}585{,}000}$
$\phantom{HN_2: 250{,}000 \times 11.334 \times 11.5\ = }\ 47{,}265{,}000$ Btu/h

or 18% of the fuel energy supplied. Some of this could be recovered by suitable waste heat equipment.

Heat Balance on a Boiler

Let us consider a further example, a heat balance on a boiler. A process steam boiler has the following specifications:

- Natural gas fuel with HHV = $1001.2\ Btu/ft^3$
- Gas firing rate = $2126.5\ ft^3/min.$
- Steam discharge at $150\ lb/in^2\,g$ saturated
- Steam capacity of $100{,}000\ lb/h$
- Condensate returned at $180°F$

The heat balance on the burner is derived from measurements made *after* the burner controls had been adjusted for an optimum air/fuel ratio corresponding to 10% excess air. All values of the heat content of the fluid streams are referred to a base temperature of $60°F$. Consequently the computations for each fluid stream entering or leaving the boiler are made by use of the equation below:

$$\dot{H} = \dot{m}\ (h-h_o) \qquad\qquad Formula\ (6\text{-}11)$$

165

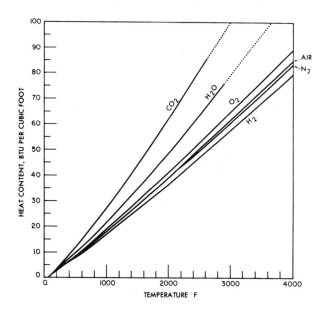

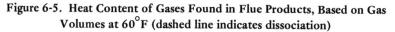

Figure 6-5. Heat Content of Gases Found in Flue Products, Based on Gas Volumes at $60°F$ (dashed line indicates dissociation)

where

$\dot{H}$ = the enthalpy rate for entrance or exit fluids

$\dot{m}$ = mass flow rates for entrance or exit fluids

h = specific enthalpy at the fluid temperature of the fluid entering or leaving the entrance or exit

h_o = specific enthalpy of that fluid at the reference temperature $T_o = 60°F$

The first law of thermodynamics for the boiler is expressed as: Sum of all the enthalpy rates of substances entering = Sum of all enthalpy rates of substances leaving + q where q is the rate of heat loss to the surroundings. This can be expressed as

$$\Sigma_{in}H_i = \Sigma_{out}H_i + q \qquad \textit{Formula (6-12)}$$

where Σ is the summation sign, and H_i is the enthalpy rate of substance i.

166

For gaseous fuels the computation for the heat content of the gases is more conveniently expressed in the form:

$$H = J_o C_{pm} \ (T - T_o)$$
Formula (6-13)

where

J_o = the volume rate of the gas stream corrected back to 1.0 atmosphere and 60° F (T_o = 60° F); and

C_{pm} = the specific heat given on the basis of a standard volume of gas averaged over the temperature range ($T - T_o$) and the gas mixture components.

$$C_{pm} = \Sigma_i \chi_i C_{pi}$$
Formula (6-14)

where

χ_i = percent by volume of a component in one of the flow paths.

C_{pi} = average specific heat over temperature range for each component.

From Formula 6-10 we derive the volume fractions of each gas component as follows:

Component	χ
CO_2	8.9
H_2O	17.2
N_2	72.2
O_2	1.7
	100.0

The average specific heat is found (using Figure 6-6).

$$
\begin{aligned}
C_{pm} = \ &0.89 \ \times \ 0.0275 = 0.00245 \\
&0.172 \times \ 0.0220 = 0.00378 \\
&0.722 \times \ 0.0186 = 0.01344 \\
&0.017 \times \ 0.0195 = \underline{0.00033} \\
& 0.02 \ \text{Btu/Scf} \cdot \text{F}
\end{aligned}
$$

167

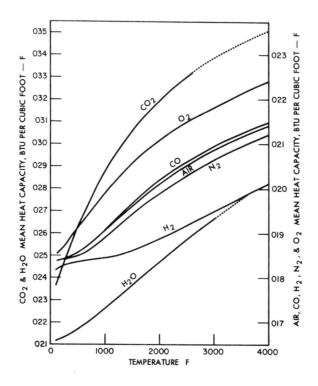

**Figure 6-6. Mean Heat Capacity of Gases Found in Flue Products from 60 to
T F (dashed line indicates dissociation)**

The combustion equation also tells that when the products are at standard conditions 11.88 ft³ fuel and air generate 11.915 ft³ of products and that for every cubic foot of gas burned, 10.880 ft³ of air is introduced. Thus for a firing rate of 2126.5 ft³/min. the air required is 2126.5 × 10.88 = 23,136 ft³/min. or 23,136 × 60 = 1,388,179 ft³/h. The total fuel and air flow rate is then almost exactly equal to 1,388,179 + 2126.5 × 60 = 1,515,769 ft³/h. This corresponds to a flue gas discharge rate of 1,519,128 ft³/h.

For a flue gas discharge rate of 1,519,128 ft³/h and a temperature of 702°F, the total exhaust heat rate is found as:

$$\dot{H}_{\text{EX GAS}} = 1,519,128 \,\frac{ft^3}{h} \times 0.02 \,\frac{Btu}{ft^3 \cdot F}$$
$$\times \,(702{-}60)F = 19,505,604 \text{ Btu/h}$$

For the steam leaving the boiler (100,000 lb/h at 150 lb/in^{2}g saturated) the energy flow rate is found using the Steam Tables in Chapter 16 and the equation below:

$$
\begin{aligned}
\dot{H}_{\text{Steam}} &= \dot{m}\,(h{-}h_o)\\
&= 100,000 \text{ lb/h }(1195.6{-}28.08) \text{ Btu/lb}\\
&= 116,752,000 \text{ Btu/h}
\end{aligned}
$$

where 1195.6 is the specific enthalpy of saturated steam at 150 lb/in^{2}g and 28.08 is the specific enthalpy of saturated liquid water at 60°F since we are using that temperature as our standard reference temperature for the heat balance. We have used 60°F as a reference temperature, this is not universal practice and in the boiler industry 70°F is more common, whereas in other areas 25°C is normal.

CHEMICAL ENERGY IN FUEL

To determine the heat content of the chemical energy in fuel, find the higher heating value (HHV) for the fuel and multiply it by the volumetric flow rate for a gaseous fuel or the mas flow rate for a liquid or solid fuel. The assumed higher heating value for the natural gas used in the boiler of our example is 1001.1 Btu/ft^3 and the heat content rate is then:

$$\dot{H} = 2126.5 \,\frac{ft^3}{min.} \times 60 \,\frac{min.}{h} \times 1001.2 \,\frac{Btu}{ft^3}$$
$$= 127,743,000 \,\frac{Btu}{h}$$

The enthalpy rates for the condensate return and make-up water are derived from data in the steam tables where the specific enthalpy of the compressed liquids are taken to be almost exactly equal to the specific enthalpy of the *saturated* liquid found at the same *temperature.*

169

The complete heat balance derived in the manner detailed above is presented in Figure 6-7.

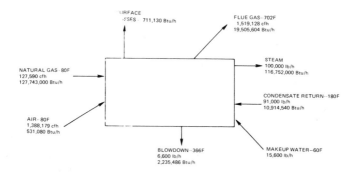

Figure 6-7. Heat Balance for a Simple Steam Generator Burning Natural Gas with 10% Excess Air

Waste heat is available from the combustion products leaving the stack. It amounts to 19,505,604 Btu/h at a temperature of 702°F. Some energy is also available from the condensers.

There may be a practical limit to the amount of heat that can be removed from the stack gases, however. Figure 6-8 gives the relationship between the percent excess air and the dew point temperature (temperature at which the water vapor in the exhaust products will condense) for the case we are considering.

If the water vapor, which is formed by burning the hydrogen in the fuel, is cooled to the point where it condenses, then part of the condensed liquid may collect on the stack liner and cause rusting or corrosion. This corrosion may increase maintenance and replacement costs for the stack to the point where additional waste heat recovery made possible by reduction of temperatures below the saturation temperature becomes uneconomical.

STACK GAS CONDENSATION

To use Figure 6-8 for the illustrative example, enter the graph at 1001.2 Btu/ft³ along one of the bottom abscissa and run a vertical

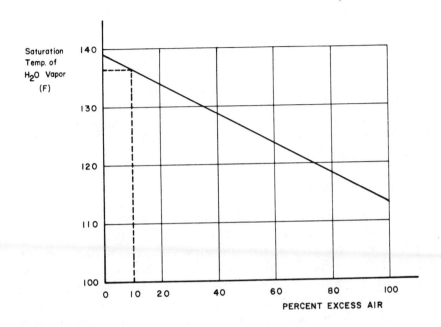

Figure 6-8. Saturation Temperature of Water Vapor in Exhaust Products from a Gas Fired Boiler as a Function of Percent Excess Air

line to the gaseous fuel curve. From the point of intersection run a horizontal line to the left ordinate and then follow the curve intersecting the ordinate at that point downward and to the right until an intersection with a vertical line representing the percent excess air is made. From that point move horizontally to the left and read the dew point temperature. For the present example enter at 1001.2 Btu/ft³, follow a vertical to the intersection of the gaseous fuel curve and then horizontally to the left to read 140°F. Follow the intersecting curve downward to the right until it intersects with the vertical from the top ordinate representing 10% excess air. Move horizontally to the left and read 137°F.

Thus the condensation temperature for the exhaust products from this boiler are 137°F when operated with 10% excess dry air.

Since the stack liner may exist at a temperature 50 to 75°F lower than the average gas stream temperature, it is prudent in this example to limit the lowest stack gas temperature to at least 212°F. The effect of water vapor in the entering air on this condensation temperature is small. If the combustion air entering the boiler were at a relative humidity of 100% at 80°F, then the saturation temperature would increase only to 140°F and the minimum stack temperature then must be held above 214°F. As mentioned before, 250 or 300°F is better for safety.

WASTE HEAT RECOVERY

The energy exhausted to the atmosphere should not be discarded. A portion of it can be recovered by using a heat exchanger. Any requirement for energy at a temperature in excess of 200°F can be satisfied. It is necessary to identify the prospective uses for the waste energy; make an economic analysis of the costs and savings involved in each of the options; and decide among those options on the basis of the economics of each. An important option in every case is that of rejecting all options if none proves economic. For the illustrative example, let us assume that the following uses for the waste heat have been identified.

- Preheating the combustion air
- Preheating the boiler feedwater
- Heating the domestic hot water supply (370 gal./h from 50 to 170°F)
- A combination of the preceding

The first two are practices which are standard in energy-intensive high technology industries (e.g., electric companies).

Two rapid calculations give the waste heat available in the exhaust gases between 702°F and 220°F and the heat requirements for the domestic hot water supply to be 14,063,000 Btu/h and 369,900 Btu/h respectively. Since the latter constitutes only 2.6% of the available waste energy, we should reject that option. The remaining options are to preheat the combustion air, the feedwater, or to divide the waste energy between those two options. For a small boiler one probably would not find the purchase of two separate heat exchangers an economic option, so that we shall limit ourselves to one of the other of the first two options. Without going into a detailed analysis, we may note that preheating the combustion air or the feedwater provides a double benefit. Since the preheated air or water requires less fuel to produce the same steam capacity, a direct fuel saving results. But the smaller quantity of fuel means a smaller air requirement, and this in turn means a smaller quantity of exhaust products and thus smaller stack loss at a given stack temperature.

The economic benefits can be estimated as follows. The air preheater is estimated to save 6% of the fuel. With an average boiler loading of 60% for the 8760 h in a year this amounts to an annual fuel saving of

$$0.06 \times 8760 \times 127,590 = 40,103,000 \ \frac{ft^3}{yr}$$

which is worth, at an average rate of $5.50/1000 ft³, an annual dollar saving of

$$\frac{40,103,000}{1000} \times 5.50 = \$220,000$$

If the costs for installing the air preheater are assumed to be

Cost of preheater	$ 52,000
Cost of installation	57,200
New burners, air piping, controls and fan	56,600
	$165,800

For the feedwater heater (or economizer) the fuel savings is 9.2% or a total annual fuel saving of:

$$0.092 \times 127,600 \times 8760 \times 0.60 = 61,696,000 \ \frac{ft^3}{yr}$$

and the economic benefit is

$$\frac{61,696,000 \, \dfrac{ft^3}{yr}}{1000} \times \$5.50 = \$340,000/yr$$

The cost of the economizer installed is estimated to be $134,000. This is clearly the best option for this particular boiler, especially as there is no need in this case for modifications to the boiler and accessories beyond the heat exchanger retrofit. There are several reasons for its superiority. The first is that in the case of air preheating one is exchanging the waste heat in the gases to the incoming air which has almost the same mass flow rate and almost the same specific heat. Thus one can expect that the final temperature of the preheater air will be almost the arithmetic mean of the ambient air temperature and the exhaust gas temperature entering the economizer. In the case of the feedwater the mass flow rate times specific heat is over four and one-quarter times that of the combustion air. Thus we can expect to transfer more energy to the water and end up with a lower flue gas temperature leaving the stack.

The second reason is that preheating the air quite often (as in this case) affects the boiler accessories which requires additional modifications and thus related capital expenditures.

In the preceding example of the process steam boiler, we are analyzing an efficient process unit. The heat available in the product (process steam) constituted a large percentage of the energy introduced in the fuel. The efficiency in percentage terms is computed as

$$\eta = \frac{\text{Useful output}}{\text{Energy input}} = \frac{Q_{steam}}{Q_{fuel}} \times 100$$

$$\eta = \frac{100,000 \text{ lb/h} \, (1195.6 - 148) \times 100}{127,590 \text{ ft}^3/\text{h} \times 1001.2 \text{ Btu/ft}^3}$$

$$= \frac{1.0476 \times 10^{10}}{1.2775 \times 10^8} = 0.82, \text{ or } 82\%$$

HEAT RECOVERY IN
STEEL TUBE FURNACE

As a second and very different example note the steel tube furnace illustrated below in Figure 6-9.

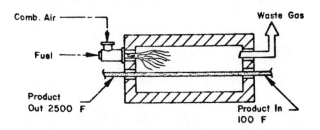

Figure 6-9. Continuous Steel Tube Furnace

The tubing enters the furnace from the right at a temperature of 100°F.

The specifications for the steel tube heating furnace are:

Product capacity — 50 ton/h
Product specifications — 0.23% carbon steel
Final product temperature — 2000°F
Air/Fuel inlet temperature — 100°F
Air/Fuel mixture —10% excess air
Fuel — No. 5 fuel oil gravity API° 16
Fuel firing rate 48.71 gpm at 240°F (factory usage)
Utilization factor — 0.62

The useful heat leaves the furnace in the steel at 2000°F. This is called the useful furnace output and equals:

$$Q_{prod} = \dot{m}_{prod} C_P (T_{out} - T_{in})$$

$$= 50 \frac{ton}{h} \times 2200 \, lb/ton$$

$$\times 0.179 \, Btu/lb - F(200-100)$$

$$= 0.3741 \times 10^8 \frac{Btu}{h}$$

175

0.179 is used as the average specific heat of steel over the 100°F–2000°F range. The heat input to the furnace is the chemical energy in the fuel oil. The heating value for No. 5 fuel oil is found from Table 6-2.

Table 6-2. Physical Properties of Fuel Oil at 60°F

Fuel oil (CS-12-48) Grade No.	Grav-ity, API	Sp gr	Lb per gal	Btu per lb	Net Btu per gal
6	3	1.0520	8.76	18,190	152,100
6	4	1.0443	8.69	18,240	151,300
6	5	1.0366	8.63	18,290	149,400
6	6	1.0291	8.57	18,340	148,800
6	7	1.0217	8.50	18,390	148,100
6	8	1.0143	8.44	18,440	147,500
6	9	1.0071	8.39	18,490	146,900
6	10	1.0000	8.33	18,540	146,200
6	11	0.9930	8.27	18,590	145,600
6	12	.9861	8.22	18,640	144,900
6, 5	14	.9725	8.10	18,740	143,600
6, 5	16	.9593	7.99	18,840	142,300
5	18	.9465	7.89	18,930	140,900
4, 5	20	.9340	7.78	19,020	139,600
4, 5	22	.9218	7.68	19,110	138,300
4, 5	24	.9100	7.58	19,190	137,100
4, 2	26	.8984	7.49	19,270	135,800
4, 2	28	.8871	7.39	19,350	134,600
2	30	.8762	7.30	19,420	133,300
2	32	.8654	7.21	19,490	132,100
2	34	.8550	7.12	19,560	130,900
1, 2	36	.8448	7.04	19,620	129,700
1, 2	38	.8348	6.96	19,680	128,500
1	40	.8251	6.87	19,750	127,300
1	42	0.8156	6.79	19,810	126,200

The relation between specific gravity and degrees API is expressed by the formula:

$$\frac{141.5}{131.5 + API} = \text{sp gr at 60 F.}$$

For each 10 F above 60 F add 0.7 API.
For each 10 F below 60 F subtract 9.7 AP

176

The heat input is determined by:

$$Q_{fuel} = 48.71 \text{ gpm} \times 142,300 \frac{\text{Btu}}{\text{gal.}}$$
$$= 4.159 \times 10^8 \text{ Btu/h}$$

The percent efficiency of the furnace is:

$$\eta = \frac{\text{output} \times 100}{\text{input}}$$

$$= \frac{\text{Enthalpy added to steel} \times 100}{\text{Enthalpy entering with fuel}}$$

$$= \frac{0.3741 \times 10^8}{4.159 \times 10^8} \times 100 = 9\%$$

This means that of the 415.9 MBtu introduced per hour to the furnace, that 378.4 million are released to the atmosphere. At the present average cost of $2/gal. for No. 5 fuel oil, the heat wasted is

$$Q_{waste} = 48.71 \text{ gpm} \times (1-0.09)\frac{\text{gal wasted}}{\text{gal used}}$$
$$\times 60 \text{ min./h} \times \$2/\text{gal.} = \$5318/\text{h}$$
$$= \$46,586/\text{yr.}$$

In order to construct the combustion equation we must note from Table 6-3 that the carbon-hydrogen ratio is about 7.3. Thus:

$$C_{6.08}H_{10} + 9.44\,O_2 + 35.49\,N_2 \rightarrow 6.08\,CO_2$$
$$+ 5\,H_2O + 0.86\,O_2 + 35.49\,N_2.$$

Again referring to Table 6-2 the density of the liquid fuel is 7.99 lb/gal. Therefore the above equation represents the reaction for 100 lb fuel/7.99 lb/gal. = 12.51 gal. and for 35.49 + 9.44 or 44.93 lb mol of air. 44.93 lb mol of air weighs 1301 lb (using the molecular weight for air found in Table 6-4). The specific volume of air at standard conditions is found from the same table as 378.5 ft^3/lb mol. Therefore the volume of air required to burn 12.51 gal. of fuel is

177

Table 6-3. Typical Properties of Commercial Petroleum Products Sold in the East and Midwest

	Premium Fuels					Fuel Oils				
	Gasolines		No. 1 fuel, light diesel, or stove oil	Diesel or "gas house" gas oil	Reduced crude, heavy gas oil or premium residuum*	No. 2	No. 3	Cold No. 5	No. 5	Bunker C or No. 6
	12 psia Reid vapor pressure natural gasoline	Straight run gasoline								
Gravity, °API	79	63	38–42	35–38	19–22	32–35	28–32	20–25	16–24	6–14
Viscosity	...	...	34–36†	34–36†	15–50‡	34–36†	35–45†	<20‡	20–40‡	50–300‡
Conradson carbon, wt %	None	None	Trace	Trace	3–7	Trace	<0.15	1–2	2–4	6–15
Pour point, °F	<0	<0	<0	<10	...	<10	<20	<15	15–60	>50
Sulfur, wt %	<0.1	<0.2	<0.1	0.1–0.6	<2.0	0.1–0.6	0.2–1.0	0.5–2.0	0.5–2.0	1–4
Water and sed., wt %	None	None	Trace	Trace	<1.0	<0.05	<0.10	0.1–0.5	0.5–1.0	0.5–2.0
Distillation, °F										
10%	110	150	390–410	400–440	550–650	420–440	450–500	<600	<600	600–700
50%	150	230	440–460	490–510	800–900	490–530	540–560	...	...	850–950
90%	235	335	490–520	580–600	...	580–620	600–670	<700	>700	...
End point	315	370	510–560	630–660	...	630–660	650–700	...	...	...
Flash point, °F	<100	<100	100–140	110–170	>150	130–170	>130	>130	>130	>150
Aniline point, °F	145	125	140–160	150–170	...	120–140	120–140	...	...	...
C to H ratio	5.2	5.6	6.1–6.4	6.2–6.5	6.9–7.3	6.6–7.1	7.0–7.3	7.3–7.7	7.3–7.7	7.7–9.0
Average MBtu recovered/gal of oil:										
Carbureted water gas	101	101	102	101	93	91	86	82	83	74
High-Btu oil gas	90	89	90	90	82	80	76	72	73	64
Tar + carbon, wt % of oil										
Carbureted water gas	...	...	20	22	35	31	36	42	42	52
High-Btu oil gas	...	...	32	34	45	42	46	52	52	60

(Data on gasolines and data below C/H ratio not given in reports.)
* The 6 wt % Conradson carbon oil used in the Hall High Btu Oil Gas Tests (A.G.A. Gas Production Research Committee. Hall High Btu Oil Gas Process, New York, 1949.) and "New England Gas Enriching Oil" are typical examples of this group.
† Saybolt Universal seconds at 100 F.
‡ Saybolt Furol seconds at 122 F.
Remarks:
No. 1 Fuel Oil—A distillate oil intended for vaporizing pot-type burners. A volatile fuel.
No. 2 Fuel Oil—For general purpose domestic heating; for use in burners not requiring No. 1 oil. Moderately volatile.
No. 3 Fuel Oil—Formerly a distillate oil for use in burners requiring low viscosity oil. Now incorporated as a part of No. 2.
No. 4 Fuel Oil—For burner installations not equipped with preheating facilities.
No. 5 Fuel Oil—A residual type oil. Requires preheating to 170–220 F.
No. 6 Fuel Oil—Preheating to 220–260 F suggested. A high viscosity oil.

$44.93 \times 378.5 = 17,000$ ft³ at standard temperature and pressure. The air input is then:

$$m = \frac{17,000 \text{ ft}^3}{12.51 \text{ gal.}} \cdot 48.71 \text{ gpm} = 66,200 \text{ ft}^3/\text{m}$$

or

$$m_{air} = 66,200 \times 60 = 3,972,000 \text{ ft}^3/\text{h}$$
$$Q_{air} = 1.2 \text{ Btu/ft}^3 \times 3,972,000$$
$$= 4,766,000 \text{ Btu/h}$$

The enthalpy of the entering air is 1.2 Btu/ft³ at 100°F.

178

Table 6-4. Gas Constants and Volume of the Pound-Mol for Certain Gases

	$R' = R/M$, specific gas constant, ft per F	M, mol wt, lb per lb-mol	R, universal gas constant, ft-lb per (lb-mol)(F)	Mv, cu ft per lb-mol *
Hydrogen	767.04	2.016	1546	378.9
Oxygen	48.24	32.000	1544	378.2
Nitrogen	55.13	28.016	1545	378.3
Nitrogen, "atmospheric" †	54.85	28.161	1545	378.6
Air	53.33	28.966	1545	378.5
Water vapor	85.72	18.016	1544	378.6
Carbon dioxide	34.87	44.010	1535	376.2
Carbon monoxide	55.14	28.010	1544	378.3
Hydrogen sulfide	44.79	34.076	1526	374.1
Sulfur dioxide	23.56	64.060	1509	369.6
Ammonia	89.42	17.032	1523	373.5
Methane	96.18	16.042	1543	378.2
Ethane	50.82	30.068	1528	374.5
Propane	34.13	44.094	1505	368.7
n-Butane	25.57	58.120	1486	364.3
iso-Butane	25.79	58.120	1499	367.4
Ethylene	54.70	28.052	1534	376.2
Propylene	36.01	42.078	1515	379.1

* At 60 F, 30 in. Hg, dry.
† Includes other inert gases in trace amounts.

The volume ratio of stack gases at standard conditions to air inlet rate is found from the combustion equation as:

$$\frac{6.08 + 5 + 0.86 + 35.49}{9.44 + 35.49} = \frac{47.43}{44.93} = 1.056$$

$$Q_{ex} = 1.056 \times 3,972,000$$
$$= 4,184,000 \text{ ft}^3/\text{h}$$

The volume fractions of the stack gas components are found from the combustion equation as:

	χ
CO_2	0.128
H_2O	0.106
O_2	0.018
N_2	0.748
	1.000

The specific enthalpy of the stack gas components may be calculated by use of Figure 6-5 for the exhaust temperature of 2200°F leading to the average specific enthalpy as indicated below:

$$
\begin{array}{llllll}
 & & & & & h \\
CO_2 & 0.128 & \times & 69.3 & = & 8.870 \\
H_2O & 0.106 & \times & 54.0 & = & 5.724 \\
O_2 & 0.018 & \times & 45.4 & = & 0.817 \\
N_2 & 0.748 & \times & 43.4 & = & \underline{32.463} \\
 & & & & & 47.874
\end{array}
$$

Therefore

$$
\begin{aligned}
Q_{ex} &= 4{,}194{,}000 \ ft^3/h \times 47{,}874 \ \frac{Btu}{ft^3} \\
&= 200{,}784{,}000 \ \frac{Btu}{h}.
\end{aligned}
$$

The heat losses from the surface equal the fuel enthalpy input less the exhaust gas enthalpy and the produce enthalpy.

$$
\begin{aligned}
Q &= 415{,}880{,}000 + 4{,}766{,}000 - 200{,}784{,}000 \\
&- 37{,}410{,}000 = 182{,}451{,}270 \ \frac{Btu}{h}.
\end{aligned}
$$

The complete heat balance diagram is shown in Figure 6-10. Now suppose that we install a metallic recuperator and preheat the combustion air to 750°F. The new heat balance is shown in Figure 6-11.

The economic analysis of this situation is as follows.

The waste heat saved is equal to the fuel savings times the heating value of the fuel.

$$
\begin{aligned}
Q_{saved} &= (48.71 - 31.66) \ gpm \times 142{,}300 \ \frac{Btu}{gal.} \\
&= 145{,}554{,}000 \ or \ 35\% \ saving
\end{aligned}
$$

which equals

$$
0.35 \times 48.71 \ gpm \times 60 \times \$2/gal. = \$2{,}045.82/h.
$$

In using Figures 6-12 and 6-13 remember that the combustion air volume (in SCF) is reduced directly proportional to the reduction

180

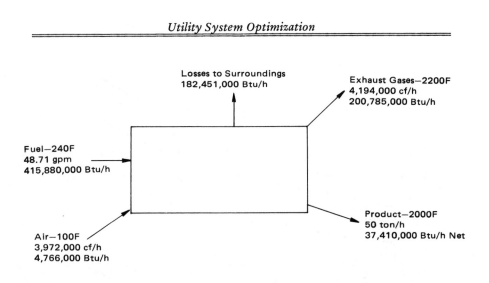

Figure 6-10. Heat Balance for a Simple Continuous Steel Tube Furnace

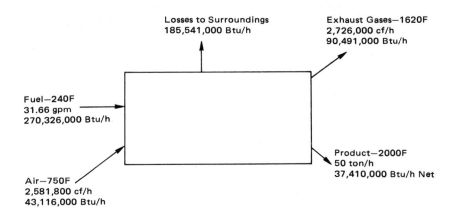

Figure 6-11. Heat Balance for a Continuous Steel Tube Furnace Equipped
with a Recuperator Preheating Combustion Air to 750°F

181

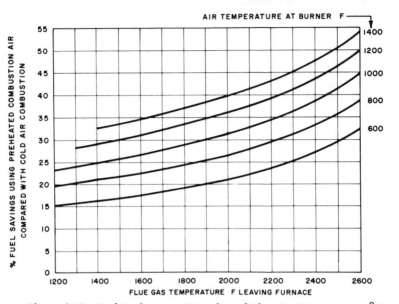

Figure 6-12. Fuel Savings as a Function of Flue Gas Temperature °F Leaving Furnace

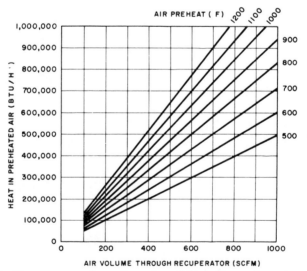

Figure 6-13. Heat Recovered as a Function of Air Volume through Recuperator (SCFM)

182

in firing rate. If we maintain the same air/fuel ratio for the furnace (10% excess air), the air rate is reduced from the initial value of 3,972,000 ft^3/h at a firing rate of 48.71 gpm to a rate of 31.66 gpm/ 48.71 gpm $\times$ 3,972,000 ft^3/h = 2,581,800 ft^3/h with the lower firing rate of 31.66 gpm when the recuperator is used.

If as in our example the furnace is used 0.62 $\times$ 8760 or 5431 h/ yr then the annual saving is 5431 $\times$ $2,045.80 = $11.10 $\times$ 10^6/yr. Since a metallic recuperator can be purchased for a fraction of the savings, the savings will allow a payoff in less than a year.

BOILERS

Boiler Configurations and Components

Industrial boiler designs are influenced by fuel characteristics and firing method, steam demand, steam pressures, firing characteristics and the individual manufacturers. Industrial boilers can be classified as either firetube or watertube indicating the relative position of the hot combustion gases with respect to the fluid being heated.

Firetube Boilers

Firetube units pass the hot products of combustion through tubes submerged in the boiler water. Conventional units generally employ from 2 to 4 passes to increase the surface area exposed to the hot gases and thereby increase efficiency. Multiple passes, however, require greater fan power, increased boiler complexity and larger shell dimensions. (Refer to Figure 6-14.) Maximum capacity of firetube units is currently limited to 25,000 lbs of steam per hour (750 boiler hp) with an operating pressure of 250 psi due to economic factors related to material strength and thickness.

Advantages of firetube units include: (1) ability to meet wide and sudden load fluctuations with only slight pressure changes, (2) low initial costs and maintenance, and (3) simple foundation and installation procedures.

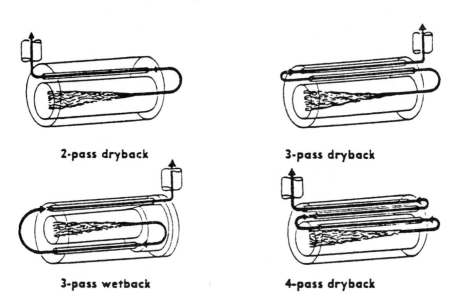

2-pass dryback   **3-pass dryback**

3-pass wetback   **4-pass dryback**

Figure 6-14. Typical Firetube Boiler Gas Flow Patterns

Watertube Boilers

Watertube units circulate the boiler water inside the tubes and the flue gases outside. Water circulation is generally provided by the density variation between cold feed water and the hot water/steam mixture in the riser as illustrated in Figure 6-15. Watertube boilers may be subclassified into different groups by tube shape, by drum number and location and by capacity. Refer to Figure 6-16.

Another important determination is "field" versus "shop" erected units. Many engineers feel that shop assembled boilers can meet closer tolerance than field assembled units and therefore may be more efficient; however, this has not been fully substantiated. Watertube units range in size from as small as 1000 lbs of steam per hour to the giant utility boilers in the 1000 MW class. The largest industrial boilers are generally taken to be about 500,000 lbs of steam per hour. Important elements of a steam generator include the firing mechanism, the furnace water walls, the superheaters, convec-

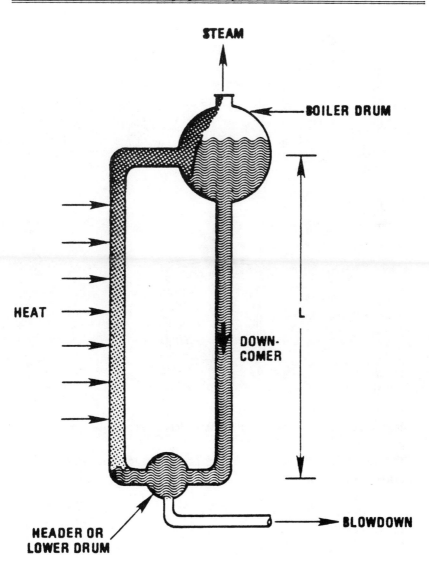

Figure 6-15. Water Circulation Pattern in a Watertube Boiler

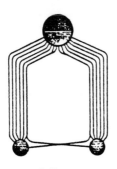

A Type

D Type

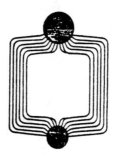

O Type

Figure 6-16. Classification of Watertube Boilers by Basic Tube Arrangement

tive regions, the economizer and air preheater and the associated ash and dust collectors.

FUEL HANDLING AND
FIRING SYSTEMS

Gas Fired

Natural gas fuel is the simplest fuel to burn in that it requires little preparation and mixes readily with the combustion air supply. Industrial boilers generally use low-pressure burners operating at a

186

pressure of 1/8 to 4 psi. Gas is generally introduced at the burner through several orifices that generate gaseous jets that mix rapidly with the incoming combustion air supply. There are many designs in use that differ primarily in the orientation of the burner orifices and their locations in the burner housing.

Oil Fired

Oil fuels generally require some type of pretreatment prior to delivery to the burner including the use of strainers to remove solid foreign material and tank and flow line preheaters to assure the proper viscosity. Oil must be atomized prior to vaporization and mixing with the combustion air supply. This generally requires the use of either air, steam or mechanical atomizers. The oil is introduced into the furnace through a gun fitted with a tip that distributes the oil into a fine spray that allows mixing between the oil droplets and the combustion air supply. Oil cups that spin the oil into a fine mist are also employed on small units. An oil burner may be equipped with diffusers that act as flame holders by inducing strong recirculation patterns near the burner. In some burners, primary air nozzles are employed.

Pulverized Coal Fired

The pulverizer system provides four functions: pulverizing, drying, classifying to the required fineness and transporting the coal to the burner's main air stream. The furnace may be designed for dry ash removal in the hopper bottom or for molten ash removal as in a slag tap furnace. The furnace is dependent on the burning and ash characteristics of the coal as well as the firing system and type of furnace bottom. The primary objectives are to control furnace ash deposits and provide sufficient cooling of the gases leaving the furnace to reduce the buildup of slag in the convective regions. Pulverized coal fired systems are generally considered to be economical for units with capacities in excess of 200,000 lbs of steam per hour.

Stoker Fired

Coal stoker units are characterized by bed combustion on the boiler grate with the bulk of the combustion air supplied through the grate. Several stoker firing methods currently in use on industrial sized boilers include underfed, overfed and spreader. In underfed and overfed stokers, the coal is transferred directly on to the burning bed. In a spreader stoker the coal is hurled into the furnace when it is partially burned in suspension before lighting on the grate. Several grate configurations can be used with overfed and spreader stokers including stationary, chain, traveling, dumping and vibrating grates. Each grate configuration has its own requirements as to coal fineness and ash characteristics for optimum operation. Spreader stoker units have the advantage that they can burn a wide variety of fuels including waste products. Underfed and overfed units have the disadvantage that they are relatively slow to respond to load variations. Stoker units can be designed for a wide range of capacities from 2,000 to 350,000 lbs of steam per hour. Spreader stoker units are generally equipped with overfire air jets to induce turbulence for improved mixing and combustible burnout. Stoker units are also equipped with ash reinjection systems that allow the ash collected that contains a significant portion of unburned carbon to be reintroduced into the furnace for burning.

COMBUSTION CONTROL SYSTEMS

Combustion controls have two purposes: (1) maintain constant steam conditions under varying loads by adjusting fuel flow, and (2) maintain an appropriate combustion air-to-fuel flow.

Classification

Combustion control systems can be classified as series, parallel and series/parallel.

In series control, either the fuel or air is monitored and the other is adjusted accordingly. For parallel control systems, changes in steam conditions results in a change in both air and fuel flow. In series/parallel systems, variations in steam pressure affect the rate of fuel input and simultaneously the combustion air flow is controlled by the steam flow.

Combustion controls can be also classified as positioning and metering controls.

Positioning controls respond to system demands by moving to a present position. In metering systems, the response is controlled by actual measurements of the fuel and/or air flows.

Application

The application and degree of combustion controls varies with the boiler size and is dictated by system costs. The parallel positioning jackshaft system has been extensively applied to industrial boilers based on minimum system costs. The combustion control responds to changes in steam pressure and can be controlled by a manual override. The control linkage and cam positions for the fuel and air flow are generally calibrated on startup.

Improved control of excess air can be obtained by substituting electric or pneumatic systems for the mechanical linkages. In addition, relative position of fuel control and combustion air dampers can be modified. More advanced systems are pressure ratio control of the fuel and air pressure, direct air and fuel metering and excess air correction systems using flue gas O_2 monitoring. Factors that have limited the application of the most sophisticated control systems to industrial boilers include cost, reliability and maintenance.

SAFETY

It is essential that energy engineers conducting boiler evaluation tests and tune-ups understand and be aware of boiler safety devices. Occasionally operators who don't understand safety have been known to bypass safety features in order to keep a unit operating.

Summary of requirements found in the NATIONAL BOILER SAFETY CODES:

189

* PREPURGE—4 to 8 air changes to insure no fuel vapors or gases remain in the boiler which could ignite or explode when the pilot light is off.

* PILOT PROVING—10 to 15 second proving period pilot must ignite and be proven before the main fuel valves open.

* MAIN FLAME TRIAL—usually 10 to 15 second trial period for natural gas and oil after main fuel valves open.

* FLAME FAILURE RESPONSE TIME—usually 3 to 4 seconds after the main flame goes out the Fuel Shut Off valves should automatically close to prevent a build up of explosive vapor concentrations within the boiler.

* INTERLOCKS—automatically shut the boiler down if certain safety features are not in the safe condition.

> Loss of atomizing means (steam/air)
> High/Low gas pressure
> Low fire for light off
> Low fuel oil temperature
> Low fuel oil pressure
> Main combustion air
> Low water level

* FUEL VALVES—Safety Shut Off Valves (SSOV). Usually two valves; closing time 1 to 5 seconds depending on fuel and size of boiler.

* FLAME DETECTORS—should be positioned and capable of detecting flame only and not sparks from the spark ignitor or hot refractory, so as not to give a false flame signal.

OPERATING GUIDES

Consult the plant engineer and the boiler manufacturers technical manuals for complete boiler operating procedures. Typical items to check before conducting efficiency checks are as follows:

Oil Burners

Make sure the atomizer is of the proper design and size and the burner is centered with dimensions according to manufacturer's drawings.

Inspect oil-tip passages and orifices for wear (use proper size drill as a feeler gage), and remove any coke or gum deposits to assure the proper oil-spray pattern.

Verify proper oil pressure and temperature at the burner.

Verify proper atomizing—steam pressure.

Make sure that the burner diffuser (impeller) is not damaged, and is properly located to the oil-gun tip.

Check to see that the oil gun is positioned properly within the burner throat, and that the throat refractory is in good condition.

Gas Burners

Inspect gas-ingestion orifices and verify that all passages are unobstructed. Also, be sure filters and moisture traps are in place, clean and operating properly, to prevent plugging of gas orifices.

Confirm proper location and orientation of diffusers, spuds etc. Look for any burned off or missing burner parts.

Combustion Controls

Inspect all fuel valves to verify proper movement, clean valve internal surfaces if necessary.

Eliminate "play" in control linkages on dampers. Any play no matter how slight will cause a loss of efficiency as a precise tune-up will be impossible.

Make sure fuel-supply inlet pressure to pressure regulators are high enough to assure constant regulator-outlet pressures for all firing rates.

Correct any control elements that fail to respond smoothly to varying steam demand. Unnecessary hunting caused by improperly adjusted regulators or automatic master controllers can waste fuel.

Check that all gages are functioning and are calibrated.

Furnace

Inspect boiler furnace and gas side surfaces for excessive deposits and fouling. These lead to higher stack temperatures and lower boiler efficiencies.

Inspect furnace refractory and insulation for cracks that may cause leaks and missing refractory.

Clean furnace inspection parts and make sure that burner throat, furnace walls, and leading connection passes are visible through them as flame observation is an essential part of efficient boiler operation and testing.

BOILER EFFICIENCY IMPROVEMENT

The boiler plant should be designed and operated to produce the maximum amount of usable heat from a given amount of fuel.

Combustion is a chemical reaction of fuel and oxygen which produces heat. Oxygen is obtained from the input air which also contains nitrogen. Nitrogen is useless to the combustion process. The carbon in the fuel can combine with air to form either CO or CO_2. Incomplete combustion can be recognized by a low CO_2 and high CO content in the stack. Excess air causes more fuel to be burned than required. Stack losses are increased and more fuel is needed to raise ambient air to stack temperatures. On the other hand, if insufficient air is supplied, incomplete combustion occurs and the flame temperature is lowered.

Boiler Efficiency

Boiler efficiency (E) is defined as:

$$\%E = \frac{\text{Heat out of Boiler}}{\text{Heat supplied to Boiler}} \times 100 \qquad \textit{Formula (6-15)}$$

For steam-generating boiler:

$$\%E = \frac{\text{Evaporation Ratio} \times \text{Heat Content of Steam}}{\text{Calorific Value of Fuel}} \times 100 \qquad \textit{Formula (6-16)}$$

For hot water boilers:

$$\%E = \frac{\text{Rate of Flow from Boiler} \times \text{Heat Output of Water}}{\text{Caloric Value of Fuel} \times \text{Fuel Rate}} \times 100$$

Formula (6-17)

The relationship between steam produced and fuel used is called the evaporation ratio.

Boilers are usually designed to operate at the maximum efficiency when running at rated output. Figure 6-17 illustrates boiler efficiency as a function of time on line.

Full boiler capacity for heating occurs only a small amount of the time. On the other hand, part loading of 60% or less occurs approximately 90% of the time.

Where the present boiler plant has deteriorated, consideration should be given to replacing with modular boilers sized to meet the heating load.

The overall thermal efficiency of the boiler and the various losses of efficiency of the system are summarized in Figure 6-18.

To calculate dry flue gas loss, Formula 6-18 is used.

$$\text{Flue gas loss} = \frac{K(T-t)}{CO_2} \qquad \textit{Formula (6-18)}$$

where

K = constant for type of fuel = 0.39 Coke
= 0.37 Anthracite
= 0.34 Bituminous Coal
= 0.33 Coal Tar Fuel
= 0.31 Fuel Oil

T = temperature of flue gases in $°F$
t = temperature of air supply to furnace in $°F$
CO_2 = percentage CO_2 content of flue gas measured volumetrically.

It should be noted that this formula does not apply to the combustion of any gaseous fuels, such as natural gas, propane, butane, etc. Basic combustion formulas or nomograms should be used in the gaseous fuel case.

193

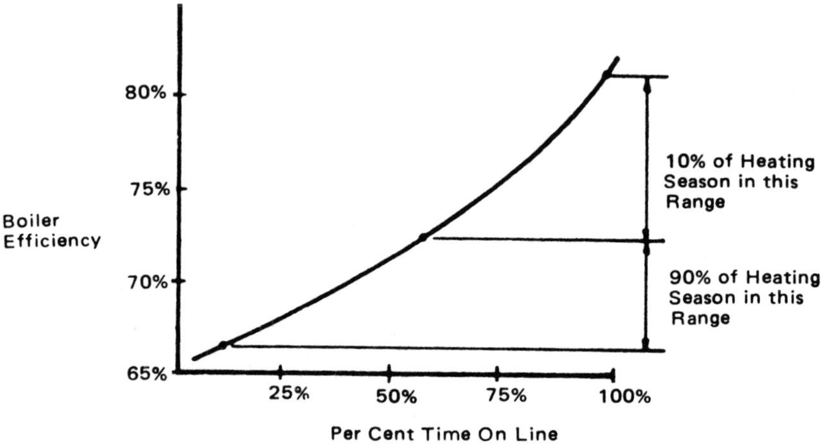

Figure 6-17. Effect of Cycling to Meet Part Loads

1. Overall thermal efficiency . . . . _____
2. Losses due to flue gases
   (a) Dry Flue Gas . . . . _____

The loss due to heat carried up the stack in dry flue gases can be determined, if the carbon dioxide (CO_2) content of the flue gases and the temperatures of the flue gas and air to the furnace are known.

   (b) Moisture % Hydrogen . . . . _____
   (c) Incomplete combustion . . . .

                                                                        _____
3. Balance of account, including radiation and other unmeasured

   losses . . . . _____
                    TOTAL 100%

Figure 6-18. Thermal Efficiency of Boiler

194

To estimate losses due to moisture, Figure 6-19 is used.

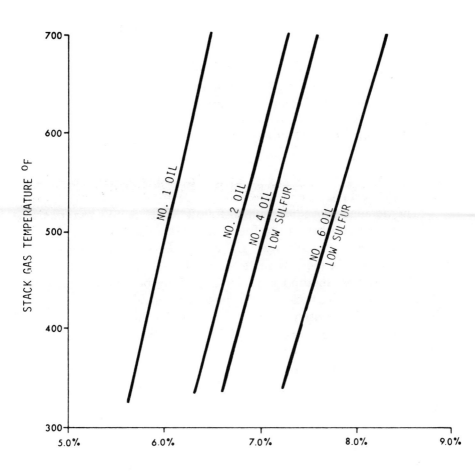

NOTE:
1. The figure gives a simple reference to heat loss in stack gases due to the formation of water in burning the hydrogen in various fuel oils.
2. The graph assumes a boiler room temperature of 80°F.

Figure 6-19. Heat Loss Due to Burning Hydrogen in Fuel
(Source: Instructions For Energy Auditors, Volume 1)

The savings in fuel as related to the change in efficiency is given by Formula 6-19.

$$\text{Savings in Fuel} = \frac{\text{New Efficiency} - \text{Old Efficiency}}{\text{New Efficiency}}$$
$$\times \text{Fuel Consumption} \qquad \textit{Formula (6-19)}$$

Figure 6-20 can be used to estimate the effect of flue gas composition, excess air, and stack temperature on boiler efficiency.

BOILER TUNE-UP TEST PROCEDURES

As illustrated by Figures 6-21 and 6-22, either % CO_2 or % O_2 can be used to determine excess air as long as the boiler is not operating on the fuel rise side of the curve.

The detailed procedures which follow illustrate how to tune-up a boiler to get the best air-to-fuel ratio. Note that for natural gas, % CO_2 must also be measured, while for fuel oil smoke spot numbers or visual smoke measurements are used.

The principal method used for improving boiler efficiency involves operating the boiler at the lowest practical excess O_2 level with an adequate margin for fuel variations in fuel properties and ambient conditions and the repeatability and response characteristics of the combustion control system.

These tests should only be conducted with a thorough understanding of the test objectives and following a systematic, organized series of tests.

Cautions

Extremely low excess O_2 operation can result in catastrophic results.

Know at all times the impact of the modification on fuel flow, air flow and the control system.

Observe boiler instrumentation, stack and flame conditions while making any changes.

When in doubt, consult the plant engineering personnel or the boiler manufacturer.

196

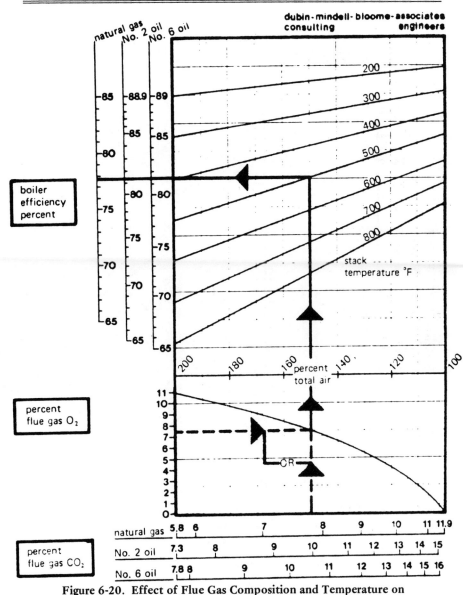

Figure 6-20. **Effect of Flue Gas Composition and Temperature on**
Boiler Efficiency
(*Source: Guidelines for Saving Energy in Existing Buildings—Engineers, Architects and*
Operators Manual, ECM-2)

197

THEORETICAL AIR CURVE
(FUEL OIL)

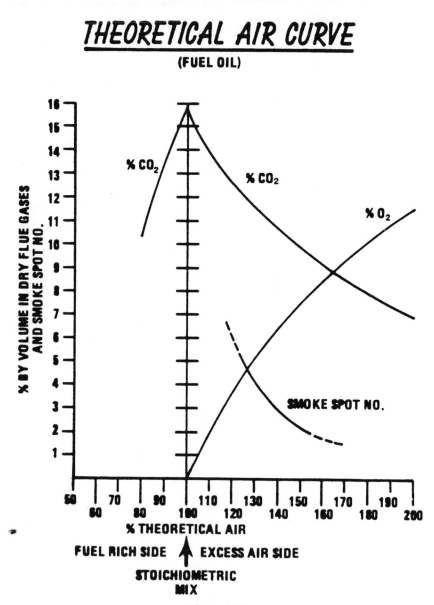

Figure 6-21

THEORETICAL AIR CURVE

(NATURAL GAS)
COMPLETE COMBUSTION

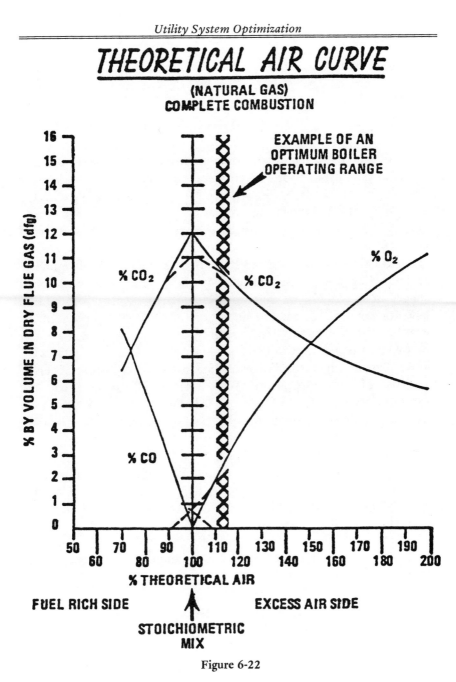

Figure 6-22

199

Consult the boiler operation and maintenance manual supplied with the unit for details on the combustion control system or methods of varying burner excess air.

Test Description

The test series begins with baseline tests that document existing "as-found conditions" for several firing rates over the boiler's normal operating range. At each of these firing rates, variations in excess O_2 level from 1 to 2% above the normal operating point to the "minimum O_2" level are made. Curves of combustibles as a function of excess O_2 level will be constructed similar to those given in Figures 6-23 and 6-24. As illustrated in these figures, high levels of smoke or CO indicating potentially unstable operation can occur with small changes in excess O_2 so that small changes in excess O_2 should be made for conditions near the smoke or CO limit. It is important to note that the boiler may exhibit a gradual smoke or CO behavior at one firing rate and a steep behavior at another. Minimum excess O_2 will be that at which the boiler just starts to smoke, or the CO emissions rise above 400 ppm or the Smoke Spot Number equals the maximum value as given in Table 6-3. Once minimum excess O_2 levels are established, an appropriate O_2 margin or operating cushion ranging from 0.5 to 2.0% O_2 above the minimum point depending on the particular boiler control system and fuels.

Repeated tests at the same firing condition approaching from both the "high side" and "low side" (i.e., from higher and lower firing rates) can determine whether there is excessive play in the boiler controls.

Record all pertinent data for future comparisons. Readings should be made only after steady boiler conditions are reached and at normal steam operating conditions.

Step-By-Step Boiler Adjustment Procedure
for Low Excess O_2 Operation

1. Bring the boiler to the test firing rate and put the combustion controls on manual.

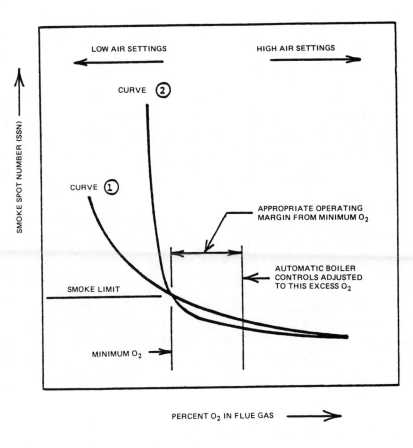

CURVE 1 – GRADUAL SMOKE/O_2 CHARACTERISTIC
CURVE 2 – STEEP SMOKE/O_2 CHARACTERISTIC

Figure 6-23. Typical Smoke-O_2 Characteristic Curves for Coal- or Oil-Fired Industrial Boilers

2. After stabilizing, observe flame conditions and take a complete set of readings.
3. Raise excess O_2 1 to 2%, allowing time to stabilize and take readings.

201

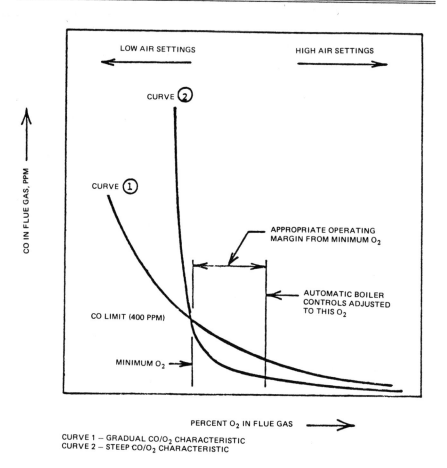

CURVE 1 — GRADUAL CO/O_2 CHARACTERISTIC
CURVE 2 — STEEP CO/O_2 CHARACTERISTIC

Figure 6-24. Typical CO–O Characteristic Curve for Gas-Fired Industrial Boilers

4. Reduce excess O_2 in small steps while observing stack and flame conditions. Allow the unit to stabilize following each change and record data.

5. Continue to reduce excess air until a minimum excess O_2 condition is reached.

Table 6-3. Maximum Desirable Smoke Spot Number

Fuel Grade	Maximum Desirable SSN
No. 2	less than 1
No. 4	2
No. 5 (light and heavy), and low-sulfur resid.	3
No. 6	4

6. Plot data similar to Figures 6-23 and 6-24.

7. Compare the minimum excess O_2 value to the value provided by the boiler manufacturers. High excess O_2 levels should be investigated.

8. Establish the margin in excess O_2 above the minimum and reset the burner controls to maintain this level.

9. Repeat Steps 1 thru 8 for each firing rate to be considered. Some compromise in optimum O_2 settings may be necessary since control adjustments at one firing rate may affect conditions at other firing rates.

10. After these adjustments have been completed, verify the operation of these settings by making rapid load pick-ups and drops. If undesirable conditions are encountered, reset controls.

11. Low fire conditions may require special consideration.

12. Perform tests on any alternate fuel used. Again, some discretion will be required to resolve differences in control settings for optimum conditions on the two fuels.

Evaluation of the New Low O_2 Settings

Extra attention to furnace and flame patterns for the first month or two following implementation of the new operating modes. Thoroughly inspect the boiler during the next shutdown. To assure high boiler efficiency, periodically make performance evaluations and compare with the results obtained during the test program.

Burner Adjustments

Adjustments to burner and fuel systems can also be made in addition to the low excess O_2 test program previously described. The approach in testing these adjustments is a "trial-and-error" procedure with sufficient organization to allow meaningful comparisons with established data.

Items that may result in lower minimum excess O_2 levels include changes in: burner register settings, oil gun tip position, oil gun diffuser position, coal spreader position, fuel oil temperature, fuel and atomizing pressure, and coal particle size. Evaluation of each of these items involves the same general procedures, precautions and data evaluation as outlined previously. The effect of these adjustments on minimum O_2 are variable from boiler to boiler and difficult to predict.

Conclusions

1. Combustion modifications are potentially catastrophic unless caution is continuously observed.
2. Test programs must be well thought out and planned in advance and anticipated results formulated and checked with obtained data.
3. Any modifications at low excess O_2 levels must be made slowly in small steps and with continuous evaluation of the flame, furnace and stack gas conditions.
4. All data must be recorded to allow for future comparisons to be made.
5. Practical operating excess O_2 levels must be given an adequate excess O_2 margin to allow for load variation and fuel property or ambient air condition changes.
6. New recommended operating conditions should be monitored for a sufficient length of time to gain confidence in their long-term use.
7. Periodic efficiency checks can indicate deviations from optimum performance conditions.

AUXILIARY EQUIPMENT FOR
INCREASING EFFICIENCY

The efficiency improvement potential of auxiliary equipment modifications is dependent on the existing boiler conditions.

Stack gas heat recovery equipment (air preheaters and economizers) are generally the most cost-effective auxiliary equipment additions. Addition of turbulators to firetube boilers can improve operating efficiencies and promote balanced gas flows between tube banks. Advanced combustion control systems and burners generally are less beneficial than stack gas heat recovery equipment on industrial sized units. Insulation and sootblowers, judiciously applied, can have beneficial effects on boiler efficiency. Significant energy saving potentials exist in waste water heat recovery from blowdown water and returned condensate.

Table 6-4 summarizes the options available to improve boiler efficiency.

Air preheaters and economizers are common equipment used. A caution should be noted in that these and other options will reduce stack temperature. In order to avoid corrosion problems exit temperatures should be as illustrated in Table 6-5. Exit gas temperature is determined by the extent of boiler convective surface or the presence of stack gas heat recovery equipment.

For boilers without heat recovery equipment, the minimum exit gas temperature is fixed by the boiler operating pressure since this determines the steam temperature.

Usual design practices result in an outlet gas temperature 150°F above the saturated steam temperature.

It becomes increasingly expensive to approach boiler saturation temperatures by simply adding convective surface area. As operating pressures increase, the stack gas temperature increases making heat recovery equipment more desirable. Economizers will permit a reduction in exit gas temperatures since the feedwater is at a lower temperature (220°F) than the steam saturation temperature. Stack gas temperatures of 300°F can be achieved with stack gas heat recovery

Table 6-4. Boiler Efficiency Improvement Equipment

DEVICE	PRINCIPLE OF OPERATION	EFFICIENCY IMPROVEMENT POTENTIAL	SPECIAL CONSIDERATIONS
Air Preheaters	Transfer energy from stack gases to incoming combustion air.	2.5 % for each 100 degree F decrease in stack gas temperature.	* Results in improved combustion condition. * Minimum flue gas temperatures limited by corrosion characteristics of the flue gas. * Application limited by space, duct orientation and maximum combustion air temperatures.
Economizers	Transfer energy from stack gases to incoming feedwater.	2.5 % increase for each 100 degree F decrease in stack gas temperature (1 % increase for each 10 degree F increase in feedwater temperature.	* Minimum flue gas temperature limited by corrosion characteristics of the flue gas. * Application limited on low pressure boilers. * Generally preferred over air preheaters for small (50,000 lbs./hr.) units.
Firetube Turbulators	Increases turbulence in the secondary passes of firetube units thereby increasing efficiency	2.5 % increase for each 100 degree F decrease in stack gas temperature.	* Limited to gas and oil fired units. * Properly deployed, they can balance gas flows through the tubes. * Increases pressure drop in the system.
Combustion Control Systems	Regulate the quantity of fuel and air flow	0.25 % increase for each 1 % decrease in excess O_2 depending on the stack gas temperature.	* Vary in complexity from the simplest jackshaft system to cross limited oxygen correction system. * Can operate either pneumatically or electrically. * Retrofit applications must be compatible with existing burner hardware.
Instrumentation	Provide operational data		* Provide records so that efficiency comparisions can be made.
Oil and Gas Burners	Promote flame conditions that result in complete combustion at lower excess air levels.	0.25 % increase for each 1 % decrease in excess O_2 depending on the stack gas temperature.	* Operation with the most elaborate low excess air burners require the use of advanced combustion control systems. * Flame shape and heat release rate must be compatible with furnace characteristics.

		* Flame scanners increase reliability. * Advanced atomizing systems are available. * Close control of oil viscosity improves atomization.
Insulation	Reduce external heat transfer	* Mass type insulation has low thermal conductivity and release heat loss by conduction. * Reflective insulation has smooth, metallic surfaces that reduce heat loss by radiation. * Insulation provides several other advantages including structural strength, reduced noise and fire protection.
Sootblowers	Remove boiler tube deposits that retard heat transfer.	* Can use steam or air as the blowing media. * Fixed position systems are used in low temperature regions whereas retractable "losses" are employed in high temperature areas. * The choice of the cleaning media will depend on the characteristics of the deposits.
Blowdown Systems	Transfer energy from expelled blowdown liquids to incoming feedwater.	* Quantity of expelled blowdown water is dependent on the boiler and makeup water quality. * Continuous blowdown operation not only decreases expelled liquids but also allows the incorporation of heat recovery equipment.
Condensate Return Systems	Reduce hot water requirements by recovering condensate.	* Quantity of condensate returned dependent on process and contamination. * Several systems available range from atmospheric (open) to fully pressurized (closed) systems.

207

Table 6-5. Minimum Exit Gas Temperatures

Oil Fuel ($>$ 2.5% S)	390°F
Oil Fuel ($<$ 1.0% S)	330°F
Bituminous Coal ($>$ 3.5% S)	290°F
Bituminous Coal ($<$ 1.5% S)	230°F
Pulverized Anthracite	220°F
Natural Gas (Sulfur-free)	220°F

equipment. Further reductions are achieved using air preheaters. Present design criteria limits the degree of cooling using stack gas heat recovery equipment to a level which will minimize condensation on heat transfer surfaces. The sulfur content of the fuel has a direct bearing on the minimum stack gas temperature as SO_3 combines with condensed water to form sulfuric acid and also the SO_3 concentration in the flue gas determines the condensation temperature.

7

Heating, Ventilating, Air Conditioning, and Building System Optimization

This chapter will review the basics of Heating, Ventilation and Air Conditioning (HVAC) and buildings as related to energy engineering.

DEGREE DAYS

Degree days are the summation of the product of the difference in temperature (ΔT) between the *average outdoor* and hypothetical *average indoor* temperatures (65°F), and the number of days *(t)* the outdoor temperature is below 65°F. Therefore:

$$DD\text{'s} = \Delta T \times t, \text{ therefore } \Delta t = DD/t \qquad Formula\ (7\text{-}1)$$

Degree Days divided by the total number of days on which Degree Days were accumulated will yield an average ΔT for the season, based on an assumed indoor temperature of 65°F. To find the average outdoor temperature of the season, this figure must be subtracted from 65°F.

Example Problem 7-1

If there are 6750 degree days recorded over a heating season of 270 days, what is the mean outdoor temperature for that season?

Answer

$$\Delta T = DD/t \quad \Delta T = \frac{6740 DD}{270 \text{ days}} \quad \Delta T = 25°F$$

The average outdoor temperature can now be found, since

$\Delta T = T$ (avg. indoor) $- T$ (avg. outdoor)
T (avg. outdoor) $= T$ (avg. indoor) $- \Delta T$
T (avg. outdoor) $= 65°F - 25°F$
T (avg. outdoor) $= 40°F$

RESISTANCE (R) TO HEAT FLOW AND CONDUCTANCE (U) AND CONDUCTIVITY (K)

The rate at which heat flows through a material depends on its characteristics. Some materials transmit heat more readily than others. This characteristic of materials which affects the flow of heat through them, can be viewed either as their *resistance* to the flow of heat or as their *conductance* allowing the flow of heat.

For a section of a building, such as a wall, the conductance is expressed as the U-value for that wall; that is, the number of Btu's that will pass through a one-square-foot section of a building in one hour with a one-degree temperature difference between the two surfaces.

U = Btu's per square foot per hour per degree Fahrenheit.

or

$$U = \text{Btu}/\text{ft}^2\,\text{h}\,°F$$

R-Value = Thermal Resistance = The unit time for a unit area of a particular body or assembly having defined surfaces with a unit average temperature difference established between the two surfaces per unit of Thermal Transmission.

$$\frac{\text{hr} \cdot \text{ft}^2 \cdot °F}{\text{Btu}}$$

$$R = 1/U \qquad \qquad \textit{Formula (7-2)}$$

The conductivity of a material as related to conductance and resistance is illustrated by Formula 7-3.

$$U = \frac{K}{d} = \frac{1}{R}$$ *Formula (7-3)*

where d is the thickness of the material.

VOLUME *(V)* OF AIR

The volume of air within a structure is constant even though the air itself changes—new air enters and old air leaves. The total volume is equal to the volume of space within the conditioned portion of the home. (Only the volume of conditioned space is considered since air entering and leaving the unconditioned part of the home does not demand energy to condition it.)

To determine the volume *(V)* of air, multiply the height *(H)* of the space times the width *(W)* of the space times the length *(L)* of the space.* While this can be done for the home as a whole, it is more accurate to calculate it for each room and then add these volumes.

AIR CHANGES PER HOUR *(AC)*

The rate at which the volume of air in a structure changes per hour differs greatly from building to building. The number of air changes per hour *(AC/h)* has wide variation due to a number of factors such as

- *The number and size of openings* in the envelope—around doors and windows and in the siding itself;
- *The average speed* of the wind blowing against the structure and the protection the structure has from this wind;
- *The number and size* of chimneys, vents, and exhaust fans and the frequency of their use;
- *The number of times* that doors and windows are opened; and
- *How the structure is used.*

---

*This is only appropriate for structures with flat ceilings.

HEATING CAPACITY OF AIR *(HC)*

Air can be heated and cooled. A certain amount of heat is necessary to change the temperature of each cubic foot (ft^3) of air one degree Fahrenheit (F). This amount of heat depends on the density of air which varies with temperature and pressure. This figure will generally be within the range of 0.018–0.022 Btu's/ft^3 °F.

BUILDING DYNAMICS

The building experiences heat gains and heat losses depending on whether the cooling or heating system is present, as illustrated in Figures 7-1 and 7-2. Only when the total season is considered in conjunction with lighting and heating, ventilation and air-conditioning (HVAC) can the energy utilization choice be decided. One way of reducing energy consumption of HVAC equipment is to reduce the overall heat gain or heat loss of a building.

CONDUCTION HEAT LOSS

The formula used to determine the amount of heat conducted through the envelope is as follows: Degree days *(DD)* is the product of the difference in temperature, ΔT, and the time *(t)* in days, providing that the days in degree days *(DD)* are converted to hours. This is accomplished by multiplying *(DD)* times 24 hours a day. This will yield the quantity of heat *(Q)* conducted through a particular section of the envelope for the entire heating season.

The formula can be written:

$$Q_{(heating\ season)} = U \times A \times DD \times 24\ \text{hours/day} \qquad \textit{Formula (7-4)}$$

or

$$Q_{(heating\ season)} = \frac{A \times DD \times 24\ \text{hrs/day}}{R} \qquad \textit{Formula (7-5)}$$

In general heat flow through a flat surface is defined as

$$Q = U A \Delta T \qquad \textit{Formula (7-6)}$$

where ΔT is the temperature difference causing the heat flow.

212

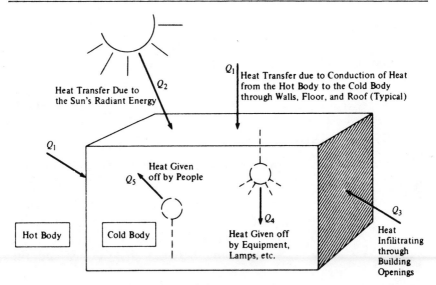

Heat Gain $= Q_1 + Q_2 + Q_3 + Q_4 + Q_5$

Figure 7-1. Heat Gain of a Building

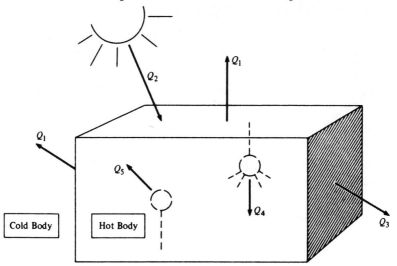

Heat Loss $= Q_1 + Q_3 - Q_2 - Q_4 - Q_5$

Figure 7-2. Heat Loss of a Building

213

For a composite wall, the heat flow is represented by Figure 7-3. To calculate the overall U or Conductance value, the resistance of each material is added in series. This is analogous to an electrical circuit.

$$R = R_1 + R_2 + R_3 + R_4 + R_5 \qquad \textit{Formula (7-7)}$$

FACTORS IN CONDUCTION

There are four factors which affect the conduction of heat from one area to another. They are:

- *The difference in temperature (ΔT)* between the warmer area and the colder area;
- *The length of time (t)* over which the transfer occurs;
- *The area (A) in common* between the warmer and the colder area; and
- *The resistance (R) to heat flow and conduction (U)* between the warmer and the colder area.

Difference in Temperature

Heat flows (much as water moves down hill) from warm areas to cold ones. The steeper the gradient between its origin and its destination the faster it will flow. In fact, the rate at which heat is conducted is directly proportionate to the difference in temperature (ΔT) between the warm area and the colder one.

Length of Time

The longer the heat is allowed to flow across the gradient, the more heat will be conducted. The amount of heat (Btu's) is directly proportionate to the time span *(t)* of the transfer.

Btu/h is the amount of heat transferred in one hour.

The Area *(A)* in Common

The larger the area common to the warmer and colder surfaces, through which the heat flows, the greater is the rate of conducted

214

heat. For the same material, for the same length of time, at the same ΔT, the amount of heat (Btu's) transferred is directly proportionate to the area (A) in common.

Example Problem 7-2

Calculate the heat loss through 20,000 ft² of building wall, as indicated by Figure 7-3. Assume a temperature differential of 17°F.

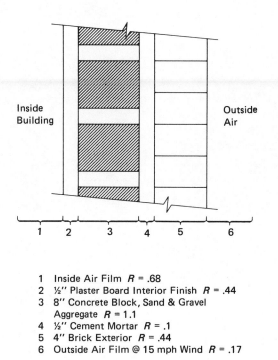

1   Inside Air Film $R = .68$
2   ½″ Plaster Board Interior Finish $R = .44$
3   8″ Concrete Block, Sand & Gravel
    Aggregate $R = 1.1$
4   ½″ Cement Mortar $R = .1$
5   4″ Brick Exterior $R = .44$
6   Outside Air Film @ 15 mph Wind $R = .17$

Figure 7-3. Typical Wall Construction

215

Answer

Description		Resistance
Outside air film at 15 mph		0.17
4" brick		0.44
Mortar		0.10
Block		1.11
Plaster Board		0.44
Inside film		0.68
	Total resistance	2.94

$U = 1/R = 0.34$

$Q = U A \Delta T$

$\quad = 0.34 \times 20,000 \times 17 = 115,600$ Btu/h.

In Figure 7-4 a surface film conductance is introduced.

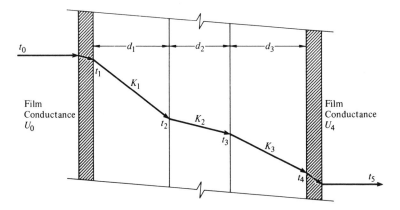

Figure 7-4. Temperature Distribution for the Composite Wall

The surface or film conductance is the amount of heat transferred in Btu per hour from a surface to air or from air to a surface per square foot for one degree difference in temperature. The flow

216

of heat for the composite material can also be specified in terms of the conductivity of the material and the conductance of the air film:

$$Q = \frac{A\,(t_0 - t_5)}{1/U_0 + d_1/K_1 + d_2/K_2 + d_3/K_3 + 1/U_4}.$$ Formula (7-8)

LATENT HEAT AND SENSIBLE HEAT

The latent heat gain of a space means that moisture is being added to the air in the space. Moisture in the air is really in the form of superheated steam. Removing sensible heat from a space through air-conditioning equipment lowers the dry-bulb temperature of the air. On the other hand, removing latent heat from a space changes the substance state from a vapor to a liquid. The latent heat gain of a space is expressed in terms of moisture, heat units (Btu) or grains of moisture per hour (7,000 grains equals one pound). The average value for the latent heat or vaporization for superheated steam in air is 1050.

Example Problem 7-3

2000 grains of moisture are released in a conditioned room each hour. Calculate the heat that must be removed in order to condense this moisture at the cooling coils.

Answer

$$\frac{2000}{7000} \times 1050 = 299.9 \text{ Btuh}$$

Example Problem 7-4

Calculate the quantity of heat (Q) required by infiltration in an 8,000 cubic foot (ft^3) home that has 1.7 air changes per hour (AC/h) when the outside temperature is 48°F and the inside temperature is 68°F ($\Delta T = 20°F$) for one day (24 hours).

217

Answer

$$Q = V \times AC/h \times 0.020^* \text{ Btu/ft}^3 \, ^\circ F \times \Delta T \times t$$
$$Q = 8{,}000 \text{ ft}^3 \times 1.7 \, AC/h \times 0.020^* \text{ Btu/ft}^3 \, ^\circ F \times 20^\circ F \times 24 \text{ hrs}$$
$$Q = 8{,}000 \times 1.7 \times 0.020^* \times 20 \times 24$$
$$Q = 130{,}560 \text{ Btu's}^{**}$$

INFILTRATION

Leakage or infiltration of air into a building is similar to the effect of additional ventilation. Unlike ventilation, it cannot be controlled or turned off at night. It is the result of cracks, openings around windows and doors, and access openings. Infiltration is also induced into the building to replace exhaust air unless the HVAC balances the exhaust. Wind velocity increases infiltration and stack effects are potential problems. Air that is pushed out the window and door cracks is referred to as exfiltration.

To estimate infiltration, the Air Change Method or Crack Method is used.

Air Change Method

The five factors which determine the amount of energy lost through infiltration can be put together in a formula that states that:

The quantity of heat (Q) equals the volume of air (V) times the number of air changes per hour (AC/h) times the amount of heat required to raise the temperature of air one degree Fahrenheit (0.018–0.022 Btu's) times the temperature difference (ΔT) times the length of time (t). This is expressed as follows:

$$Q = V \times AC/h \times 0.020 \text{ Btu's} \cdot /\text{ft}^3 \, ^\circ F \times \Delta T \times t \qquad \textit{Formula (7-9)}$$

The Air Change Method is considered to be a quick estimation method and is not usually accurate enough for air-conditioning design. A second method used to determine infiltration is the Crack Method.

---

*This is a regional variable.
** Once again, all of the units in the formula cancel except Btu's, leaving the units for Q as Btu's.

Crack Method

When infiltration enters a space it adds sensible and latent heat to the room load. To calculate this gain the following equations (7-10) and (7-11) are used.

Sensible Heat Gain

$$Q_S = 1.08 \text{ CFM } \Delta T \qquad \qquad Formula\ (7\text{-}10)$$

Latent Heat Gain

$$Q_L = .7 \text{ CFM } (HR_o - HR_i) \qquad Formula\ (7\text{-}11)$$

where

Q_S = Sensible heat gain Btuh
Q_L = Latent heat gain Btuh
CFM = Air Flow Rate
ΔT = Temperature differential between outside and inside air, F
HR_o = Humidity ratio of outside air, grains per lb
HR_i = Humidity ratio of room air, grains per lb

BODY HEAT

The human body releases sensible and latent heat depending on the degree of activity. Heat gains for typical applications are summarized in Table 7-1.

EQUIPMENT, LIGHTING
AND MOTOR HEAT GAINS

It is important to include heat gains from equipment, lighting systems and motor heat gain in the overall calculations.

For a manufacturing facility, the major source of heat gains will be from the process equipment. Consideration must be given to all equipment including motors driving supply and exhaust fans.

To convert motor horsepower to heat gain in Btuh, equation 7-12 is used.

Table 7-1. Heat Gain from Occupants

Activity	Sensible Heat Btuh	Latent Heat Btuh	Total Heat Gain Btuh
Very Light Work — Seated (Offices, Hotels, Apartments)	215	185	400
Moderately Active Work (Offices, Hotels, Apartments)	220	230	450
Moderately Heavy Work (Manufacturing)	330	670	1000
Heavy Work (Manufacturing)	510	940	1450

Source: ASHRAE—Guide & Data Book

$$Q = \frac{hp \times .746}{\eta} \times 3412 \qquad \textit{Formula (7-12)}$$

where

hp is the running motor horsepower

η is the efficiency of the motor

Q is the heat gain from the motor Btuh

Similarly, the kilowatts of the lighting system can be converted to heat gain.

$$Q = (KW_F + KW_B) \times 3412 \qquad \textit{Formula (7-13)}$$

where

KW_F is the kilowatts of the lighting fixtures

KW_B is the kilowatts of the ballast

Q is the heat gain from the lighting system Btuh

RADIANT HEAT GAIN

Heat from the sun's rays greatly increases heat gain of a building. If the building energy requirements were mainly due to cooling, then this gain should be minimized. Solar energy affects a building in the following ways:

1. *Raises the surface temperature:* Thus a greater temperature differential will exist at roofs than at walls.

220

2. A large percentage of direct solar radiation and diffuse sky radiation *passes through* transparent materials, such as glass.

SURFACE TEMPERATURES

The temperature of a wall or roof depends upon:

(a) the angle of the sun's rays
(b) the color and roughness of the surface
(c) the reflectivity of the surface
(d) the type of construction.

When an engineer is specifying building materials, he should consider the above factors. A simple example is color. The darker the surface, the more solar radiation will be absorbed. Obviously, white surfaces have a lower temperature than black surfaces after the same period of solar heating. Another factor is that smooth surfaces reflect more radiant heat than do rough ones.

In order to properly take solar energy into account, the angle of the sun's rays must be known. If the latitude of the plant is known, the angle can be determined.

SUNLIGHT AND GLASS CONSIDERATIONS

A danger in the energy conservation movement is to take steps backward. A simple example would be to exclude glass from building designs because of the poor conductance and solar heat gain factors of clear glass. The engineer needs to evaluate various alternate glass constructions and coatings in order to maintain and improve the aesthetic qualities of good design while minimizing energy inefficiencies. It should be noted that the method to reduce heat gain of glass due to conductance is to provide an insulating air space.

To reduce the solar radiation that passes through glass, several techniques are available. Heat absorbing glass (tinted glass) is very popular. Reflective glass is gaining popularity, as it greatly reduces solar heat gains.

To calculate the relative heat gain through glass, a simple method is illustrated below:

$$Q = \pm U A (t_0 - t_1) + A \times S_1 \times S_2 \qquad \textit{Formula (7-14)}$$

where

Q is the total heat gain for each glass orientation (Btuh)

U is the conductance of the glass (Btu/h-ft^2-°F)

A is the area of glass; the area used should include framing, since it will generally have a poor conductance compared with the surrounding material. (ft^2)

$t_0 - t_1$ is the temperature difference between the inside temperature and outside ambient. (°F)

S_1 is the shading coefficient; S_1 takes into account external shades, such as venetian blinds and draperies, and the qualities of the glass, such as tinting and reflective coatings.

S_2 is the solar heat gain factor. This factor takes into account direct and diffused radiation from the sun. Diffused radiation is basically caused by reflections from dust particles and moisture in the air.

THE PSYCHROMETRIC CHART

Just as the steam table and the Mollier Diagram are used to relate the properties of steam, the psychrometric chart is used to illustrate the properties of air. The psychrometric chart is a very important tool in the design of air-conditioning systems.

PROPERTIES OF AIR

Air expands and contracts with temperature. If pressure is held constant, then air expands or contracts at a specified rate with change in temperature as defined by equation 7-15.

$$V_2 = V_1 \frac{T_2}{T_1} \qquad \textit{Formula (7-15)}$$

222

where

P_2 = initial pressure, psia
P_1 = final pressure, psia

The change in the volume occupied by air at any temperature can be found by first using Formula 7-15 to calculate the change in volume with pressure and then using Formula 7-14 to calculate the change in volume with temperature.

The temperature at which the water vapor in the atmosphere begins to condense is the *Dew Point Temperature*. It should be noted that the weight of moisture per pound of dry air in a mixture of air and water vapor depends on the dew point temperature alone. If there is no condensation of moisture, the dew point temperature remains constant.

Humidity Ratio is defined as the weight of water vapor mixed with one pound of dry air.

Degree of Saturation is defined as the actual humidity ratio divided by the humidity ratio at saturation.

Relative Humidity is defined as the vapor pressure of air divided by the saturation pressure of pure water at the same temperature.

Sensible Heat of an Air-Vapor Mixture is defined as the heat which affects the dry-bulb temperature of the mixture only.

Wet-Bulb Temperature can be determined by covering the bulb of a thermometer with a wet wick and holding it in a stream of swiftly moving air. At first the temperature will drop quickly and then reach a stationary point referred to as the wet-bulb temperature. The wet-bulb temperature is lower than the dry thermometer. The amount of water which evaporates from the wet wick into the air depends on the amount of water vapor initially in the air flowing past the wet bulb. A sling psychrometer is a convenient instrument used to measure wet-bulb temperatures.

Total Heat is defined as the sum of the sensible and latent heat. Sensible heat depends only upon the dry-bulb temperature, while the latent heat content depends only upon its dew point.

The *Enthalpy of an Air-Water Vapor Mixture* can be calculated by equation 7-16.

223

$$h_{(mix)} = h_{(dry\ air)} + h_{(water\ vapor)} \qquad \textit{Formula (7-16)}$$

or

$$h_{(mix)} = Cp \times T_{DB} + HR + hg$$

where

$h_{(mix)}$ = enthalpy of the mixture of dry air and water vapor, Btu per lb

Cp = specified heat, Btu per LvF

T_{DB} = dry bulb temperature

HR = humidity ratio of the mixture

hg = enthalpy of saturated vapor (steam) at the dew point temperature

To determine the properties of air such as the humidity ratio, relative humidity, enthalpy, the psychrometric chart is frequently used. Figure 7-4 illustrates the psychrometric chart.

Example Problem 7-4

Given air at 70° DB and 50% relative humidity, for the air vapor mixture find:

- Wet-Bulb Temperature
- Enthalpy
- Humidity Ratio
- Dew Point Temperature
- Specific Volume
- Vapor Pressure
- Percentage Humidity

Answer

From Figure 7-4 find the intersection of 70° DB and 50% *HR*, point "A."

The WB temperature is found as 58.6 WB, point "B."

The Enthalpy is found as 25.5 Btu/lb, point "C."

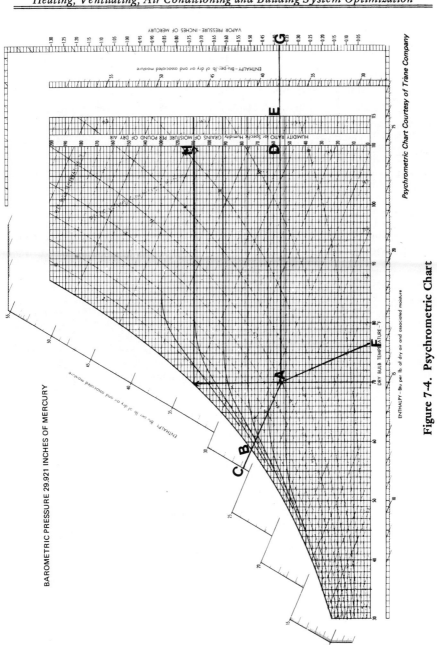

Figure 7-4. Psychrometric Chart

225

The Humidity Ratio is found to be 56 grains of moisture per pound of dry air, point "D" and the Dew Point temperature is 53°F, point "E."

The Specific Volume is found to be 13.5 cubic feet per pound of air, point "F," with a Vapor Pressure of .38 inches of mercury, point "G."

The percentage humidity equals the actual humidity 56, point "D" divided by the humidity ratio at saturation (100% *RH)* which is found to be 110, point "H." Thus % humidity = 56/110 = .50.

Example Problem 7-5

Given 8000 CFM of chilled air at 55°F DB and 50°F WB mixed with 3000 CFM of outside air at 90°F DB and 80°F WB, compute the properties of the mixture.

Answer

From Figure 7-5, the intersection of 55°F DB and 50°F WB is point "A." The specific volume is then 13.1 cubic feet/lb, point "B."

Similarly, for the outside air, the specific volume is 14.3 cubic feet/lb, point "D."

The total weight and dry-bulb temperature of the mixture can be found by the following ratios:

$$\frac{8{,}000}{13.1} = 610.6 \text{ lb/min.}$$

$$\frac{3{,}000}{14.3} = 209.7 \text{ lb/min.}$$

$$\underline{ \;\; 820.3 \text{ lb/min. for total weight}}$$

The dry-bulb temperature is:

$$\frac{610.6}{820.3} \times 55 = 40.93°F$$

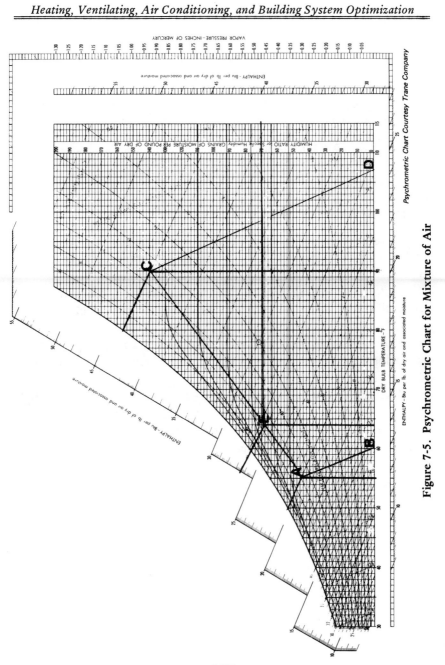

Psychrometric Chart Courtesy Trane Company

Figure 7-5. Psychrometric Chart for Mixture of Air

$$\frac{209.7}{820.3} \times 90 = \underline{23.0^\circ F}$$
$$\underline{63.9^\circ F \text{ DB}}$$

The properties of the mixture, point "E" can now be determined from the chart.

WB $\quad = 59.8^\circ F$
$h \quad = 26.6$ Btu/lb
humidity ratio $\quad = 70$ gr of moisture/lb of dry air.

BASICS OF FAN DISTRIBUTION SYSTEMS

In order to distribute conditioned or ventilated air, fans are the chief vehicle used. Several basic types of fans commonly used in industry are illustrated in Figure 7-6 and are listed below:

Centrifugal, airfoil blade—used on large heating, ventilating, and air-conditioning systems. Airfoil fans are used where clean air is handled.

Centrifugal, backward curved blade—used for general heating, ventilating, and air-conditioning systems. Air handled need not be as clean as above.

Centrifugal, radial blade—a rugged, heavy duty fan for high pressure applications. It is designed to handle sand, wood chips, etc.

Centrifugal, forward curved blade—ideal for low pressure applications, such as domestic furnaces or room and packaged air conditioners.

The brake horsepower of a fan is illustrated by Formula 7-17.

$$\text{Brake Horsepower} = \frac{\text{CFM} \times \text{Fan PS}}{6356 \times \eta_F} \qquad \textit{Formula (7-17)}$$

where

CFM $\quad$ is the quantity of air in CFM
$\eta \quad$ is the fan static efficiency
Fan PS $\quad$ is the Fan Static Pressure in inches

To compute the fan static pressure

228

Figure 7-6. Fan Types: (A) Vaneaxial; (B) Backward Curved Blade;
(C) Tubeaxial; (D) Radial; (E) Radial Tip Blade; (F) Airfoil Blade
Courtesy of Buffalo Forge Company

$$\text{Fan PS} = P_T\ (0) - P_T\ (i) - P_v\ (O) \qquad \textit{Formula (7-18)}$$

where

$P_T\ (O)$ is the total pressure at fan outlet
$P_T\ (i)$ is the total pressure at fan inlet
$P_v\ (O)$ is the velocity pressures at fan outlet

The excess pressure above the static pressure is known as the velocity pressure, and is computed by Formula 7-19 for standard air having a value of 13.33 cu ft per lb as

$$PV = \left(\frac{V}{4005}\right)^2 \qquad \textit{Formula (7-19)}$$

where V is the velocity of air in FPM.

Note that the pressure of air in sheet metal ducts is so low that ordinary pressure gauges (Bourdon type) cannot be used, thus a V-tube or manometer is used, which measures pressure in inches of water. A pressure of 1 psi will support a column of water 2.31 ft high or 27.7 inches.

Fan Laws

The performance of a fan at varying speeds and air densities may be predicted by certain basic fan laws as illustrated in Table 7-2.

Example Problem 7-6

An energy audit indicates that the ventilation requirements of a space can be reduced from 15,000 CFM to 12,000 CFM. Comment on the savings in brake horsepower if the fan pulley is changed to reduce the fan speed accordingly.

Answer

From the Fan Laws:

Table 7-2. Fan Laws

1.	Fan Law for variation in fan speed at constant air density with a constant system
	1.1 Air volume, CFM varies as fan speed
	1.2 Static velocity or total pressure varies as the square of fan speed
	1.3 Power varies as cube of fan speed
2.	Fan Law for variation in air density at constant fan speed with a constant system
	2.1 Air volume is constant
	2.2 Static velocity or pressure varies as density
	2.3 Power varies as density

$$hp_1 = hp_2 \times \left(\frac{\text{CFM new}}{\text{CFM old}}\right)^3$$

$$= hp_2 \times \left(\frac{12,000}{15,000}\right)^3 = hp_2 \, (.8)^3 = .512 \, hp_2$$

or a 48.8% savings.

The fan performance is affected by the density of the air that the fan is handling. All fans are rated at standard air with a density of .075 lb per cu ft and a specific volume of 13.33 cu ft per lb. When a fan is tested in a laboratory at different than standard air, the brake horsepower is corrected by using the Fan Laws.

Fan Performance Curves

Fan performance curves are used to determine the relationship between the quantity of air that a fan will deliver and the pressure it can discharge at various air quantities. For each fan type, the manufacturer can supply fan performance curves which can be used in design and as a tool of determining the fan efficiency.

As illustrated by problem 7-6, one energy engineering technique to reduce fan horsepower is to reduce fan speed. An alternate way is to throttle the air flow by a damper. The fan performance curves can be used to illustrate the best choice of these options. The system characteristics can be plotted on the fan curves to show the static pressure required to overcome the friction loss in the duct system. From the Fan Laws the system friction loss varies with the square of fan speed; thus, as the air quantity increases, the friction loss will vary as illustrated by Figure 7-7.

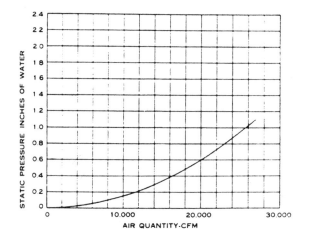

Figure 7-7. System Characteristic Curve

For a detailed analysis, the fan performance curve should be used to predict how a specific fan will perform in a desired application.

Example Problem 7-7

Given Fan Performance Curve Figure 7-8. The fan delivers 21,500 CFM at 600 rpm at a brake horsepower of 12.3. Comment on the savings in brake horsepower by reducing air flow to 14,400 CFM by each of the following methods: (a) reducing fan speed to 400 rpm, (b) throttling the air flow by a damper.

232

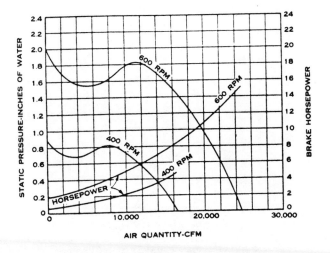

Figure 7-8. Fan Performance Curves

(a) From the Fan Laws the brake horsepower is reduced as follows:

$$hp = 12.3 \ \frac{400}{600}^{\ 3} = 3.64$$

Using the fan performance curve, Figure 7-9, the system characteristic curve "A" is plotted.

By reducing the rpm from 600 to 400, the system operates at point 1 and then moves to point 2. The brake horsepower is found from Figure 7-8 to be 3.7.

(b) By closing the air damper, the air flow is reduced to 14,400 CFM while still running the fan at 600 rpm. Using the fan performance curve Figure 7-9, the system operates at point 1 and then moves to point 3. The power to operate the fan at point 3 is 7.2 hp from Figure 7-8. Thus, if the fan speed can be reduced it is more efficient than throttling the air flow damper.

233

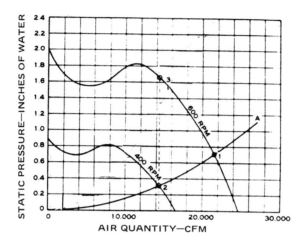

Figure 7-9. Fan Performance and System Characteristic Curves

FLUID FLOW

Pump and piping considerations are extremely important due to the fact that energy transport losses are a part of any distribution system. Losses occur due to friction and that lost energy must be supplied by pump horsepower.

Centrifugal pumps are commonly used in heating, ventilation and air-conditioning applications as well as utility systems. The output torque for the pump is supplied by a driver such as a motor. Liquid enters the eye of the impeller which rotates. Pressure energy builds up by the action of centrifugal force, which is a function of the impeller vane peripheral velocity.

As with fan systems, Pump Laws and curves can be used to predict system responsiveness. The affinity laws of a pump are illustrated in Table 7-3.

The horsepower required to operate a pump is illustrated by Formula 7-20.

$$hp = \frac{\Delta P \text{ GPM}}{17.5\eta}$$

Formula (7-20)

234

Table 7-3. Affinity Laws for Pumps

Impeller Diameter	Speed	Specific Gravity (SG)	To Correct for	Multiply by
Constant	Variable	Constant	Flow	$\left(\dfrac{\text{New Speed}}{\text{Old Speed}}\right)$
			Head	$\left(\dfrac{\text{New Speed}}{\text{Old Speed}}\right)^2$
			BHP (or kW)	$\left(\dfrac{\text{New Speed}}{\text{Old Speed}}\right)^3$
Variable	Constant		Flow	$\left(\dfrac{\text{New Diameter}}{\text{Old Diameter}}\right)$
			Head	$\left(\dfrac{\text{New Diameter}}{\text{Old Diameter}}\right)^2$
			BHP (or kW)	$\left(\dfrac{\text{New Diameter}}{\text{Old Diameter}}\right)^3$
	Constant	Variable	BHP (or kW)	$\dfrac{\text{New SG}}{\text{Old SG}}$

where

ΔP is the differential pressure across a pump in psi
GPM is the required flow rate in gallons per minute
η is the pump efficiency

To convert psi to read in feet use Formula 7-21.

$$\text{Head in Feet} = \frac{\text{psi} \times 2.31}{\text{Specific Gravity of Fluid}} \qquad \textit{Formula (7-21)}$$

235

Basically, the size of discharge line piping from the pump determines the friction loss through the pipe that the pump must overcome. The greater the line loss, the more pump horsepower required. If the line is short or has a small flow, this loss may not be significant in terms of the total system head requirements. On the other hand, if the line is long and has a large flow rate, the line loss will be significant.

To calculate the pressure loss for water system piping and the corresponding velocity, Formulas 7-22 and 7-23 are used.

$$\Delta P = \frac{.055 \, CF^{1.85}}{d^{4.87}} \qquad \text{Formula (7-22)}$$

$$V = \frac{.41 \, F}{d^2} \qquad \text{Formula (7-23)}$$

where

ΔP = pressure loss per 100 feet of pipe, psi
V = velocity of fluid, ft/sec.
C = roughness factor
    1 for copper tubing
    1.62 for steel pipe
    .77 for plastic pipe
F = flow rate in gallons per minute
d = inside diameter of pipe, inches

The pressure loss due to fittings is determined by Formula 7-24.

$$\Delta P = .0067 \, KV^2 \qquad (Formula \, 7-24)$$

where K is the loss coefficient.

Options to Reduce Pump Horsepower

There are several alternates that will significantly reduce pump horsepower. The below summarizes some of the options available.

1. Many pumps are oversized due to very conservative design practices. If the pump is oversized, install a smaller impeller to match the load.

2. In some instances heating or cooling supply flow rates can be reduced. To save on pump horsepower either reduce motor speed or change the size of the motor sheave.

3. Check economics of replacing corroded pipe with a large pipe diameter to reduce friction losses.

4. Consider using variable speed pumps to better match load conditions. Motor drive speed can be varied to match pump flow rate or head requirements.

5. Consider adding a smaller auxiliary pump. During part load situations a larger pump can be shut down and a smaller auxiliary pump used.

HVAC SYSTEMS

The below summary illustrates the types of systems frequently encountered in heating air-conditioning systems.

Single Zone System

Single zone systems consist of a mixing, conditioning and fan section. The conditioning section may have heating, cooling, humidifying or a combination of capabilities. Single zone systems can be factory assembled roof-top units or built up from individual components and may or may not have distributing duct work.

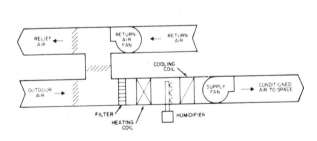

Terminal Reheat System

Reheat systems are modifications of single zone systems. Fixed cold temperature air is supplied by the central conditioning system and reheated in the terminal units when the space cooling load is less than maximum. The reheat is controlled by thermostats located in each condition space.

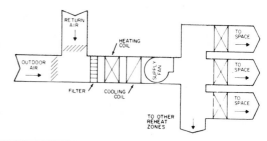

Multizone Systems

Multizone systems condition all air at the central system and mix heated and cooled air at the unit to satisfy various zone loads as sensed by zone thermostats. These systems may be packaged roof-top units or field-fabricated systems.

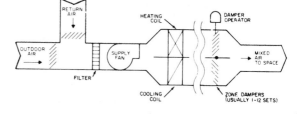

Dual Duct Systems

Dual duct systems are similar to multizone systems except heated and cooled air is ducted to the conditioned spaces and mixed as required in terminal mixing boxes.

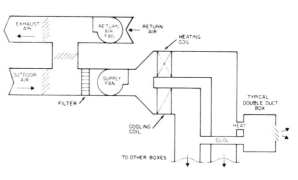

238

Variable Air Volume Systems

A variable air volume system delivers a varying amount of air as required by the conditioned spaces. The volume control may be by fan inlet (vortex) damper, discharge damper or fan speed control. Terminal sections may be single duct variable volume units with or without reheat, controlled by space thermostats.

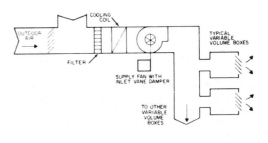

Induction Systems

Induction systems generally have units at the outside perimeter of conditioned spaces. Conditioned primary air is supplied to the units where it passes through nozzles or jets and by induction draws room air through the induction unit coil. Room temperature control is accomplished by modulating water flow through the unit coil.

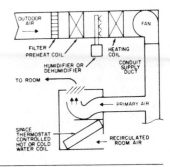

Fan Coil Units

A fan coil unit consists of a cabinet with heating and/or cooling coil, motor and fan and a filter. The unit may be floor or ceiling mounted and uses 100% return air to condition a space.

Unit Ventilator

A unit ventilator consists of a cabinet with heating and/or cooling coil, motor and fan, a filter and a return air—outside air mixing section. The unit may be floor or ceiling mounted and uses return and outside air as required by the space.

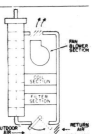

Unit Heater

Unit heaters have a fan and heating coil which may be electric, hot water or steam. They do not have distribution duct work but generally use adjustable air distribution vanes. Unit heaters may be mounted overhead for heating open areas or enclosed in cabinets for heating corridors and vestibules.

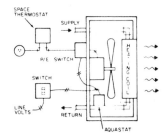

Perimeter Radiation

Perimeter radiation consists of electric resistance heaters or hot water radiators usually within an enclosure but without a fan. They are generally used around the conditioned perimeter of a building in conjunction with other interior systems to overcome heat losses through walls and windows.

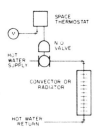

Hot Water Converters

A hot water converter is a heat exchanger that uses steam or hot water to raise the temperature of heating system water. Converters consist of a shell and tubes with the water to be heated circulated through the tubes and the heating steam or hot water circulated in the shell around the tubes.

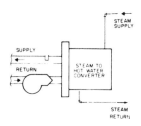

Source: "Energy Conservation with Comfort," Honeywell

240

Reducing Energy Consumption
in HVAC Systems

Variable Air Volume System—A variable volume system provides heated or cooled air at a constant temperature to all zones served. VAV boxes located in each zone or in each space adjust the quantity of air reaching each zone or space depending on its load requirements. Methods for conserving energy consumed by this system include:

1. Reduce the volume of air handled by the system to that point which is minimally satisfactory.

2. Lower hot water temperature and raise chilled water temperature in accordance with space requirements.

3. Lower air supply temperature to that point which will result in the VAV box serving the space with the most extreme load being fully open.

4. Consider installing static pressure controls for more effective regulation of pressure bypass (inlet) dampers.

5. Consider installing fan inlet damper control systems if none now exist.

Constant Volume System—Most constant volume systems either are part of another system—typically dual duct systems—or serve to provide precise air supply at a constant volume. Opportunities for conserving energy consumed by such systems include:

1. Determine the minimum amount of airflow which is satisfactory and reset the constant volume device accordingly.

2. Investigate the possibility of converting the system to variable (step controlled) constant volume operation by adding the necessary controls.

Induction System—Induction systems comprise an air handling unit which supplies heated or cooled primary air at high pressure to induction units located on the outside walls of each space served. The high pressure primary air is discharged within the unit through nozzles inducing room air through a cooling or heating coil in the unit. The resultant mixture of primary air and induced air is discharged to the room at a temperature dependent upon the cooling and heating load of the space. Methods for conserving energy consumed by this system include:

1. Set primary air volume to original design values when adjusting and balancing work is performed.

2. Inspect nozzles. If metal nozzles, common on most older models, are installed, determine if the orifices have become enlarged from years of cleaning. If so, chances are that the volume/pressure relationship of the system has been altered. As a result, the present volume of primary air and the appropriate nozzle pressure required must be determined. Once done, rebalance the primary air system to the new nozzle pressures and adjust individual induction units to maintain airflow temperature. Also, inspect nozzles for cleanliness. Clogged nozzles provide higher resistance to air flow, thus wasting energy.

3. Set induction heating and cooling schedules to minimally acceptable levels.

4. Reduce secondary water temperatures during the heating season.

5. Reduce secondary water flow during maximum heating and cooling periods by pump throttling or, for dual-pump systems, by operating one pump only.

6. Consider manual setting of primary air temperature for heating, instead of automatic reset by outdoor or solar controllers.

Dual-Duct System—The central unit of a dual-duct system provides both heated and cooled air, each at a constant temperature.

Each space is served by two ducts, one carrying hot air, the other carrying cold air. The ducts feed into a mixing box in each space which, by means of dampers, mixes the hot and cold air to achieve that air temperature required to meet load conditions in the space or zone involved. Methods for improving the energy consumption characteristics of this system include:

1. Lower hot deck temperature and raise cold deck temperature.

2. Reduce air flow to all boxes to minimally acceptable level.

3. When no cooling loads are present, close off cold ducts and shut down the cooling system. Reset hot deck according to heating loads and operate as a single duct system. When no heating loads are present, follow the same procedure for heating ducts and hot deck. It should be noted that operating a dual-duct system as a single-duct system reduces air flow, resulting in increased energy savings through lowered fan speed requirements.

Single Zone System—A zone is an area or group of areas in a building which experiences similar amounts of heat gain and heat loss. A single zone system is one which provides heating and cooling to one zone controlled by the zone thermostat. The unit may be installed within or remote from the space it serves, either with or without air distribution ductwork.

1. In some systems air volume may be reduced to minimum required therefore reducing fan power input requirements. Fan brake horsepower varies directly with the cube of air volume. Thus, for example, a 10% reduction in air volume will permit a reduction in fan power input by about 27% of original. This modification will limit the degree to which the zone serviced can be heated or cooled as compared to current capabilities.

2. Raise supply air temperatures during the cooling season and reduce them during the heating season. This procedure

reduces the amount of heating and cooling which a system must provide, but, as with air volume reduction, limits heating and cooling capabilities.

3. Use the cooling coil for both heating and cooling by modifying the piping. This will enable removal of the heating coil, which provides energy savings in two ways. First, air flow resistance of the entire system is reduced so that air volume requirements can be met by lowered fan speeds. Second, system heat losses are reduced because surface area of cooling coils is much larger than that of heating coils, thus enabling lower water temperature requirements. Heating coil removal is not recommended if humidity control is critical in the zone serviced and alternative humidity control measures will not suffice.

Multizone System—A multizone system heats and cools several zones—each with different load requirements—from a single, central unit. A thermostat in each zone controls dampers at the unit which mix the hot and cold air to meet the varying load requirements of the zone involved. Steps which can be taken to improve energy efficiency of multizone systems include:

1. Reduce hot deck temperatures and increase cold deck temperatures. While this will lower energy consumption, it also will reduce the system's heating and cooling capabilities as compared to current capabilities.

2. Consider installing demand reset controls which will regulate hot and cold deck temperatures according to demand. When properly installed, and with all hot deck or cold deck dampers partially closed, the control will reduce hot and raise cold deck temperature progressively until one or more zone dampers is fully open.

3. Consider converting systems serving interior zones to variable volume. Conversion is performed by blocking off the hot deck, removing or disconnecting mixing dampers, and

adding low pressure variable volume terminals and pressure bypass.

Terminal Reheat System—The terminal reheat system essentially is a modification of a single-zone system which provides a high degree of temperature and humidity control. The central heating/cooling unit provides air at a given temperature to all zones served by the system. Secondary terminal heaters then reheat air to a temperature compatible with the load requirements of the specific space involved. Obviously, the high degree of control provided by this system requires an excessive amount of energy. Several methods for making the system more efficient include:

1. Reduce air volume of single zone units.

2. If close temperature and humidity control must be maintained for equipment purposes, lower water temperature and reduce flow to reheat coils. This still will permit control, but will limit the system's heating capabilities somewhat.

3. If close temperature and humidity control are not required, convert the system to variable volume by adding variable volume valves and eliminating terminal heaters.

THE ECONOMIZER CYCLE

The basic concept of the economizer cycle is to use outside air as the cooling source when it is cold enough. There are several parameters which should be evaluated in order to determine if an economizer cycle is justified. These include:

- Weather
- Building occupancy
- The zoning of the building
- The compatibility of the economizer with other systems
- The cost of the economizer

What Are the Costs of Using
the Economizer Cycle?

Outside air cooling is accomplished usually at the expense of an additional return air fan, economizer control equipment, and an additional burden on the humidification equipment. Therefore, economizer cycles must be carefully evaluated based on the specific details of the application.

Using outside air to cool a building can result in lower mechanical refrigeration cost whenever outdoor air has a lower total heat content (enthalpy) than the return air. This can be accomplished by an "integrated economizer" or enthalpy control. See Figure 7-10 for a comparison of controls.

Dry Bulb Economizer

Operation of the "integrated economizer" can be made automatic by providing (1) dampers capable of providing 100% outdoor air, and (2) local controls that sequence the chilled water or DX (direct expansion) coil and dampers so that during economizer operation, on a rise in discharge (or space) temperature, the outdoor damper opens first; then on a further rise, the cooling coil is turned "on."

Economizer operation is activated by outside air temperature, say 72°F DB*. If outside air is below 72°F, the above described economizer sequence occurs. Above 72°F, outside air cooling is not economical, and the outdoor air damper closes to its minimum position to satisfy ventilation requirements only.

Enthalpy Control

If an economizer system is equipped with enthalpy control, savings will accrue due to a more accurate changeover point. The load on a cooling coil for an air handling system is a function of the *total heat* of air entering the coil. Total heat is a function of *two*

---

*This varies according to location.

246

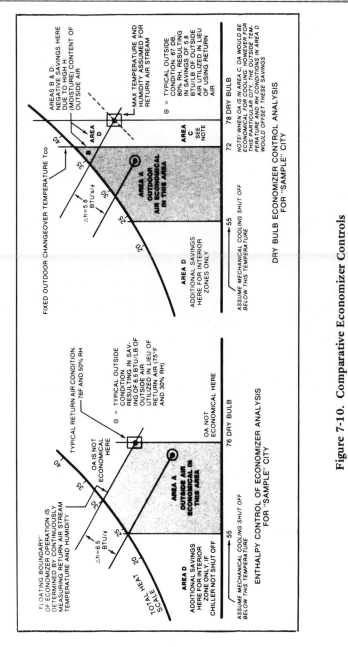

Figure 7-10. Comparative Economizer Controls

247

measurements, dry bulb (DB) and relative humidity (RH) or dew point (DP). The enthalpy control measures both conditions (DB and RH) in the return air duct and outdoors. It then computes which air source would impose the lowest load on the cooling system. If outside air is the smallest load, the controller enables the economizer cycle. (Dry Bulb Economizer control savings will be less than those shown for enthalpy control, except in dry climates.)

Enthalpy Control Savings Calculations

Savings are based on the assumption that the system previously had either (a) no 100% O.A. damper or (b) a fixed minimum outdoor air setting (whenever the fan operates) sufficient for ventilation purposes; it is also assumed that (c) minimum outdoor air has already been reduced, and the minimum damper opening will be at the *new value*.

Step 1. Determine minimum cfm of outdoor air to be used during occupied hours.

Step 2. Calculate annual savings.

$$A \frac{ft^3}{min.} \left(1 - \frac{B\%}{100}\right) \times K \frac{10^6 \ Btu}{yr \ 1000 \ cfm} \times \frac{operating \ hrs/wk}{50} \times J \frac{\$}{10^6 \ Btu}$$

$$= \$ \ SAVED \ PER \ YEAR \qquad \qquad Formula \ (7\text{-}25)$$

where

A = air handling capacity $\left(\dfrac{ft^3}{min.}\right)$

B = present ventilation air (%)

J = cost of cooling $\left(\dfrac{\$}{10^6 \ Btu}\right)$

K = seasonal cooling savings $\left(\dfrac{10^6 \ Btu}{yr \ 1000 \ cfm}\right)$

Formula 7-25 is used to calculate the savings resulting from enthalpy control of outdoor air. The calculated savings generally will be greater than the savings resulting from a dry bulb economizer. To estimate dry bulb economizer savings, multiply the enthalpy savings by .93.

8

HVAC Equipment

PERFORMANCE RATIOS

Measuring Efficiency by Using the Coefficient of Performance

The coefficient of performance *(COP)* is the basic parameter used to compare the performance of refrigeration and heating systems. *COP* for cooling and heating applications is defined as follows:

$$COP \text{ (Cooling)} = \frac{\text{Rate of Net Heat Removal}}{\text{Total Energy Input}} \qquad \textit{Formula (8-1)}$$

$$COP \text{ (Heating, Heat Pump*)} = \frac{\text{Rate of Useful Heat Delivered*}}{\text{Total Energy Input}}$$
$$\textit{(Formula (8-2)}$$

Measuring System Efficiency Using the Energy Efficiency Ratio

The energy efficiency ratio *(EER)* is used primarily for air-conditioning systems and is defined by Formula 8-3.

$$EER = \frac{COP \,(3412)}{1000} \text{ Btu/watt-hr} \qquad \textit{Formula (8-3)}$$

---

*For Heat Pump Applications, exclude supplemental heating.

249

THE HEAT PUMP

The heat pump has gained wide attention due to its high potential *COP.* The heat pump in its simplest form can be thought of as a window air conditioner. During the summer, the air on the room side is cooled while air is heated on the outside air side. If the window air conditioner is turned around in the winter, some heat will be pumped into the room. Instead of switching the air conditioner around, a cycle reversing valve is used to switch functions. This valve switches the function of the evaporator and condenser, and refrigeration flow is reversed through the device. Thus, *the heat pump is heat recovery through a refrigeration cycle.* Heat is removed from one space and placed in another. In Chapter 7, it was seen that the direction of heat flow is from hot to cold. Basically, energy or pumping power is needed to make heat flow "up hill." The mechanical refrigeration compressor "pumps" absorbed heat to a higher level for heat rejection. The refrigerant gas is compressed to a higher temperature level so that the heat absorbed by it, during the evaporation or cooling process, is rejected in the condensing or heating process. Thus, the heat pump provides cooling in the summer and heating in the winter. The source of heat for the heat pump can be from one of three elements: air, water or the ground.

Air to Air Heat Pumps

Heat exists in air down to 460°F below zero. Using outside air as a heat source has its limitations, since the efficiency of a heat pump drops off as the outside air level drops below 55°F. This is because the heat is more dispersed at lower temperatures, or more difficult to capture. Thus, heat pumps are generally sized on cooling load capacities. Supplemental heat is added to compensate for declining capacity of the heat pump. This approach allows for a realistic first cost and an economical operating cost.

An average of two to three times as much heat can be moved for each Kw input compared to that produced by use of straight resistance heating. Commercially available heat pumps range in size from 2 to 3 tons for residences up to 40 tons for commercial and industrial users. Figure 8-1 illustrates a simple scheme for determining the supplemental heat required when using an air–air heat pump.

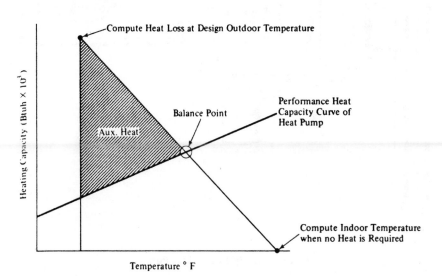

Figure 8-1. Determining Balance Point of Air to Air Heat Pump

Hydronic Heat Pump

The hydronic heat pump is similar to the air to air unit, except the heat exchange is between water and refrigerant instead of air to refrigerant, as illustrated in Figure 8-2. Depending on the position of the reversing valve, the air heat exchanger either cools or heats room air. In the case of cooling, heat is rejected through the water-cooled condenser to the building water. In the case of heating, the reversing valve causes the water to refrigerant heat exchanger to become an evaporator. Heat is then absorbed from the water and discharged to the room air.

251

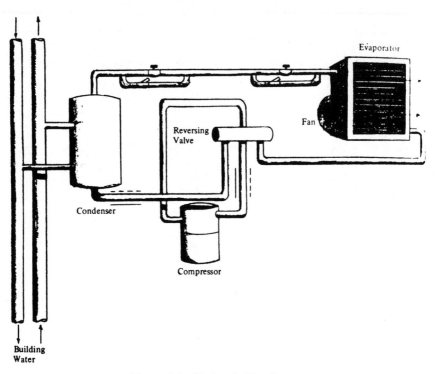

Figure 8-2. Hydronic Heat Pump

Imagine several hydronic heat pumps connected to the same building water supply. In this arrangement, it is conceivable that while one unit is providing cool air to one zone, another is providing hot air to another zone; the first heat pump is providing the heat source for the second unit, which is heating the room. This illustrates the principle of energy conservation. In practice, the heat rejected by the cooling units does not equal the heat absorbed. An additional evaporative cooler is added to the system to help balance the loads. A better heat source would be the water from wells, lakes, or rivers which is thought of as a constant heat source. Care should be taken to insure that a heat pump connected to such a heat source does not violate ecological interests.

252

Liquid Chiller

A liquid chilling unit (mechanical refrigeration compressor) cools water, brine, or any other refrigeration liquid, for air conditioning or refrigeration purposes. The basic components include a compressor, liquid cooler, condenser, the compressor drive, and auxiliary components. A simple liquid chiller is illustrated in Figure 8-3. The type of chiller usually depends on the capacity required. For example, small units below 80 tons are usually reciprocating, while units above 350 tons are usually centrifugal.

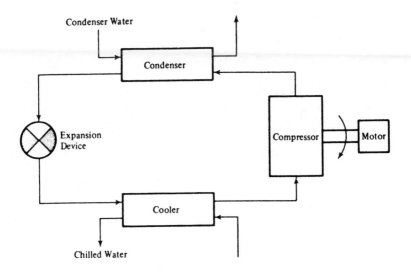

Figure 8-3. Liquid Chiller

A factor which affects the power usage of liquid chillers is the percent load and the temperature of the condensing water. *A reduced condenser water temperature saves energy.* In Figure 8-4, it can be seen that by reducing the original condenser water temperature by 10 degrees, the power consumption of the chiller is reduced. Likewise, a chiller operating under part load consumes less power. The

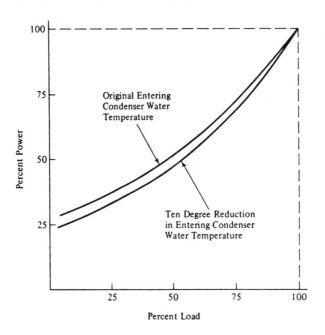

Figure 8-4. Typical Power Consumption Curve for Centrifugal Liquid Chiller

"ideal" coefficient of performance *(COP)* is used to relate the measure of cooling effectiveness. Approximately 0.8 Kw is required per ton of refrigeration (0.8 Kw is power consumption at full load, based on typical manufacturer's data).

$$COP = \frac{1 \text{ ton} \times 12{,}000 \text{ Btu/ton}}{0.8 \text{ Kw} \times 3412 \text{ Btu/Kw}} = 4.4$$

Chillers in Series and in Parallel

Multiple chillers are used to improve reliability, offer standby capacity, reduce inrush currents and decrease power costs at partial loads. Figure 8-5 shows two common arrangements for chiller staging; namely, chillers in parallel and chillers in series.

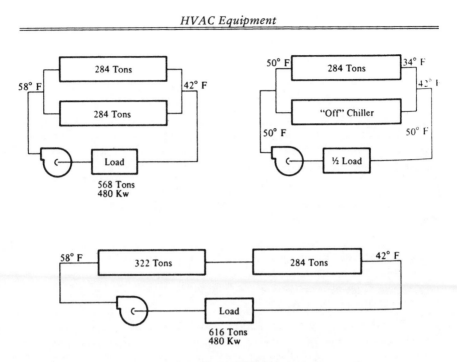

Figure 8-5. Multiple Chiller Arrangements

In the parallel chiller arrangement, liquid to be chilled is divided among the liquid chillers and the streams are combined after chilling. Under part load conditions, one unit must provide colder than designed chilled liquid so that when the streams combine, including the one from the off chiller, the supply temperature is provided. The parallel chillers have a lower first cost than the series chillers counterparts but usually consume more power.

In the series arrangement, a constant volume of flow of chilled water passes through the machines, producing better temperature control and better efficiency under part load operation; thus, the upstream chiller requires less Kw input per ton output. The waste of energy during the mixing aspect of the parallel chiller operation is avoided. The series chillers, in general, require higher pumping costs. The energy conservation engineer should evaluate the best arrangement, based on load required and the partial loading conditions.

The Absorption Refrigeration Unit

Any refrigeration system uses external energy to "pump" heat from a low temperature level to a high temperature. Mechanical refrigeration compressors pump absorbed heat to a higher temperature level for heat rejection. Similarly, absorption refrigeration changes the energy level of the refrigerant (water) by using lithium bromide to alternately absorb it at a low temperature level and reject it at a high level by means of a concentration-dilution cycle.

The single-stage absorption refrigeration unit uses 10 to 12 psig steam as the driving force. Whenever users can be found for low pressure steam, energy savings will be realized. A second aspect for using absorption chillers is that they are compatible for use with solar collector systems. Several manufacturers offer absorption refrigeration equipment which uses high temperature water (160°–200°F) as the driving force.

A typical schematic for a single-stage absorption unit is illustrated in Figure 8-6. The basic components of the system are the evaporator, absorber, concentrator, and condenser. These components can be grouped in a single or double shell. Figure 8-6 represents a single-stage arrangement.

Evaporator—Refrigerant is sprayed over the top of the tube bundle to provide for a high rate of transfer between water in the tubes and the refrigerant on the outside of the tubes.

Absorber—The refrigerant vapor produced in the evaporator migrates to the bottom half of the shell where it is absorbed by a lithium bromide solution. Lithium bromide is basically a salt solution which exerts a strong attractive force on the molecules of refrigerant (water) vapor. The lithium bromide is sprayed into the absorber to speed up the condensing process. The mixture of lithium bromide and the refrigerant vapor collects in the bottom of the shell; this mixture is referred to as the dilute solution.

Concentrator—The dilute solution is then pumped through a heat exchanger where it is preheated by hot solution leaving the

256

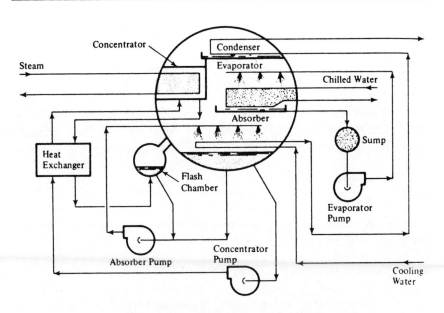

Figure 8-6. One-Shell Lithium Bromide Cycle Water Chiller
(Source: Trane Air Conditioning Manual)

concentrator. The heat exchanger improves the efficiency of the cycle by reducing the amount of steam or hot water required to heat the dilute solution in the concentrator. The dilute solution enters the upper shell containing the concentrator. Steam coils supply heat to boil away the refrigerant from the solution. The absorbent left in the bottom of the concentrator has a higher percentage of absorbent than it does refrigerant, thus it is referred to as concentrated.

Condenser—The refrigerant vapor boiling from the solution in the concentrator flows upward to the condenser and is condensed. The condensed refrigerant vapor drops to the bottom of the condenser and from there flows to the evaporator through a regulating orifice. This completes the refrigerant cycle.

The single-stage absorption unit consumes approximately 18.7 pounds of steam per ton of capacity (steam consumption at full load based on typical manufacturers data). For a single-state absorption unit,

$$COP = \frac{1 \text{ ton} \times 12{,}000 \text{ Btu/ton}}{18.7 \text{ lb} \times 955 \text{ Btu/lb}} = 0.67$$

The single-stage absorption unit is not as efficient as the mechanical chiller. It is usually justified based on availability of low pressure steam, equipment considerations, or use with solar collector systems.

Example Problem 8-1

Compute the energy wasted when 15 psig steam is condensed prior to its return to the power plant. Comment on using the 15 psig steam directly for refrigeration.

Answer

From Steam Table 16-14 for 30 psia steam, hfg is 945 Btu per pound of steam; thus, 945 Btu per pound of steam is wasted. In this case where *excess low pressure* steam cannot be used, absorption units should be considered in place of their electrical-mechanical refrigeration counterparts.

Example Problem 8-2

2000 lb/hr of 15 psig steam is being wasted. Calculate the yearly (8000 hr/yr) energy savings if a portion of the centrifugal refrigeration system is replaced with single-stage absorption. Assume 20 Kw additional energy is required for the pumping and cooling tower cost associated with the single-stage absorption unit. Energy rate is $.09 Kwh and the absorption unit consumes 18.7 lb of steam per ton of capacity.

The centrifugal chiller system consumes 0.8 Kwh per ton of refrigeration.

258

Answer

Tons of mechanical chiller capacity replaced = 2000/18.7 = 106.95 tons. Yearly energy savings = 2000/18.7 × 8000 × 0.8 × $.09 = $61,602.

Two-Stage Absorption Unit

The two-stage absorption refrigeration unit (Figure 8-7) uses steam at 125 to 150 psig as the driving force. In situations where excess medium pressure steam exists, this unit is extremely desirable. The unit is similar to the single-stage absorption unit. The two-stage absorption unit operates as follows:

Medium pressure steam is introduced into the first-stage concentrator. This provides the heat required to boil out refrigerant from the dilute solution of water and lithium bromide salt. The liberated refrigerant vapor passes into the tubes of the second-stage concentrator, where its temperature is utilized to again boil a lithium bromide solution, which in turn further concentrates the solution and liberates additional refrigerant. In effect, the concentrator frees an increased amount of refrigerant from solution with each unit of input energy.

The condensing refrigerant in the second-stage concentrator is piped directly into the condenser section. The effect of this is to reduce the cooling water load. A reduced cooling water load decreases the size of the cooling tower which is used to cool the water. The remaining portions of the system are basically the same as the single-stage unit.

The two-stage absorption unit consumes approximately 12.2 pounds of steam per ton of capacity; thus, it is more efficient than its single-stage counterpart. The associated COP is:

$$COP = \frac{1 \text{ ton} \times 12{,}000 \text{ Btu/ton}}{12.2 \text{ lb} \times 860 \text{ Btu/lb}} = 1.14$$

Either type of absorption unit can be used in conjunction with centrifugal chillers when it is desirable to reduce the peak electrical demand of the plant, or to provide for a solar collector addition at a later date.

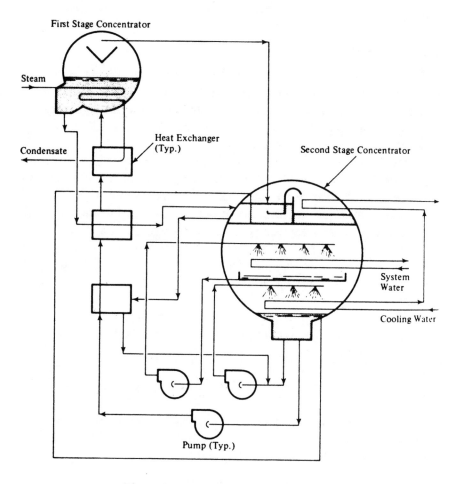

Figure 8-7. Two-Stage Absorption Unit

9

Cogeneration: Theory and Practice

CARL E. SALAS, P.E.
Vice President, O'Brien and Associates, Inc.

Because of its enormous potential, it is important to understand and apply cogeneration theory. In the overall context of Energy Management Theory, cogeneration is just another form of the conservation process. However, because of its potential for practical application to new or existing systems, it has carved a niche that may be second to no other conservation technology.

This chapter is dedicated to development of a sound basis of current theory and practice of cogeneration technology. It is the blend of theory and practice, or praxis of cogeneration, that will form the basis of the most workable conservation technology during the remainder of the 20th century.

INTRODUCTION

There are a myriad of definitions for cogeneration. One of the more acceptable definitions is:

Cogeneration is the sequential production of electricity and heat, steam or useful work from the same fuel source.

This means that rather than using the energy in the fuel for a single function, as typically occurs, the available energy is cascaded through at least two *useful* cycles.

To put it in more simple terms: cogeneration is a very efficient method of making use of *all* of the available energy expended during any process generating electricity (or shaft horsepower) and then utilizing the waste heat.

A more subjective definition of cogeneration calls upon current practical applications of power generation and process needs. Nowhere, more than in the United States is an overall system efficiency of only 30% tolerated as "standard design." In the name of limited *initial* capital expenditure, all of the waste heat from most process is rejected to the atmosphere.

In short, present design practices dictate that of the useful energy in one gallon of fuel, only 30% of that fuel is put to useful work. The remaining 70% is rejected randomly.

If one gallon of fuel goes into a process, the designer may ask, "how much of that raw energy can I make use of within the constraints of the overall process?"

In this way, cogeneration may be taken as a way to use a maximum amount of available energy from any raw fuel process. Thus, cogeneration may be thought of as: *just good design.*

AN OVERVIEW OF
COGENERATION THEORY

As discussed in the introduction, and as may be seen from Figure 9-1, standard design practices make use of, at best, 30% of available energy from the raw fuel source (gas, oil, coal).

Of the remaining 70% of the available energy, approximately 30% of the heat is rejected to the atmosphere through a condenser (or similar) process. An additional 30% of the energy is lost directly to the atmosphere through the stack, and finally, approximately 7% of the available energy is radiated to the atmosphere because of the high relative temperature of the process system.

262

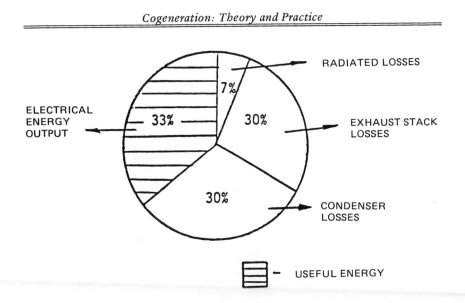

Figure 9-1. Energy Balance Without Heat Recovery

With heat recovery, however, potential useful application of available energy more than doubles. Although in a "low quality" form, *all* of the condenser-related heat may be used, and 40% of the stack heat may be recovered. This optimized process is depicted (in theory) as Figure 9-2.

Thus, it may be seen that effective use of all available energy may more than double the "worth" of the raw fuel. System efficiency is increased from 30% to 75%.

This efficiency may be taken advantage of by a clear understanding of the available uses of low grade energy and the various cogeneration cycles available to the designer.

Example Problem 9-1

A cogeneration system vendor recommends the installation of a 20-megawatt cogeneration system for a college campus. Determine the approximate *range* of useful *thermal* energy. Use the energy balance of Figure 9-2.

263

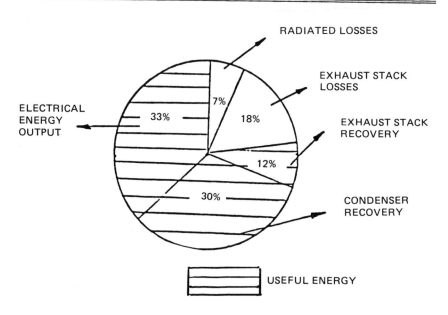

Figure 9-2. Energy Balance With Heat Recovery

Analysis

Step 1: "Range" is defined by the best and worst operating *times* of the installed system:

At best, system will operate 365 days per year, 24 hours per day.

At worst, system will operate 5 days per week, 10 hours per day.

Step 2: Perform heat balance. See diagram on the next page.

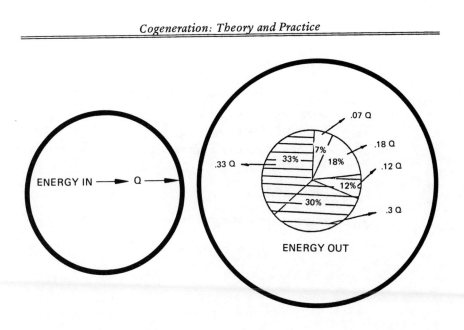

$$Q = E_{out} = E_{electricity} + E_{condenser} + E_{stack} + E_{radiated} \quad Formula\ (9\text{-}1)$$

Step 3: Calculate *available thermal* energy.

From Figure 9-2; available energy equals condenser energy and 40% stack energy.

$$E_{available} = E_{condenser} + .4\ E_{stack} \qquad Formula\ (9\text{-}2)$$

$$E_{condenser} = .3Q \qquad Formula\ (9\text{-}3)$$

$$E_{stack} = .3Q \qquad Formula\ (9\text{-}4)$$

$$E_{electrical} = .33Q \qquad Formula\ (7\text{-}5)$$

$$Q = \frac{E_{electrical}}{.33} \qquad Formula\ (9\text{-}6)$$

By substitution

$$E_{available} = 1.272\ \text{Electrical} \qquad Formula\ (9\text{-}7)$$

265

Calculate available energy

$$Q = Q_{1t} = K_1 \, E_{available} \times K_2 \times t$$

K_1 = 3413 Btu/Kwh

K_2 = 1000 Kilowatts per megawatt

t = equipment hours of operation

Q_1 = 3413 Btu/Kwh $\times$ (1.272 $\times$ 20 Mwatt)

$\times$ 1000 $\dfrac{\text{Kw}}{\text{Mwatt}}$

Q_1 = 86.82 $\times$ 10^6 Btu/hr

Worst Case: System operates

5 days/wk $\times$ 52 wks/yr $\times$ 10 hrs/day = 2600 hrs/yr

Best Case: System operates

365 days/yr $\times$ 24 hrs/day = 8670 hrs/yr

Thus range is

Q = 86.82 $\times$ 10^6 $\times$ 2600 = 225.7 $\times$ 10^9 Btu/yr

Q = 86.82 $\times$ 10^6 $\times$ 8760 = 760.5 $\times$ 10^9 Btu/yr

APPLICATION OF THE
COGENERATION CONSTANT

The *cogeneration constant* may be used as a fast check on any proposed cogeneration installation. Notice from the sample problem which follows, the ease with which a thermal vs. electrical comparison of end needs may be made.

Example Problem 9-2

A cogeneration system vendor recommends a 20 megawatt installation. Determine the approximate rate of *useful* thermal energy.

Analysis

$$Q = E \times K_c \qquad\qquad \textit{Formula (9-9)}$$

E is the cogeneration system electrical rated capacity

K_c is the cogeneration constant

Q = 20 Mw $\times$ 1.272

    = 25.4 megawatt of *useful* heat

    or

$$25.4 \text{ Mw} \times 1000\,\frac{\text{Kw}}{\text{Mw}} \times \frac{3413 \text{ Btu}}{\text{Kwh}} \times \frac{\text{Therm}}{100{,}000 \text{ Btu}}$$

Q = 866.9 therms/hour!

APPLICABLE SYSTEMS

To ease the complication of matching power generation to load, and because of newly established laws, it is most advantageous that the generator operate in parallel with the utility grid which thereby "absorbs" all generated electricity. The requirements for "qualify facility (QF) status," and the consequent utility rate advantages which are available when "paralleling the grid," is that a significant portion of the thermal energy produced in the generation process must be recovered. Specifically, the equation 9-10 must be satisfied:

$$\frac{\text{Power Output} + \tfrac{1}{2} \text{ Useful Thermal Output}}{\text{Energy Input}} \geqslant 42.5\% \text{ (for any calendar year)}$$

$$\textit{Formula (9-10)}$$

Note that careful application of the *cogeneration constant* will generally assure that the qualifying facility status is met.

BASIC THERMODYNAMIC CYCLES

A sound understanding of basic cogeneration principles dictates that the energy manager be familiar with two standard thermodynamic cycles. These cycles are:

1. Brayton Cycle
2. Rankine Cycle

The Brayton Cycle is the basic thermodynamic cycle for the simple gas turbine power plant. The Rankine Cycle is the base cycle for a vapor-liquid system typical of steam power plants. An excellent theoretical discussion of these two cycles appears in Reference 5.

The Brayton Cycle

In the open Brayton Cycle plant, energy input comes from the fuel that is injected into the combustion chamber. The gas/air mixture drives the turbine with high temperature waste gases exiting to the atmosphere. (See Figure 9-3.)

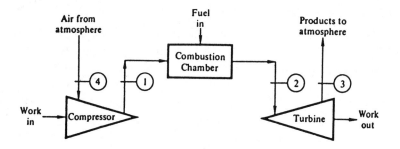

Figure 9-3. Open Gas-Turbine Power Plant

The basic Brayton Cycle, as applied to cogeneration, consists of a gas turbine, waste heat boiler and a process or "district" heating load. This cycle is a *full-load* cycle. At part loads, the efficiency of the gas turbine goes down dramatically. A simple process diagram is illustrated in Figure 9-4.

Because the heat rate of this arrangement is superior to that of all other arrangements at full load, this simple, standard Brayton Cycle merits consideration under all circumstances. Note that to complete the loop in an efficient manner, a deaerator and feedwater pump are added.

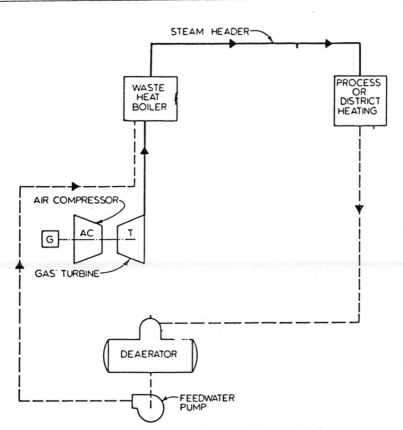

Figure 9-4. Process Diagram — Brayton Cycle

The Rankine Cycle

The Rankine Cycle is illustrated in the simplified process diagram, Figure 9-5. Note that this is the standard boiler/steam turbine arrangement found in many power plants and central facility plants throughout the world.

The Rankine Cycle, or steam turbine, provides a real-world outlet for waste heat recovered from any process or generation situation. Hence, it is the steam turbine which is generally referred to as the topping cycle.

269

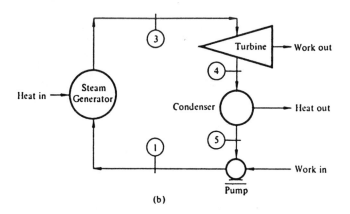

Figure 9-5. Simple Rankine Cycle

Combined Cycles

Of major interest and import for the serious central plant designer is the Combined Cycle. This cycle forms a hybrid which includes the Brayton Cycle on the "bottoming" portion; and a standard Rankine Cycle on the "topping" portion of the combination. A process diagram with standard components is illustrated in Figure 9-6.

The Combined Cycle, then, greatly approximates the cogeneration Brayton Cycle but makes use of a knowledge of the plant requirements and an understanding of Rankine Cycle theory. Note also, that the ideal mix of power delivered from the Brayton and Rankine portions of the Combined Cycle is 70% and 30%, respectively.

Even within the seemingly limited set of situations defined as "Combined Cycle" many, many variations and options become available. These options are as much dependent on any local plant requirements and conditions as they are on available equipment.

Some examples of Combined Cycle variations are:

1. Gas turbine exhaust used to produce 15 psi steam for Rankine Cycle turbine with no additional fuel burned. This situation is shown in Figure 9-6.

270

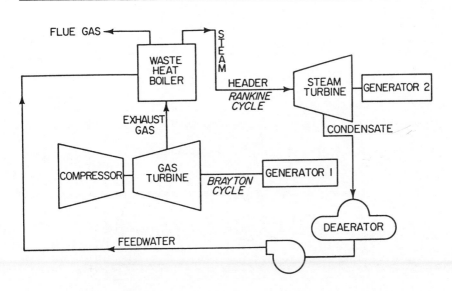

Figure 9-6. Combined Cycle Operation

2. Gas turbine exhaust fired in the duct with additional fuel. This provides a much greater amount of produced power with a correspondingly greater amount of fuel consumption. This situation generally occurs with a steam turbine pressure range of 900–1259 psig.

3. Gas turbine exhaust fired directly and used directly as combustion air for a conventional power boiler. Note here that the boiler pressure range may vary between 200 and 2600 psig. See Figure 9-7.

One other note to keep in mind is that in any Combined Cycle case, the primary or secondary turbine may supply direct mechanical energy to a refrigerant compressor. As discussed, the variations are endless. However, a thorough understanding of the end process generally will end in a final, and best, cogeneration system selection.

271

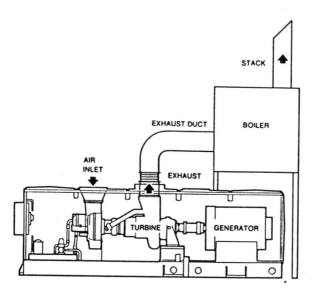

Figure 9-7. Gas Turbine Used As Combustion Air

SUMMARY

During the remainder of the 20th century, the one power technology that has the greatest potential toward making maximum use out of present wasteful plant and facility conditions is cogeneration.

The following guidelines should be considered in cogeneration design:

- The *Cogeneration Constant* must be applied to the rated electrical capacity of any cogeneration system design. If *local* conditions do not necessitate a use for thermal energy equivalent to 1.272 X electrical output, the overall system sizing should be decreased.

- Rankine, Brayton and Combined Cycle scenarios have typical site-specific advantages and disadvantages. It is only by understanding the local site and facility conditions, and further blending site needs with system strengths that a truly viable system may be designed.

272

10

Passive Solar Energy Systems*

INTRODUCTION

Passive solar energy systems use natural, nonmechanical forces—including sunshine, shading, and breezes—to help heat and cool homes with little or no use of electrical controls, pumps, or fans.

It should not be assumed that passive solar building heating measures are less effective or their application less widespread than the active measures. In fact there are few, if any, buildings that would not benefit in comfort and energy cost savings by the use of one or more passive measures.

Passive heating systems often depend on "thermal storage mass" to absorb and store the sun's energy for use within a building. Storage mass includes materials such as rock, stone, concrete, or water, which store large quantities of heat as they absorb the sun's energy. This storage is accompanied by a moderate rise in temperature—the larger the temperature rise, the greater the energy stored. Later, as the thermal storage mass cools, the stored heat is released to heat the living space. "Phase change" materials may also be used for storage.

---

*Source: Residential Conservation Service Auditor Training Manual.

New passive buildings are generally integrated into their sites. Their design takes advantage of the natural environment to heat and cool the building by natural means. For example, evergreen trees are planted on the north side for winter protection, and deciduous trees on the south side to heighten winter sun access. Hillsides are used as earth berms that insulate the building. Indigenous materials, such as adobe or rocks are used for thermal mass and insulation. Southerly orientation and glazing contribute direct solar gain for heating. Thermal mass on or near south glazing contributes to heating, as well. Shading devices and overhangs are used to keep out the high summer sun when cooling is needed, and the thermal mass absorbs heat, effectively cooling the structure. Many of these passive features can be retrofitted to appropriately oriented existing structures.

Passive systems may require some attention from household members to perform at their peak. For example, someone may have to slide insulation into place to close off windows at night or during periods of extended cloud cover to reduce heat losses. Also, vents may have to be opened to control heat flow.

Well-designed passive buildings incorporate energy-saving features beyond standard improvements (insulation, weatherstripping, and caulking). Depending on the specific location and design of the building, a variety of methods are suitable for additional energy savings in retrofit situations. For example, vestibules, now called airlock entries, are reappearing as a popular way to keep incoming blasts of cold or hot outside air to a minimum. Effective placement of vegetation such as trees, shrubs, and vines is a natural, simple way to protect a building from temperature extremes.

This chapter discusses five passive solar systems.

- thermosyphon domestic hot water systems;
- direct gain glazing systems;
- indirect gain systems;
- sunspaces; and
- window heat gain retardant devices.

THERMOSYPHON DOMESTIC HOT WATER SYSTEMS

Thermosyphon systems are generally considered to be passive measures most suitable for warm climates. Chapter 12 describes this system.

DIRECT GAIN GLAZING SYSTEMS

In direct gain systems, the building itself is a solar collector. Direct gain in its simplest form is probably the easiest way to apply solar energy to any building. It simply involves letting the sun shine in through added south-facing windows, thereby heating space directly. Nearly all of the sunlight entering a room is immediately converted to heat.

Thermal mass for storing excess heat (such as a concrete floor) may be located with direct exposure to the sunlight or in some other part of the building. All materials can store heat, but some store more than others. Table 10-1 lists the heat capacities (ability to store heat) of some common materials. Of the materials listed, air holds the least amount of heat and water the most.

To reduce heat loss, and thus increase overall thermal performance, insulation may be placed next to the glass at night, either inside or outside. During the heating season, south-facing glass takes advantage of the sun's low position in the sky; in the summer when the sun is high in the sky, the glass may be shaded by overhangs or foliage. Both east and west glass, even in cold climates, can admit somewhat more solar energy than they lose if nighttime insulation is used. North glazing should be kept at a minimum.

Vertical glass admits almost as much heat during the winter as tilted glass (skylights) and is much easier to insulate and keep clean; also, vertical glass does not break as easily. Many codes require tilted glass to be tempered, which is more expensive than regular window glass.

South-facing windows that are designed to distribute heat to as much of the building as possible are preferred. Clerestory windows with adequate overhangs should be considered before skylights.

275

Table 10-1. Heat Capacities of Common Materials

Material	Heat Capacity (Btu/ft³–°F)
Air (75°F)	0.018
Clay	13.9
Sand	18.1
Gypsum	20.3
Limestone	22.4
Wood, oak	26.8
Glass	27.7
Brick	28
Concrete	28
Asphalt	29
Aluminum	36.6
Marble	38
Copper	51.2
Iron	55
Water	62.5

If possible, south-facing windows that are exposed to the sun should be located so that the sunlight falls directly on thermal mass (heavy masonry fireplaces, masonry floors, water walls, etc.). Structures built with concrete floors (e.g., slab on grade) are frequently carpeted, eliminating the exposure of thermal mass. Replacing the carpeting with heavy ceramic tiles could result in greatly increased solar storage, reduced building heat load, and lower heating bills.

Thermal mass provides nighttime heating by reradiating stored heat to living space. Thermal mass also tempers temperature swings that can occur. Some people will not want nighttime heating if it means altering their homes to provide storage; others will not mind the changes necessary to provide some nighttime heating, to prevent overheating, and to moderate temperature variations.

Applications

Addition of south-facing glazing should be considered as a potential passive solar retrofit wherever south-facing walls are available and an outside view is desired. Direct gain systems help to heat buildings.

Advantages and Disadvantages

Glass is relatively inexpensive, widely available, and thoroughly tested. The overall direct gain system can be one of the least expensive means of solar heating, the simplest solar energy system to conceptualize, and the easiest to build. In many instances, it can be achieved by simply enlarging existing windows. Direct gain systems, besides heating the interior, provide natural lighting and a view. To meet a small fraction of the heating needs of a building, direct gain systems do not necessarily need thermal storage.

On the other hand, ultraviolet radiation in the sunlight can degrade fabrics and photographs. If the desire is to achieve large energy savings, then relatively large glazing areas and correspondingly large amounts of thermal mass are required to decrease temperature swings. Thermal mass can be expensive, unless it serves a structural purpose. Interior daily temperature swings of 15° to 20°F are common even with thermal mass.

Construction Terms

Clerestory—Vertical window placed high in wall near eaves; used for light, heat gain, and ventilation.

Double Glazed—A frame with two panes of transparent glazing with space between the panes.

Fenestration—The arrangement, proportioning, and design of windows or doors in a building.

Glazing—Transparent or translucent material, generally glass or plastic, used to cover a window opening in a building.

Header—A horizontal structural member over an opening used to support the load above the opening.

277

Lite—A single pane of glazing.

Movable Insulation—Insulation (such as shutters, panels, curtains, or reflective foil draperiss) that can be moved manually or by mechanical means.

Shading Coefficient—The ratio of the solar heat gain through a specific glazing system to the total solar heat gain through a single layer of clear double-strength glass.

Tempered Glazing—Glazing that has been specially treated to resist breakage.

Thermal Mass—Any material used to store the sun's heat or the night's coolness. Water, concrete, and rock are common choices for thermal mass. In winter, thermall mass stores solar energy collected during the day and releases it during sunless periods (nights or cloudy days). In summer, thermal mass absorbs excess daytime heat, and ventilation allows it to be discharged to the outdoors at night.

Thermal Storage Floors, Ceilings, and Interior Walls—Floors, ceilings, and interior walls that contain thermal mass and are used to collect and/or store heat in solar energy systems. They can be exposed to sun directly or receive only indirect solar heat to be effective.

INDIRECT GAIN GLAZING SYSTEMS

Any passive heating system that uses some intermediary material for heat collection and/or storage before passing that heat on to its desired place of use can be called an indirect passive system. The term "indirect gain systems" means the use of panels of insulated glass, fiberglass or other transparent substances that direct the sun's rays onto specially constructed thermal walls, ceilings, rockbeds, or containers of water or other fluids where heat is stored and radiated.

Trombe Wall

The Trombe wall uses a heat storage mass placed between glass and the space to be heated. Per unit of thermal storage mass used, the Trombe wall makes the best use of the material. While the temperature swing in the material is great, the temperature variation in the heated space is small. See Figure 10-1 for some features of a Trombe wall.

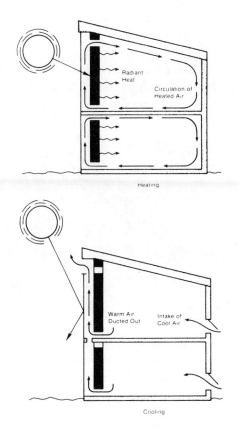

Figure 10-1. Trombe Wall: Heating and Cooling

Water Wall

The water wall uses the same principle as the Trombe wall and involves replacing the existing wall, or parts of it, with containers that hold water. The water mass then stores heat during the day and releases the heat as needed.

279

Thermosyphon Air Panels (TAP)

TAP systems are often called "day heaters" because of their effective use during the day, since they have no storage. They are similar in appearance to active flat-plate collectors, and are often mounted vertically. A TAP system has one or more glazings of glass or plastic, an air space, an absorber, another air space, and (often) an insulated backing. Air flows naturally up in front of, behind, or through the absorber and re-enters the building through a vent at the top. Figure 10-2 displays structural features of TAP systems.

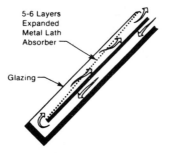

5-6 Layers
Expanded
Metal Lath
Absorber

Glazing

Figure 10-2. Airflow Through a TAP System

Applications

Trombe walls, water walls, and thermosyphon air panel systems all provide heat for residential living space. Each system requires south-facing opaque walls with solar access.

Advantages and Disadvantages

Trombe and water walls have several advantages:

- Glare and ultraviolet degradation of fabrics are not problems.
- Temperature swings in the living space are lower than with direct gain or convective loop systems.
- The time delay between the absorption of solar radiant energy by the surface and the delivery of the resulting heat to

280

the space provides warmth in the evening when most residences need it.

Trombe and water walls have the following disadvantages:

- Two south walls (a glazed wall and a mass wall) are needed.
- Massive walls are not often found in residential construction (although thermal storage walls may be the least expensive way to achieve the required thermal storage since they are compactly located behind the glass).
- In cold climates, considerable heat is lost to the outside from the warm wall through the glazing unless the glazing is insulated at night.

Thermosyphon Air Panels (TAP) systems have the following advantages:

- Glare and ultraviolet degradation of fabrics are not problems.
- TAPs provide one of the least expensive ways to collect solar heat.
- To provide a small fraction of the heating needs of a building, thermal storage is not needed.
- TAPs are easily incorporated onto south facades.
- TAPs are readily adaptable to existing buildings, and require less skill for resident installation.
- Because the collector can be thermally isolated from the building interior, night heat losses are lower than for any other uninsulated passive system.

TAP systems have the following disadvantages:

- The collector is an obvious add-on device.
- Both careful engineering and construction are required to ensure proper airflows, air seals, and adequate thermal isolation at night.
- The thermal energy is delivered as warmed air—it is difficult to store this heat for later retrieval because air transfers heat poorly to other mass.

281

- It's often impractical to add much collector area.
- Occupant lifestyle, room use patterns, and natural airflows within the residence will strongly affect the usability of energy delivered; siting should thus be thoughtful and creative.

Construction Terms

Absorber—The surface in a collector that absorbs solar radiation and converts it to heat energy. Generally, matte black surfaces are good absorbers and emitters of thermal radiation, while white and metallic or shiny surfaces are not.

Backdraft Damper—A damper designed to allow air flow in only one direction.

Damper—A device used to vary the volume of air passing through an air outlet, inlet, or duct.

Drum Wall—A type of thermal storage wall in which the thermal mass is large metal drums filled with a storage medium, usually water.

Masonry—Stone, brick rammed earth, adobe, ceramic, hollow tile, concrete block, gypsum block, or other similar building units or materials, or a combination thereof, bonded together with mortar to form a wall, pier, floor, roof, or similar form.

Solid Masonry—Masonry in which there are no voids; for instance, concrete block with filled cores.

Thermal Lag—The ability of materials to delay the transmission of heat; can be used interchangeably with time lag.

Trombe Wall (or Solar Mass Wall)—A massive wall that absorbs collected solar heat and holds it until it is needed to heat the house interior.

SOLARIA/SUNSPACE SYSTEMS

The terms "sunspaces" and "greenhouses" are both commonly used to refer to this measure; they are used interchangeably in this discussion. Sunspaces could be considered a form of direct gain system. Sunlight is absorbed in the sunspace by thermal storage mass (such as bricks or water in containers), which then radiates heat. Existing doors and windows are frequently used to allow heated greenhouse air to flow into interior living spaces. The sunspace can be sealed off from the rest of the house when too much heat is gathered or when there is insufficient sunlight to contribute to home heating. Closing doors and windows between interior space and the sunspace at night or on particularly cloudy days may also be necessary.

Although single glazing for sunspaces will result in a maximum light transmission for plant growth and solar gain, single glass will permit a large amount of heat loss at night. In northern climates, double glazing will retard this heat loss. Additionally, movable insulation can be applied to prevent nighttime heat loss.

Thermal mass can be expensive. Therefore, if it can be reduced or eliminated and wider temperature fluctuations allowed, the cost of a sunspace will be substantially less. However, system efficiency will be reduced.

A greenhouse can be thought of as a buffer zone between the outside environment and the inside of the building. In this buffer zone, direct solar gain causes wide temperature fluctuations and a higher temperature indoors than outdoors at virtually all times of the day. These higher indoor temperatures buffer the adjacent living area, reduce building heat loss, and can be used immediately to help heat living area.

The sunspace is a versatile passive solar measure. It can be added to many different architectural designs with pleasing results. Sunspaces are equally compatible with expensive and inexpensive homes. Figure 10-3 illustrates attached retrofit sunspaces. Figure 10-4 illustrates a retrofit greenhouse for a mobile home.

The four basic methods for transferring thermal energy from the greenhouse into interior living space are:

- direct solar transmission;
- direct air exchange;
- conduction through common walls; and

• storage in and transfer from gravel beds or other thermal
  mass.

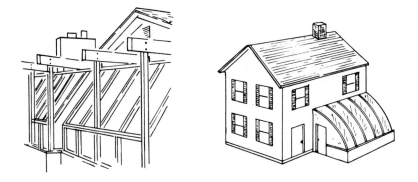

Attached
Pit Greenhouse

Figure 10-3. Attached Retrofit Greenhouses

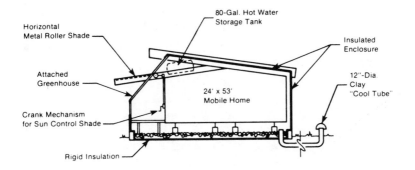

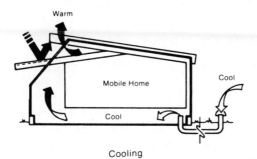

Cooling

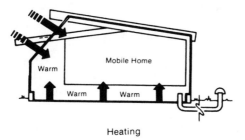

Heating

Figure 10-4. Attached Greenhouse on a Mobile Home

285

Figure 10-5 illustrates an example of thermal mass in the form of water drums. These storage methods are often used in combination. For example, in addition to a common heat storage wall to conduct heat from the greenhouse to the building, forced or natural air flow (direct air exchange) can also be used.

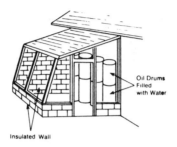

Insulated Wall

Figure 10-5. Example of Storage in an Attached Greenhouse

Applications

A sunspace can provide heat and additional living space, as well as the opportunity to grow vegetables during most of the year, in nearly all climates. Building a sunspace is a good do-it-yourself project, especially for small groups of neighbors and friends.

Advantages and Disadvantages

Greenhouses have several advantages:

- Temperature swings in adjacent living spaces are moderated.
- They provide space for growing food and other plants.
- They reduce heat loss from buildings by acting as a buffer zone.
- They are readily adaptable to most existing residential buildings.
- Since the greenhouse serves more than one function, it can be a natural and integrated part of the building design.

286

Because sunspaces are a form of direct gain system, their greatest disadvantage is that a poor design can lead to wide indoor temperature fluctuations. To combat this problem, indirect gain elements are often combined with direct gain, e.g., Trombe walls or water walls, to provide the tempering effect of thermal mass. In general, greenhouses are most economical when they have purposes in addition to providing heat and are built to a standard of quality that will enhance the functional and appraised value of the home.

A second need is to have maximum sunlight penetration in the winter (for heat) and less penetration in the summer (to prevent overheating). To achieve this, a rule of thumb of latitude $\pm 15°$ is applicable for optimal greenhouse glazing tilt, with appropriate shading.

Aesthetics can also be a concern with sunspaces. The architectural lines of a house need to be aesthetically matched. Sometimes this leads to an extreme tilt, and sometimes no tilt. In each case, interior design (addition or subtraction of thermal mass, outside venting, movable insulation, etc.) has to compensate for poor design from a passive solar energy point of view.

A fourth concern is movable insulation, which is necessary in most cases for retention of winter heat gain.

WINDOW HEAT GAIN RETARDANTS

Shading is the most effective method of reducing heat gain through transparent materials, and, ideally, a shading device should allow sunlight in during the winter. The term "window heat gain retardants" means mechanisms that significantly reduce summer heat gain through south-facing windows ($\pm 45°$ of true south) by use of devices such as awnings, insulated rollup shades, metal or plastic solar screens, or movable rigid insulation. Included as window heat gain retardants (WHGR) are any devices that provide shading. Other examples of these devices are sunscreens, exterior roll blinds, exterior shutters, venetian blinds, film shades, opaque rolls, and insulating shutters and blinds.

Internal shading can reduce the amount of heat dispersed within a space. The most common internal shading devices are venetian blinds, vertical blinds, shades, draperies, reflective films, and shutters. These devices can reject up to 65% of the solar radiation that strikes the glass directly.

External shading is more effective than internal shading; it can keep up to 95% of the solar radiation from entering the building (100% if opaque shutters are used). Many devices are available for exterior shading. Horizontal overhangs, with fixed and movable elements, are very effective on south windows, because the sun is highest as it approaches due south during midday. The sun also is higher in summer than in winter, and the overhang can be positioned to screen the sun in summer but admit it in winter. On east or west elevations, however, the sun's angle is too low to be blocked out by horizontal overhangs, and properly oriented vertical louvers have proven more beneficial. If the louvers are movable, the user can control them to provide a better view or greater admission of light when the sun is located on the opposite face of the building. Trees, vines, and adjacent buildings may also provide shade, depending on their proximity, height, and orientation. Many summer shading devices can also provide window *insulation* in the winter. Care should be taken in their application so that beneficial winter solar heat gain is not lost. As an example, heat reflecting material may work best year-round if it is mounted on a retractable shade.

Applications

Window heat gain retardants help keep buildings cool in summer. They should be considered wherever summer solar heat gain through windows causes overheating. More particularly, substantial dollar (and resource) savings are possible when retardants are used to diminish expenses for air conditioning.

288

Advantages and Disadvantages

Any heat gain retardant device has advantages and disadvantages and to list all of them for the wide range of shading devices would be impractical. Instead, we list a few key ones for each device.

For sunscreens, visibility can remain as high as 86% while providing daytime privacy. Sunscreens block solar radiation reflected from nearby more effectively than other measures. However, drapes or blinds must be used for privacy at night.

With an exterior roll blind, energy-conserving effects can be achieved year-round. In summer, shading is provided by day and ventilation by night (with the blind raised). In winter, insulation is provided by night with the blind lowered, and solar gain by day with the blind raised. However, the blind cannot be tilted to provide a view.

For exterior shutters and blinds, protection of windows is provided from storm damage, vandalism, or intrusion. However, operation and maintenance of exterior devices can be time-consuming.

For awnings, ease of installation and attractiveness are advantages. However, they are subject to wind damage and they must be maintained; e.g., new fabric may be necessary every four to eight years, if fabric awnings are used instead of more permanent wood or metal awnings.

Construction Terms

Awning—A shading device, usually movable, used over the exterior of a window.

Heat Reflective and Heat Absorbing Window or Door Materials —Glazing, films or coating applied to existing windows or doors, that have exceptional heat absorbing or heat reflecting properties.

Overhang—A horizontal or vertical projection over or beside a window used to selectively shade the window or door on a yearly basis.

289

Shading Device—A covering that blocks the passage of solar radiation; common shading devices consist of awnings, overhangs, or trees.

Shutter—Movable cover or screen for a window or door.

11

Wind Energy Technology

The Department of Energy defines "large" wind turbines as 100 Kw or more, and "small" wind turbines as under 100 Kw.

Off-grid units will generally be small scale (e.g. less than 100 Kw) devices, frequently located at remote sites. They will most commonly be used to recharge batteries, to pump water, and to heat water. Remote locations will minimize the exposure of the public to physical injury, so that liability will generally be accommodated with traditional self-insurance and/or general liability insurance. For these reasons, liability issues will not strongly impact this class of wind units.

Where off-grid wind units are remotely sited, the principal land use issue will be the visual, aesthetic impact of the wind unit itself. There are a variety of ways in which the visual impacts of such structures are controlled. In urban areas, the principal tools are zoning and building regulations. In rural areas, where most off-grid wind devices will be located, building codes and regional land use commissions are the most important.

Wind systems which are integrated into a utility grid, but which are not owned by that utility, will be exposed to the same type of nontechnical issues that were discussed for off-grid systems. In addition, the connection with a utility system will introduce major issues in determining equitable rate structures for the transfer of electrical power between the owner of the wind unit and the utility.

A utility-owned wind system will probably be quite different, physically, from the previously discussed nonutility-owned systems. These will undoubtedly be larger machines (of over one megawatt) located at sites which are specifically selected because of their high wind speeds. Since such sites are finite, there will be an incentive to locate multiple wind turbines (e.g. tens of units) at each site. This grouping of numerous wind turbines at a single site has been referred to as a "wind farm."

Such large wind farms would have impacts quite distinct from those of smaller wind units. For example, the radio and television interference which large metallic blades can cause could restrict the sites for wind farms. The aesthetic impact of a rural wind farm would be quite different from that of smaller units on urban homes and business.

Compared to other types of power plants that utilities operate, the problems that wind farms will face should be small. Utilities are accustomed to dealing with the regulatory systems which oversee aesthetic, safety, and land use matters. Thus not only will many of these factors be mitigated by the remote siting of wind farms, but also because many of the institutional channels are already established, both within the utility and within the regulatory/public interest system, to manage such wind development.

BASICS

Wind is the movement of air caused by forces acting on the atmosphere. These forces include (1) the uneven heating of the earth's surface by the sun and (2) the earth's rotation.

The uneven heating of the earth depends on the amount of energy absorbed by its surface, which varies with the amount of energy striking the surface. This amount depends on the angle of the sun (which varies seasonally) and on the type of surface (land or water) being struck. Uneven heating of the earth's surface causes the atmosphere above the surface to be unevenly heated. Because air that is hot is at a higher pressure than air that is cold, an imbalance is created by the different pressures: to even out this difference in pressure, air flows from a hot, high-pressure region to a cool, lower-pressure region. This is wind.

Figure 11-1 is a simplified version of airflow, showing two columns of air (one warm one cold), with airflow (arrows) moving from high- to low-pressure regions.

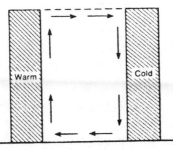

Figure 11-1. Simplified Version of Airflow

The second major force causing wind is the rotation of the earth. This rotation has two effects: a direct acceleration of the air and a deflection (caused by the Coriolis force) of the wind's direction. In the temperate zones (30°–60° north or south of the equator), the Coriolis force causes the "Westerlies"—a general flow of air from west to east. The general flow of the wind on the earth is shown in Figure 11-2.

Since the United States lies mainly in the temperate zone, the general airflow across the country is, predictably, from west to east. (An especially vivid example of this flow occurred when the Mt. St. Helens volcano in Washington first erupted. Its ash, thrown into the high atmosphere, moved across the United States and was approaching the East Coast within two days, driven by the Westerlies.)

Other factors account for variability in the flow of wind near the ground. Two of these factors are *terrain* and *surface roughness.*

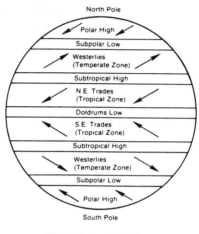

Wind and Pressure Distribution

Figure 11-2. General Flow of Wind on the Earth

Together with the uneven solar heating of the atmosphere, the terrain and roughness can force the air to change direction and to flow at extremely varied speeds. It is important to be aware of the wind's variability (and its causes) to understand the variability of the wind resource.

WIND CHARACTERISTICS

Most information about how wind behaves in relation to the various terrain and surface roughness encountered in this country comes from the work done by Battelle at the Pacific Northwest Laboratory (PNL). PNL has developed a data base of 1200 wind measuring stations across the country. From these data and the techniques that relate it to local conditions, a wind atlas has been prepared giving the expected wind resource in a series of grids across the USA. Use of this atlas, and specifics based on the terrain and surface roughness at the site, provides the best possible estimate of the wind resource at a specific site.

Terrain

Terrain has a major effect on the speed and direction of wind as it follows a course of least resistance. PNL has developed generalized models describing airflow (speed and direction) as it relates to the terrain of the surrounding land and the surface roughness of the immediate site. Examples of terrain effects on wind are shown in Figure 11-3. Note that the shape of the cliff determines the magnitude and location of the turbulent flow area, which changes dramatically in the examples in the figure.

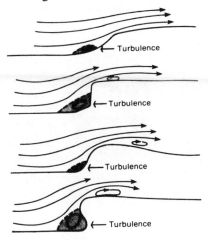

Figure 11-3. Airflow Over Cliffs Having Differently Sloped Faces

Figure 11-4 shows that speed can be enhanced by a topographical feature like a cliff, if the area of turbulence is avoided. In Figure 11-5, the sides of a hill can be regions of accelerated airflow as the winds are divided by the hill. Since winds seek the path of least resistance, gaps and gorges also tend to affect wind direction and velocity.

Because the land surface gives off its heat more quickly than a water mass, a small-scale "uneven heating of the earth's surface" takes place daily near a body of water (Figure 11-6). Since air tends to flow from an area of high pressure to an area of low pressure, this

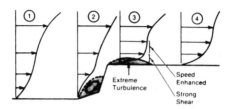

Figure 11-4. Vertical Profiles of Air Flowing Over a Cliff

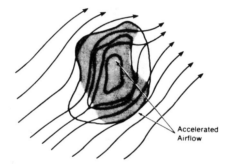

Figure 11-5. Airflow Around an Isolated Hill

sets up local-scale winds that usually blow in one of two directions only, depending on the time of day.

Surface Roughness

"Surface roughness" is the term applied to site-specific conditions that affect the wind. Surface friction causes a wind speed to decrease (Figure 11-7).

The greater the surface roughness, the greater the impact on wind speed and turbulence (Figure 11-8).

296

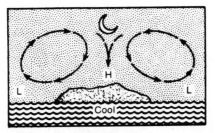

Night (Winter)

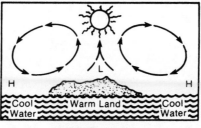

Day (Summer)

Figure 11-6. Uneven Heating of the Earth's Surface

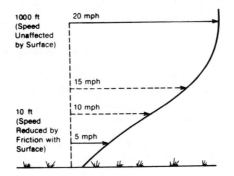

Figure 11-7. Effective Surface Friction on Low-Level Wind

297

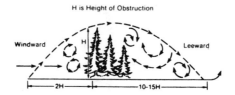

Figure 11-8. Airflow Near a Shelterbelt

In addition to the turbulence caused by natural obstruction such as trees, manmade objects can affect a potential small wind energy conversion system.

Energy in the Wind

The energy in the wind varies as the cube of the wind speed. Therefore, if you double the wind speed, the power available in the wind is represented by $2^3 = 8$; stated in another way, the power of a windstream increases by a factor of eight for each doubling of the wind speed.

The energy potential of the wind is also related to the air. This means that at higher altitudes, where the air is thinner (for example, in Denver), less energy is available at a certain wind speed than in Atlanta. If local wind speeds are used, adjustments for altitude may be necessary in some regions. The PNL data that will be used most places have already been adjusted for the reduction in energy potential with altitude, and can be used without change.

Another factor affecting the energy extracted from wind is the area that the rotor sweeps as it turns. Since that swept area is proportional to the square of the rotor's diameter, the swept area is increased by $2^2 = 4$ when the rotor diameter is doubled. Usable energy in the wind is illustrated in Figure 11-9.

298

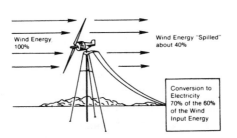

Figure 11-9. Usable Energy in the Wind

CONVERSION OF WIND ENERGY

To use the energy available in the wind, it is necessary to convert its kinetic energy into mechanical or electrical energy that can do work. This energy is captured by rotor blades, which rotate as wind flows around them or strikes them. The blades convert the kinetic energy of the wind into mechanical energy. The rotor is connected to a shaft, and the torque on the shaft, created by the rotating blades, can do mechanical work or can generate electricity.

In general, wind acts upon rotor blades to produce rotation by either causing *lift* or *drag*, depending upon the aerodynamic design of the blade. Drag is the principle by which the wind physically strikes the rotor and drags it with that impinging force. Lift, an aerodynamic principle, results from a properly designed airfoil: the flow of air over the airfoil (rotor blade) causes a different pressure on each surface of the rotor, causing the airfoil to "lift," or basically to fly. Lift is more efficient than drag for extracting energy from the wind, and most modern machines use the principle in their operation. Even os, the maximum theoretical limit (called "the Betz limit") for extraction is only about 60% of the energy available in the windstream. One reason this is so involves the question of what to do with the air after it passes through the wind machine's blades. If all the moving energy of the wind could be removed, you'd have a large amount of still air behind the blades, blocking the access through the blades. Then everything would stop until the still air was removed, which would take energy.

299

An electrical generator can be set up by using the rotating shaft of the wind machine as the armature of a generator and applying an electric field from an external source. The mechanical energy of the rotating shaft is converted into electrical energy as the armature rotates in the field. The output is electricity in a form determined by the characteristics of the generator, and the maximum available, after losses, is 70% of the energy of the moving blades.

Many electronic and mechanical controls are necessary to control the orientation of the rotor in the wind, the flow of electricity from the generator, and the several safety features that are needed because the wind turbine generator is subject to greatly varied weather conditions.

There are many different designs for rotors to extract energy from the wind, and you should become familiar with them. Also, realize that although the theory behind their operation is similar, their different shapes cause them to operate at different efficiencies. Some of the types of rotor designs with which you should be familiar are shown in Figures 11-10 through 11-12.

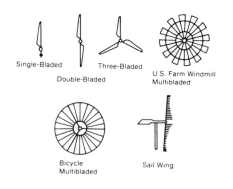

Figure 11-10. Horizontal Axis Blade Design

300

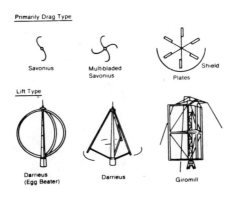

Primarily Drag Type

Savonius

Multibladed
Savonius

Shield

Plates

Lift Type

Darrieus
(Egg Beater)

Darrieus

Giromill

Figure 11-11. Vertical Axis Blade Design

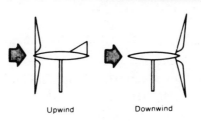

Upwind   Downwind

Figure 11-12. Upwind and Downwind Design

TOWERS

Wind turbine generators should be located in the least-turbulent, highest-speed wind regime possible. Towers are an appropriate means of doing this. A wind turbine generator should be installed on a tower whose height and construction have been determined by the manufacturer to match it safely and cost-effectively. The two main types of towers available commercially are freestanding and guyed towers, in heights from 40 to over 100 feet.

301

12

Active Solar Energy Systems

ACTIVE SOLAR ENERGY SYSTEM COMPONENTS

An active solar energy system includes three major components:

- the solar collectors;
- the circulation system (or loop), including pumps, controls and heat exchangers; and
- the storage tank.

This system is illustrated in a simple schematic shown in Figure 12-1. The system is integrated with the existing conventional space and/or water heating system. The conventional system acts as a back-up when solar energy systems cannot fully supply needed heat and hot water.

Collectors

Many types of active solar collectors are currently on the market. These include flat-plate, evacuated tube, and flexible black plastic collectors.

Most flat-plate collectors available on the market have the same general components. Figure 12-2 shows a typical flat-plate collector.

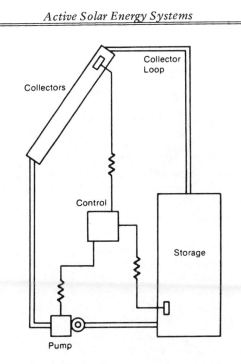

Figure 12-1. Simple Schematic of an Active Solar System

Typically, collectors have blackened copper or aluminum absorber plates with copper tubing that is either part of their structure or continuously attached (for most effective heat transfer by conduction). The absorber plate may be packaged as a modular collector panel, complete with glazing covers, insulation, and enclosure box, or it may be a structural (built-in) part of a roof or other structure. The collector itself has one or two layers of glazing material (e.g., glass), held by a frame, protecting the blackened absorbing plate that receives the sun's energy and turns it into heat. The plate has an integral or attached array of tubes through which liquid circulates. This liquid carries the heat from the collector to storage so that it can be used. The liquid inlet connection shown at the bottom of the collector is diagonally opposite the liquid outlet connection at the

303

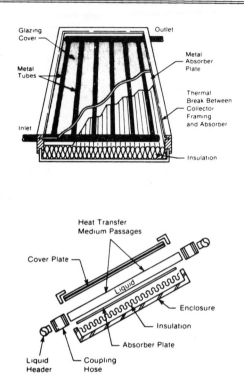

Figure 12-2. Major Components of a Flat-Plate Collector

top, which means that the flow length is the same through every passage. Thus, a balanced flow of liquid through the collector is achieved, which sustains a relatively uniform temperature over the plate. The heated outlet fluid from the collector is transferred to storage for later use. The inlet fluid has been cooled by leaving its heat in the storage tank; it is then recirculated back to the collector.

Beneath the absorber plate and tube array is an insulating material to retard the loss of the absorbed heat through the back of the collector panel. Heat losses from the sides of the collector panel are minimized by maintaining a space, or "thermal break," between the metal absorber plate and the collector framing. To further reduce heat loss, insulation is often used on the sides of the collector.

304

It is helpful to maintain a moisture-free condition in the collector to retard heat loss, which tends to be greater in moist than in dry air. A second reason is to prevent condensation of water on the glazing covers, which would reflect incoming solar raidation from the collector.

One method for keeping air dry within the collector is to use spacers made of a moisture-absorbing (desiccant) material. These spacers are inserted between cover and absorber plate and, also, between layers of multiple-glazing covers. Another method is to use "weep holes," which allow moisture to escape from within the collector box. Liquid flat-plate collectors are capable of delivering water as hot as $140°-160°F$; however, most systems operate in the range of $120°F$ to $140°F$. Hot water demand temperatures range from $110°F$ to $160°F$.

Evacuated tube collectors can reach higher temperatures and since they are under vacuum they have a higher efficiency. These collectors have been used in industrial drying processes. Their main limitation is their high initial first cost.

The Circulation System

The circulation system (or loop) performs the function of *transporting* heat from the collectors (where it is captured) to a place where it can be stored and used. A fluid, called a "heat transfer fluid," carries the heat absorbed in the collectors through the circulation system to the storage medium. The circulation system includes devices to control fluid movement.

The fluid chosen for an active system will be the result of a compromise between several ideal features. It should:

- be an excellent heat conductor;
- resist degrading;
- cost little or nothing;
- be lightweight;
- not damage pipes or ducts by freezing, boiling, or corroding;
- be nontoxic.

The most common transport fluids used today in active systems are air, water, and antifreeze solutions. Compare their qualities to the list above. The main advantages of air are that it will not freeze, boil, or cause corrosion. The main drawback to air is that it is a relatively poor heat conductor, thus requiring more pumping power per unit of energy gathered. Water is a better heat conductor than air, and generally requires less pumping power per unit of energy gathered. However, water can freeze in extreme weather conditions, thus damaging the system. The antifreeze solutions used to prevent this problem are nonpotable (toxic), and the antifreeze solution must be separated from the domestic water supply by a double-walled heat exchanger to ensure a safe potable water supply.

Several terms should be understood relative to system circulation. There are "closed-loop" and "open-loop" systems. In a "closed-loop" system, the piping is sealed off from the atmosphere (air) and from the water in the storage tank. An "open-loop" system is open to the atmosphere during operation. Open-loop systems are nonpressurized; closed-loop systems are pressurized.

There are "direct" and "indirect" system designs. "Indirect" systems generally use a heat exchanger to transfer the collected heat to the storage medium. Indirect systems are commonly used to protect the potable water supply from a toxic circulatory antifreeze mixture used for system freeze protection. In a "direct" system, potable water is circulated through the collector loop without using a heat exchanger in the storage tank.

In active systems, heat exchangers transfer heat across an enlarged surface area, while maintaining separation of the heat transfer fluid in the collector loop from the domestic water supply in the storage tank. This keeps the domestic water from being contaminated. Also, with a heat exchanger only a small portion of the total amount of system fluid must be treated to prevent system corrosion or freezing. Although heat exchangers are usually built into or around the hot water storage tank, they also exist as separate units.

Solar domestic hot water heat exchangers can be either single- or double-walled. Systems using transfer fluids other than potable water (as defined by HUD's intermediate Minimum Property Standards (MPS) or by the local health authority) in the collector loop are required by the HUD/MPS to have double-walled heat exchangers. Extra thick single walls are not considered acceptable protection, nor are single-walled configurations that depend on potable water pressures to prevent contamination. Several different heat exchanger designs can be used in an active system.

In active systems employing a liquid heat transfer medium, the liquid must be protected from freezing and boiling. This is done by chemicals in the heat transfer fluid or by mechanical control.

In climates where freezing occurs frequently, an antifreeze solution is commonly used (ethylene or propylene/glycol in the appropriate proportion with water). Glycols can still boil under stagnant conditions in a closed loop. This may occur when controls or pump operation fails or when a high temperature limit (around 190°F) is reached in the storage tank. While the boiling action itself should cause no harm (pressure relief valves will handle the overpressure situation), the liquid's ability to act as an effective antifreeze is diminished once it has boiled because of chemical breakdown in the antifreeze solution. This can be a maitenance problem. Silicone oils offer both boil and freeze protection and are essentially inert. However, because of their higher viscosities, they often require larger piping and pumping power as compared to antifreeze.

When water is used as the heat transfer fluid, system freeze protection is generally offered in three mechanical ways: (1) draindown systems; (2) drainback systems; (3) recirculation systems.

In the *draindown* system, potable water is circulated from the storage tank through the collector loop. Freeze protection is provided by solenoid valves opening and draining all the water in the collectors at a preset low temperature. Collectors and piping must be pitched (or tilted) so that the system will naturally drain down, even in a power failure. This type of system is exposed to city water line pressures and must be assembled carefully to withstand pressures as high as 100 psi. Pressure-reducing valves are used when city water pressure is greater than the working pressure of the system. System

307

protection is dependent on the successful operation of solenoid valves during freezing conditions.

In the *drainback* system, the solar heat transfer fluid automatically drains *into a tank* by gravity each time the pump shuts off, leaving no liquid to freeze in the collectors (see Figure 12-3). A heat exchanger is necessary, because the city water inlet pressure would prevent draining of pipes directly attached to the city's mains. The heat transfer fluid in the collector loop may be distilled or city water if the loop plumbing is copper. If the plumbing is galvanized pipe, an inhibitor may be added to prevent corrosion. Most inhibitors are nonpotable and require a double-walled heat exchanger.

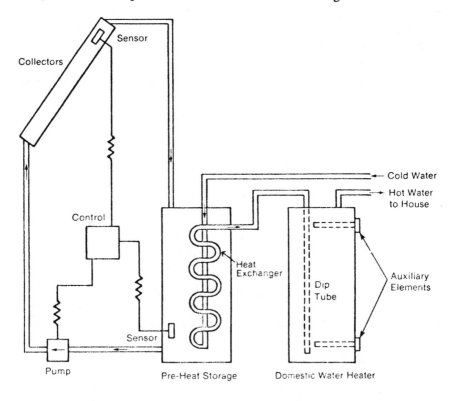

Figure 12-3. Simple Schematic of a Drainback System

In the *recirculation* system, the controller turns the pump on during freezing situations to circulate warmed, stored water through the collectors to prevent freezing. This system is acceptable only in warmer climates where freezing seldom occurs, because, as will be discussed later, it is obviously not desirable to lose collected heat in addition to using energy to run the pump to heat the collectors.

Storage Systems

The storage tank serves the function of storing the heat, which is often needed at a time other than when it is collected, until it is used. An ideal storage material:

- has a high capacity to store heat,
- is lightweight,
- costs little or nothing,
- is a good conductor,
- is nontoxic,
- causes no damage to storage location, and
- is functionally durable.

Water, rocks and phase-change eutectic salts are the most common storage materials in use today. Eutectic salts are not made of table salt, but are chemical mixtures that change from solid phase to liquid phase and back again at certain temperatures, storing or releasing energy in the process. Eutectic salts have been compounded with phase changes in the 75°F to 120°F temperature range for solar heating applications. When eutectic salts change phase from solid to liquid, 90–100 Btu/lb are stored (or absorbed).

Water is also, of course, a phase-change material. While the heat absorbed in a phase change from solid (ice) to liquid is good (144 Btu/lb), the temperature (32°) is obviously not in the usable range for solar storage. It still has excellent properties, though, for thermal storage without phase change.

When comparing storage space requirements for the three materials, the specific heat of the material, its density, and the air space it requires for heat transfer must be considered.

Table 12-1 compares the properties of water, rock and a specific eutectic salt. Considering the need for an air space of 25% in the case of rock and salt storage (since neither can be packed in so close that no air gap exists), water is able to store about 2.5 times as many Btu/ft^3 as rock, with eutectic salt storing 3.7 times as much energy as water per ft^3 when each material changes temperature by 40°F (80°F to 120°F). Rocks require the most space of the three materials to store the same amount of Btu, and eutectic salts require the least. Eutectic salts, however, are by far the most expensive storage material.

Table 12-1. Properties of Selected Storage Materials

	Water	Rock	Eutectic Salt*
Specific heat (Btu/lb°F)	1	0.2	0.35
Applicable heat fusion (Btu/lb)	—	—	108
Density (lb/ft^3)	62.4	170	100
Storage air space (%)	—	25	25
Effective density (lb/ft^3)	62.4	127.5	75
Effective storage capacity (Btu/ft^3 at 80°F–120°F)	2496	1020	9150

*Sodium sulfate decahydrate; phase change occurs at 90°F.

Other considerations are also important in selecting storage materials. Research on eutectic salt packaging and installation is presently in progress, although various firms are currently marketing the salts commercially.

AVAILABLE ENERGY

The earth travels around the sun in an elliptical orbit. Thus, the solar radiation beyond the earth's atmosphere varies, depending on the season and time of day. The sun provides energy at a rate of about 429.2 Btuh/ft^2 or about 135.3 mW/cm^2 on a surface normal to the sun's rays and outside the earth's atmosphere. This rate of energy is referred to as the solar constant.

However, the energy available at the earth's surface is only approximately 50% of incident radiation because the solar radiation is scattered and absorbed by dust, gas molecules, ozone, and water vapor as it passes through the earth's atmosphere. The total solar radiation reaching a surface on earth is thus greatly reduced.

COMPUTING THE SOLAR ENERGY
RECEIVED BY A SOLAR COLLECTOR

The total shortwave radiation that reaches a solar collector is comprised of three components.

1. Direct solar radiation
2. Diffused sky radiation
3. Reflected energy

The amount of radiant energy or insolation received by a solar collector is of primary importance. There are several sources of insolation data. Data from different sources vary due to the following:

1. Atmospheric effects are complex and difficult to model
2. Variances in the elipicity of the earth's orbit (3.4%)
3. Calibration and type of measuring instruments
4. Number of years in which measurements were compiled
5. Limited coverage

The total shortwave radiation that reaches the earth's surface is comprised of three components:

$$E_T = A(E_d + E_2 + E_3) \qquad\qquad Formula\ (12\text{-}1)$$

311

where

E_T is the total solar radiation reaching a surface Btuh/ft^2.

E_d is the direct component of the solar energy radiated on the surface, Btuh/ft^2.

E_2 is the diffuse sky radiation, Btuh/ft^2.

E_3 is the solar energy reflected from the surrounding surfaces, Btuh/ft^2.

A is the clearness factor which takes into account areas which will have a radiation above or below average, due to the relative dryness of the area.

Computing the Direct Energy
Component—E_d

To compute the solar energy radiated on a surface, first determine the angle of incidence θ on the surface in question.

Figure 12-4 shows a simplified solar angle diagram where the surface "A" is facing near south. The position of the sun in the sky is expressed in terms of the solar altitude angle β and the solar azimuth angle ϕ. β represents the sun's direction above the horizontal, while ϕ is a measure of the direction of the sun from the south. Note these angles will change during the time of day and period of the year. The tilt angle Σ indicates the angle of surface "A" with respect to the horizontal plane. If the surface were facing in a direction other than south, the angle γ would be used to reflect this.

The general solar energy equation for the direct rays incident on any surface is

$$E_d = E_1 \cos \theta \qquad\qquad \textit{Formula (12-2)}$$

where

E_d is the direct component of the solar energy radiated on the surface Btuh/ft^2.

E_1 is the solar energy radiated directly from the sun.

θ is the incident angle with the surface.

Referring to Figure 12-4, when the surface is horizontal $\Sigma = 0°$ and the angle of incidence is $\theta = \theta_H$.

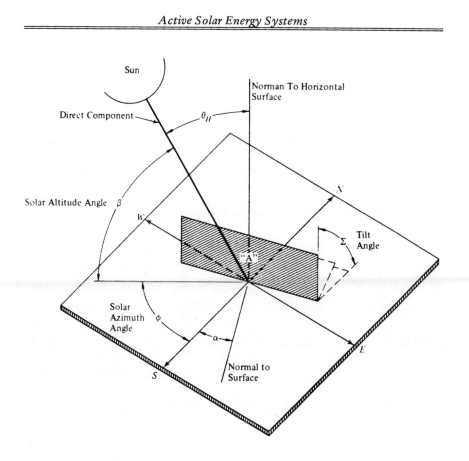

Figure 12-4. Angle of Incidence from Solar Rays

In general: $\cos \theta = \cos \beta \times \cos (\phi + \gamma) \sin \Sigma$ *Formula (12-3)*

Example Problem 12-1

Determine the solar energy related on a surface *"A,"* when it faces due south.

Answer

From Formulas 12-1 and 12-2,

313

$$E_d = E_1 \ (\cos \beta \cos \phi \sin \Sigma + \sin \beta \cos \Sigma) \qquad \textit{Formula (12-4)}$$

Example Problem 12-2

Determine the solar energy radiated on horizontal surfaces in terms of the solar altitude β.

Answer

When

$$\Sigma = 0; \cos \theta = \sin \beta$$

Thus:

$$E_d = E_1 \sin \beta \qquad \textit{Formula (12-5)}$$

Example Problem 12-3

Determine the solar energy radiated on a vertical surface.

Answer

$$\Sigma = 90°$$

Thus:

$$\cos \theta = \cos \beta \cos (\phi + \gamma)$$
$$E_d = E_1 \cos \beta \cos (\phi + \gamma) \qquad \textit{Formula (12-6)}$$

Example Problem 12-4

Determine the solar energy radiated on a $45°$ tilted surface, facing south.

Answer

$$
\begin{aligned}
\Sigma &= 45° \\
\cos 45° &= 0.707 \\
\sin 45° &= 0.707 \\
\cos \theta &= 0.707 \cos \phi \cos \beta + 0.707 \sin \beta \\
E_d &= E_1 \ 0.707 \ (\cos \phi \cos \beta + \sin \beta)
\end{aligned}
$$

$$\textit{Formula (12-7)}$$

314

Table 12-2 is used to find the solar position β and ϕ for various solar times during the 21st day of each month. The direct normal radiation component E_1 is also found, along with the constant C, which is used to compute diffuse solar radiation. The table is based on ASHRAE Handbook of Fundamentals, with a 0% ground reflectance and a clearness factor of 1. Solar time differs from local Standard time, but for most calculations, this difference is insignificant. This table is useful when detailed calculations are required to compute energy absorbed by a collector for various times, tilts, and orientations.

Example Problem 12-5

An existing building (40° N latitude) is to be retrofitted with solar collector. The building design does not permit the collector to be located due south. Comment on the effect of E_d by locating the collectur due north or due west. The collector is tilted 45°.

Answer

Collector Facing North—In Formula 12-3,

$$\gamma = 180° \text{ and } \Sigma = 45°$$

Thus: $E_d = E_1 \; 0.707 \; (\cos (\phi + 180°) \cos \beta + \sin \beta)$.

When E_d is negative, no direct component of radiation is received by the collector. Until $\cos (\phi + 180)$ is positive, little or no direct energy is received.

From Table 12-2, it is seen that the collector performs extremely poorly in the winter months and only marginally during the summer.

Collector Facing West—A far better choice would be to locate the collector facing west. In this case, $\cos (\phi + 270°)$ is a positive number throughout the year. Thus, a good portion of solar energy will be received by the collector when it faces due west.

Table 12-2. Solar Position and Irradiation

Date	Solar Time AM	Solar Time PM	32 Deg. N. Latitude Solar Position β	φ	E₁ Btuh/sq.ft	40 Deg. N. Latitude Solar Position β	φ	E₁' Btuh/sq.ft	48 Deg. N. Latitude Solar Position β	φ	E₁ Btuh/sq.ft
Jan. 21	7	5	1.4	65.2	1						
C = .058	8	4	12.5	56.5	202	8.1	55.3	141	3.5	54.6	36
	9	3	22.5	46.0	269	16.8	44.0	238	11.0	42.6	185
	10	2	30.6	33.1	295	23.8	30.9	274	16.9	29.4	239
	11	1	36.1	17.5	306	28.4	16.0	289	20.7	15.1	260
		12	38.0	0.0	309	30.0	0.0	293	22.0	0.0	267
Feb. 21	7	5	6.7	72.8	111	4.3	72.1	55	1.8	71.7	3
C = .060	8	4	18.5	63.8	244	14.8	61.6	219	10.9	60.0	180
	9	3	29.3	52.8	287	24.3	43.7	271	19.0	47.3	247
	10	2	38.5	38.9	305	32.1	35.4	293	25.5	33.0	275
	11	1	44.9	21.0	314	37.3	18.6	303	29.7	17.0	288
		12	47.2	0.0	316	39.2	0.0	306	31.2	0.0	291
Mar. 21	7	5	12.7	81.9	184	11.4	80.2	171	10.0	78.7	152
C = .071	8	4	25.1	73.0	260	22.5	69.6	250	19.5	66.8	235
	9	3	36.8	62.1	289	32.8	57.3	281	28.2	53.4	270
	10	2	47.3	47.5	304	41.6	41.9	297	35.4	37.8	287
	11	1	55.0	26.8	310	47.7	22.6	304	40.3	19.8	295
		12	58.0	0.0	312	50.0	0.0	306	42.0	0.0	297
Apr. 21	6	6	6.1	99.9	66	7.4	98.9	89	8.6	97.8	108
C = .097	7	5	18.8	92.2	206	18.9	89.5	207	18.6	86.7	205
	8	4	31.5	84.0	256	30.3	79.3	253	28.5	74.9	247
	9	3	43.9	74.2	278	41.3	67.2	275	37.8	61.2	269
	10	2	55.7	60.3	290	51.2	51.4	286	45.8	44.6	281
	11	1	65.4	37.5	296	58.7	29.2	292	51.5	24.0	287
		12	69.6	0.0	298	61.6	0.0	294	53.6	0.0	289

May 21
C = .121

5	7				1.9	114.7	1	5.2	114.3	41
6	6	10.4	107.2	118	12.7	105.6	143	14.7	103.7	162
7	5	22.8	100.1	211	24.0	96.6	216	24.6	93.0	218
8	4	35.4	92.9	249	35.4	87.2	249	34.6	81.6	248
9	3	48.1	84.7	269	46.8	76.0	267	44.3	68.3	264
10	2	60.6	73.3	279	57.5	60.9	277	53.0	51.3	274
11	1	72.0	51.9	285	66.2	37.1	282	59.5	28.6	279
12		78.0	0.0	286	70.0	0.0	284	62.0	0.0	280

June 21
C = .134

5	7				4.2	117.3	21	7.9	116.5	77
6	6	12.2	110.2	130	14.8	108.4	154	17.2	106.2	172
7	5	24.3	103.4	209	26.0	99.7	215	27.0	95.8	219
8	4	36.9	96.8	244	37.4	90.7	246	37.1	84.6	245
9	3	49.6	89.4	263	48.8	80.2	262	46.9	71.6	260
10	2	62.2	79.7	273	59.8	65.8	272	55.8	54.8	269
11	1	74.2	60.9	278	69.2	41.9	276	62.7	31.2	273
12		81.5	0.0	280	73.5	0.0	278	65.4	0.0	275

July 21
C = .136

5	7				2.3	115.2	2	5.7	114.7	42
6	6	10.7	107.7	113	13.1	105.1	137	15.2	104.1	155
7	5	23.1	100.6	203	24.3	97.2	208	25.1	93.5	211
8	4	35.7	93.6	241	35.8	87.8	241	35.1	82.1	240
9	3	48.4	85.5	261	47.2	76.7	259	44.8	68.8	256
10	2	60.9	74.3	271	57.9	61.7	269	53.5	51.9	266
11	1	72.4	53.3	277	66.7	37.9	274	60.1	29.0	271
12		78.6	0.0	278	70.6	0.0	276	62.6	0.0	272

Aug. 21
C = .122

6	6	6.5	100.5	59	7.9	99.5	80	9.1	98.3	98
7	5	19.1	92.8	189	19.3	90.0	191	19.1	87.2	189
8	4	31.8	84.7	239	30.7	79.9	236	29.0	75.4	231
9	3	44.3	75.0	263	41.8	67.9	259	38.4	61.8	253
10	2	56.1	61.3	275	51.7	52.1	271	46.4	45.1	265
11	1	66.0	38.4	281	59.3	29.7	277	52.2	24.3	271
12		70.3	0.0	283	62.3	0.0	279	54.3	0.0	273

Table 12-2. Solar Position and Irradiation (Continued)

Date	Solar Time AM	Solar Time PM	32 Deg. N. Latitude β	32 Deg. N. Latitude φ	E₁ Btuh/sq. ft	40 Deg. N. Latitude β	40 Deg. N. Latitude φ	E₁ Btuh/sq. ft	48 Deg. N. Latitude β	48 Deg. N. Latitude φ	E₁ Btuh/sq. ft
Sept. 21 C = .092	7	5	12.7	81.9	163	11.4	80.2	149	10.0	78.7	131
	8	4	25.1	73.0	240	22.5	69.6	230	19.5	66.8	215
	9	3	36.8	62.1	272	32.8	57.3	263	28.2	53.4	251
	10	2	47.3	47.5	287	41.6	41.9	279	35.4	37.8	269
	11	1	55.0	26.8	294	47.7	22.6	287	40.3	19.8	277
		12	58.0	0.0	296	50.0	0.0	290	42.0	0.0	280
Oct. 21 C = .073	7	5	6.8	73.1	98	4.5	72.3	48	2.0	71.9	3
	8	4	18.7	64.0	229	15.0	61.9	203	11.2	60.2	165
	9	3	29.5	53.0	273	24.5	49.8	257	19.3	47.4	232
	10	2	38.7	39.1	292	32.4	35.6	280	25.7	33.1	261
	11	1	45.1	21.1	301	37.6	18.7	290	30.0	17.1	274
		12	47.5	0.0	304	39.5	0.0	293	31.5	0.0	278
Nov. 21 C = .063	7	5	1.5	65.4	1						
	8	4	12.7	56.6	196	8.2	55.4	136	3.6	54.7	36
	9	3	22.6	46.1	262	17.0	44.1	232	11.2	42.7	178
	10	2	30.8	33.2	288	24.0	31.0	267	17.1	29.5	232
	11	1	36.2	17.6	300	28.6	16.1	283	20.9	15.1	254
		12	38.2	0.0	303	30.2	0.0	287	22.2	0.0	260
Dec. 21 C = .057	8	4	10.3	53.8	176	5.5	53.0	88			
	9	3	19.8	43.6	257	14.0	41.9	217	8.0	40.9	140
	10	2	27.6	31.2	287	20.7	29.4	261	13.6	28.2	214
	11	1	32.7	16.4	300	25.0	15.2	279	17.3	14.4	242
		12	34.6	0.0	304	26.6	0.0	284	18.6	0.0	250

Reprinted by permission from *ASHRAE Handbook of Fundamentals*, 1972.

Calculating the Diffuse Solar Radiation and Reflection—E_2 and E_3

To calculate E_2 and E_3, the following equations are used:

$$E_2 = CE_1(1-F) \qquad \text{Formula (12-8)}$$

$$E_3 = E_1 F(C + \sin \beta)R \qquad \text{Formula (12-9)}$$

where

E_1 is the direct radiation of solar energy

E_2 is the diffuse radiation from the sky falling directly on the surface Btuh/ft^2

C is the diffuse radiation factor from Table 12-2. (This factor takes into account the dust and moisture content of the atmosphere, and varies throughout the year.)

E_3 is the ground reflected diffuse radiation from the sky falling on the surface

β is the solar altitude from Table 12-2

R is the reflectance factor of the ground or lower surface; use 0.20 unless otherwise known

F is the angle factor between the surface and the ground

For the case where the reflection is from the ground to an inclined surface:

$$F = (1-\cos \Sigma)/2 \qquad \text{Formula (12-10)}$$

where Σ is the angle of tilt of the surface.

Table 12-2 is based on computer programs for estimating solar energy received on surfaces based on cloudless days, 0% ground reflectance, and a clearness factor of 1. A clearness factor of 1 assumes an average atmospheric water content at sea level. From this table, solar insolation for any area can be estimated. The actual amount of solar energy received will be less, due to cloud coverage. Table 12-2 contains hour by hour data for solar position and direct normal solar energy received. From this data, detailed calculations can be made to estimate total energy received for any location. This table will

319

usually give conservative values of insolation since ground reflectance has been neglected.

Example Problem 12-6

Calculate the diffuse solar energy contribution for a $45°$ roof-mounted solar collector facing due south for the following conditions:

Plant location at $40°$ N latitude, during December 21 at 12:00 Noon. Assume a clearness factor of 1.

Answer

From Table 12-2,

$$\beta = 26.6° \quad \sin\beta = 0.447$$
$$C = 0.057$$
$$E_1 = 284$$
$$\Sigma = 45°$$

$$F = \frac{(1-\cos\Sigma)}{2} = 0.1465$$
$$E_2 = 0.057 \times 284 \times (1-.1465) = 13.8 \text{ Btuh/ft}^2$$
$$E_3 = 284 \times 0.1465 \ (0.057 + 0.447) \times 0.20 = 4.19 \text{ Btuh/ft}^2.$$

Example Problem 12-7

Calculate the total energy reaching the solar collector of Problem 12-6. Assume a clearness factor of 1.

Answer

$$E_T = E_1 \cos\theta + E_2 + E_3$$
$$E_d = 0.707 \ E_1 \ (\cos\theta \ \cos\beta + \sin\beta)$$

From Table 12-2,

$$\theta = 0°$$
$$\beta = 26.6° \quad \cos\beta = 0.894$$

320

$$E_1 = 284$$
$$E_d = 0.707 \times 284 \, (0.894 + 0.447)$$
$$= 269.2 \text{ Btuh/ft}^2$$

Combining Problems 12-6 and 12-7,

$$E_T = 269.2 + 13.8 + 4.19 = 287.2 \text{ Btuh/ft}^2 .$$

INSOLATION DATA

The U.S. Weather Bureau information represents average values over several years of measurement. A complete reference to weather data is found in the *Weather Atlas of the United States,* Gale Research Co. This book is a reprint of the *Climatic Atlas of the U.S.,* U.S. Dept. of Commerce. Included in the *Atlas* are maps of monthly and annual sunshine, total hours of sunshine, mean solar radiation, wind speed, and wind direction. Another good reference is presented by Liu and Jordon in the ASHRAE publication, *Low Temperature Engineering Application of Solar Energy.* The material therein is based on Weather Bureau data for 80 cities in the United States.

Information on annual mean daily insolation is sometimes expressed in the energy unit Langleys per day. To convert Langleys per day to Btu/ft^2 per day, multiply the values by 3.69.

The percentage of possible sunshine greatly affects the amount of energy received by a collector. The direct component of solar energy is proportionately reduced by cloud cover while the diffuse radiation prevails on cloudy days.

SOLAR SYSTEM APPLICATION

Domestic Hot Water

A conventional water heater uses electricity, gas, or oil, and can work day or night to produce hot water.

In a solar water heater, the heating source is a solar collector mounted in a sunny location. Water or other fluids moving through the collector are heated by the sun. The solar-heated water circulates

321

into an insulated storage tank when water is the fluid in the collector. For nonpotable collector fluids, a heat exchanger is used to transfer heat from these fluids to the domestic water in the storage tank.

Six major types of domestic hot water systems are currently in use. The following domestic hot water systems are described:

- draindown, closed-loop, water;
- drainback, open-loop, water;
- double tank, closed-loop, antifreeze;
- single tank, closed-loop, antifreeze;
- thermosyphon; and
- air system.

Draindown Systems—The draindown system operation is based on the success of three solenoid valves draining the collectors when freezing conditions occur. The draindown system circulates potable water (under city water main pressure) through the collectors. It may be a one- or two-tank system. In any system type, a separate solar storage tank (two-tank system) allows the collectors to work more efficiently; cooler water is introduced into the collectors than in the case of a single-tank system where the electric element always keeps water at the top of the tank hot. A general disadvantage of two-tank systems over single tanks are the greater heat losses and extra expense of the second tank. Well-insulated tanks reduce heat losses.

A draindown system uses potable water only, with no antifreeze. Only small pumps are needed owing to city water supply pressure. Since no heat exchangers are required, heat transfer to storage is more efficient. The system drains down only for freeze protection, and then loses less than one gallon of water per panel.

The three fail-safe solenoids use control wiring more complex than that used with other systems. Pipe corrosion can also occur. Draindown systems are not best in climates where it freezes most winter nights.

Drainback Systems—A two-tank drainback system for solar domestic hot water systems uses pure water (usually deionized water obtained from a drug store or chemical supply house) and completely drains water from the pipes each time the pump shuts off. Freeze sensing is not necessary because the collectors are void of water except during times when the sun is shining on them. The collector loop is nonpressurized, with water falling back by gravity to the storage tank from the top of the collector loop. Downward pitch of the piping is important in both draindown and drainback systems, so that water will not remain in the collectors to freeze. Because the drainback system is nonpressurized, a larger pump is required to push the water to the top of the collectors (overcome the system head) before gravity returns the water to the storage tank. The drainback system requires a heat exchanger between the storage tank and delivery to the hot water supply.

A drainback system uses no antifreeze. Its controls are simpler than those for a draindown system, and it offers more efficient heat transfer than do antifreeze systems. Collectors must be positioned above the storage tank so that gravity can drain them for freeze protection. A large heat exchanger is needed in a preheating tank, and a pure water system must be maintained.

A popular method of freeze protection in warm climates is to recirculate warm water from the storage tank through the collectors when freezing conditions occur. Some collected heat is thus lost back to the atmosphere. The advantages are that antifreeze solutions and heat exchangers are not necessary. Greater system efficiency (owing to lack of heat exchangers) more than offsets the energy lost in the seldom-used recirculation mode.

Double-Tank Closed-Loop Systems—Both double- and single-tank closed-loop systems use an antifreeze mixture in the collector loop for freeze protection. Some common antifreeze solutions used in collectors (and their freezing points) are: 50% water/ethylene glycol ($-33°F$), 50% water/propylene glycol ($-28°F$), and silicone oils ($-120°F$). Antifreeze liquids are used in climates where freezing occurs frequently.

323

As already noted, antifreeze systems generally require double-walled heat exchange for consumer protection against toxic fluids. However, not all antifreeze solutions are toxic; state and local codes dictate heat exchange requirements. Greater system efficiency can be expected from a single-walled heat exchanger than from a double-walled one.

The major advantage of a double-tank system, when compared to a single-tank system, is that cooler liquids are brought to the collector, increasing its operating efficiency. Warmer liquids are brought to the collector from a single-tank system, since the back-up heating element is contained in the single tank.

In addition to greater collector efficiency and freeze protection, double-tank systems require a smaller pump than draindown or drainback systems. However, the two tanks and associated plumbing are more expensive; more insulation is required; heat transfer is not as efficient; and extra components such as expansion tank, air eliminator, pressure relief valves and gauges are needed.

Single-Tank Closed-Loop Systems—The back-up heating source is usually electricity. As noted earlier, tank heat losses are less from a single-tank system than from a double-tank system. A single-tank system costs less to install and requires a smaller pump than open-loop systems. A double-walled heat exchanger may be needed; heat transfer is not as efficient.

Thermosyphon Systems—Thermosyphon hot water systems are generally considered to be passive solar energy systems because they do not use pumps or valves. They are described here, with the chapter on active systems, because the audit procedure for thermosyphon hot water systems is the same as for active domestic hot water systems. In addition, comparison of the different systems is made easier.

Figure 12-5 shows the components of a thermosyphon hot water system. This type of system is used in warm climates. Potable water circulates through the collector loop. Thermosyphon systems rely on the principle that water warmed in the collector is less dense than the cooler water entering the collector; the warmed water rises

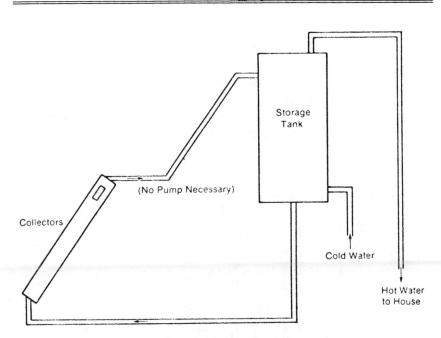

Figure 12-5. Simple Schematic of a Thermosyphon System

to the top of the storage tank, generally located above the collector. The bottom of the storage tank is 1 to 1½ feet higher than the bottom of the collectors, and the top of the tank is similarly higher than the top of the collectors. This placement of system components permits the "convective (thermosyphon) loop" to operate effectively. Locating the storage tank can be more of a problem with thermosyphon systems than with other types of systems. Rarely-needed freeze protection is provided by "reverse thermosyphon" or heat tape on the collectors, among other methods.

Thermosyphon systems require no antifreeze and produce highly efficient heat transfer. They are relatively simple, inexpensive systems to install. No pumps or controls are required.

325

Air Systems—In the five system types just described, flat-plate liquid-cooled collectors (circulation loops contained liquids) were used. Air collectors for domestic hot water heating can also be used. Figure 12-6 is a simple diagram of an air system. A fan is used to blow air through collectors. The heated air blows over a large air-to-liquid heat exchanger through which the domestic water supply is being circulated through the storage tank.

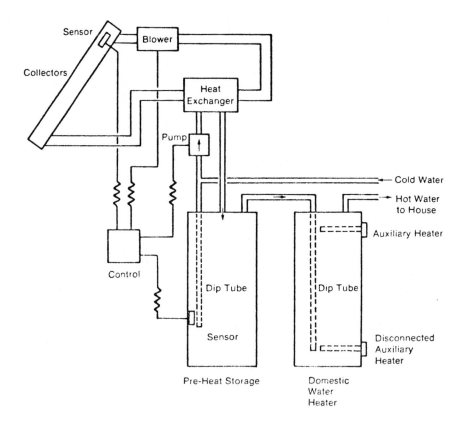

Figure 12-6. Simple Schematic of an Air System

326

Air obviously does not need freeze protection, and it is free. Corrosion is not a problem. In general, air systems require less maintenance than liquid systems. However, air ducts and air handling units require greater space than piping in liquid systems. In addition, air leaks are difficult to detect. Both air fans and liquid pumps are required for the air system.

Space Heating

A solar space heating system is essentially the same as one for solar water heating except that it is larger in scale, creating additional design and installation considerations. Since space heating requires much more heat than water heating, there are more collectors, more valves, more piping, larger storage, and more controls. All solar space heating systems include a collector array, a storage system, the piping between these two, a distribution system to the house, an auxiliary heater, a heat exchanger between storage and auxiliary heater, and a control system to make the components of the system work together smoothly.

Figure 12-7 is a simple schematic of connection to forced air delivery system.

- A disadvantage of the forced air delivery system is the high initial cost, particularly when expensive prefabricated collectors are employed. With the use of large areas of lower-efficiency collectors, the total system cost could be lowered considerably. However, system lifetime is an important consideration.
- Care must be taken to prevent the occurrence of corrosion, scale, or freeze-up capable of causing damage or blockage.
- Leakage anywhere in the system can cause a considerable amount of damage to the system and the dwelling.
- Contamination of the domestic hot water supply is possible if a leak allows a treated heat transfer liquid to enter the domestic water system.

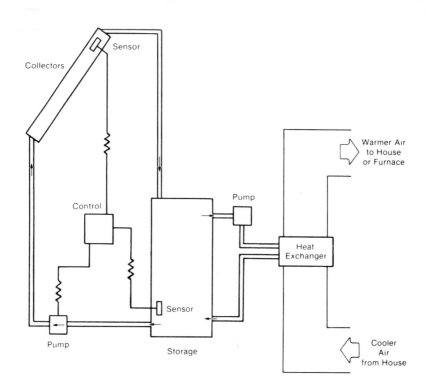

Figure 12-7. Simple Schematic of Connection to Forced Air Delivery System

Air Flat-Plate Systems

- There is no problem with corrosion, rust, clogging, or freezing.
- Air leakage does not have the serious consequences of water leakage.
- There is no danger of contamination by leakage from heat storage, as in the water system.
- A disadvantage is that ductwork risers occupy usable floor space and must be aligned from floor to floor.
- Also, air, having a lower thermal storage capacity than water, requires correspondingly more energy to transfer a given

328

amount of heat from collector to storage and from storage to occupied spaces.

- Air collectors and storage may need frequent cleaning to remove deposits of dust (filters may solve this problem).
- Air systems require a much larger heat exchange surface than liquid systems.
- Mildew is possible in rock storage.

13

Biomass Technology

The amounts of biomass wastes which are already collected, or are readily collectible, are summarized in this chapter. The characteristics of crop wastes, animal manures, municipal solid wastes and forestry residues are discussed, as they relate to collection and utilization as fuels. Processing techniques for municipal solid waste to permit separation of a refuse derived fuel which can be co-combusted with coal, or otherwise used as a fuel, are discussed also. Advanced thermal processes including pyrolysis and gasification, are summarized as they relate to preferred means of preparing wood wastes for such treatment.

Organic solid wastes and residues from biomass are now generally recognized as renewable fuels, which must increasingly supplement fossil nonrenewable fuels. However, utilization of such wastes and residues requires some modifications in procedures, many of which have been developed and tested only in the last 5 to 10 years. Only a few of the processes can be considered to have been proven, but many appear promising as development work continues.

This chapter discusses agricultural crop residues, municipal solid wastes, and forestry wastes and residues as they might provide supplements to fossil fuels in the near term.

Source: "Energy from Biomass Wastes," by D. J. Lohuis, World Energy Engineering Congress, Association of Energy Engineers.

AMOUNTS OF BIOMASS WASTES

The amounts of such wastes which are produced are impressive. Crop wastes are estimated at about 600 million dry tons annually, municipal solid wastes at about 140 million tons annually and forestry residues including logging wastes left in the forest at 130 million dry tons.

In addition, there are about 1,740 million tons of animal and poultry manure. It is to be noted that much of the crop wastes and manures are not collected, and that the quantity of manure cited is not on a moisture free basis.

The most extensive study of the amounts of crop wastes and manures which might be collected was done by Stanford Research Institute[1]* on a county by county basis, and was published in July 1976. The totals were in millions of dry tons per year:

	Total	Available	Collected
Crop Residues	322	278	7
Manures	36	26	26

It is to be noted that about 75% of the crop residues are now returned to the soil, and that the manure figures cited include only those from barns and feedlots on the basis that only these are economically collectible.

The municipal solid waste figure of 140 million tons is that which is now collected or is readily collectible. About 80% of that weight is combustible. It is also to be noted that there is presently a cost for disposal of municipal waste, estimated at about $6 billion annually.[2]

SRI estimates the forest and wood processing mill residues as 116 million dry tons annually of which 76 million tons are now collected. 38 million tons are sold for nonfuel uses; 19 million used as fuel, and 59 million tons wasted. The latter are principally logging wastes.

---

*See References at the end of the chapter.

Other estimates of the amounts of material left at logging sites are higher. Zerbe of the U.S. Forest Service[3] estimates 130 million dry tons per year, not including unharvested trees of commercially useful species, or of noncommercial species. If "rough, rotten but salvable" wood is included, it could bring the total to more than 1000 million dry tons, but much of this would be on a one time basis, and some realizable only with several years' intervening for growth.

COLLECTION AND TRANSPORTATION

The ability to use any of the cited biomass materials efficiently depends on the moisture content, physical form, and the costs of collection. Therefore, in the case of wood residues, the uniform, finely divided and especially the dry materials arising from wood-working operations are already almost completely used. There is much more bark unused because it is often wet, contaminated and of nonuniform character; however, the use of even these less desirable materials is increasing rapidly as the very substantial energy needs of the pulp and paper industry become ever more costly.

Costs of collection of bark are also increasing, as more of the wood is debarked in remote locations and only wood chips, not logs are brought to the pulp and paper mills. Costs of collection of the "rough, rotten but salvable dead wood" which constitutes the bulk of the unused wood residue resource are also high. This is in part because much of such material is on mountainous terrain. While harvesting equipment for trees for fuel can be lighter and more mobile than harvesting for timber, it remains difficult to operate on steep slopes.

Transportation costs could be reduced by at least partial drying and densification at the wood-for-fuel harvesting site or nearby. Drying must eventually be done anyway, either in the combustion chamber, if direct fired, or in the gasification or pyrolysis unit.

An alternate to use of "rough, rotten and salvable dead wood" for fuel, is the growing of trees or crops specifically for fuel in "bio-

332

mass plantations." Studies done for the Department of Energy by Intertechnology Corporation and Mitre have included examination of short rotation forestry using species that sprout from stumps, grow rapidly when young and reproduce vegetatively. Poplar, cottonwood, alder, locust and soft maple have been considered.

Growing of corn, sorghum, sugarcane and certain perennial grasses in "plantations" for fuel have also been considered. One disadvantage of annual crops is the necessity to harvest at specific times of the year, while harvesting of trees can proceed year round. The annual crops are also projected to have greater need for fertilization.

Questions have also been raised as to loss of soil fertility by growing trees for frequent harvesting. It is proposed that the ash resulting from combustion be applied to the harvested area, thus returning most of the potassium and phosphorus.

Collection costs are projected to be high for most crop residues. Most crop residues are returned to the land; storing and transporting them to a central point for use as fuel will be more expensive. However, certain crops produce residues which are collected at central processing sites; for example, cotton gin trash, and corn cobs; or produce troublesome residues in the field. A notable example in the latter category is rice straw.

Crop residues are generally low in density, adding to storage and transportation costs. Pelletization has been tested, but has usually been judged not economic for fuel preparation, because of high electrical energy usage, the need to control moisture content, and high replacement rates for the forming dies because of the presence of abrasive sand. Cubing to a lower density is under test and seems more promising with some materials. Straw is not readily cubable.

PROBLEMS IN USE OF
CROP RESIDUES

Besides collection and storage costs, there are some special problems with certain crop residues, when used as fuels. As noted, densification of straws is not readily accomplished. Straws also tend to be high in silica content, and also have relatively large amounts of sodium and potassium salts. Their presence leads to low ash fusion temperatures, thus the problem of ready formation of slags on combustion grates. Disposal of rice straw is a particular problem; 4 million dry tons annually are wasted.[1] Most of it has been burned in the field; this will be prohibited for environmental reasons.

36 million dry tons of corn cobs are produced annually, but only small amounts are collected at central locations, as in production of seed corn. Two companies producing seed corn are using cobs as fuel to dry the corn. Gasification is being used. General Motors has evaluated corn cobs and cubed corn stalks as fuel for some of their plants in the major corn producing states. Corn cobs and corn stalk cubes have shapes which facilitate gasification but potassium salt carryover is a problem.

MANURES AS FUEL

Crop residues in the form of animal manures present a particular disposal problem in the case of dairy cattle feeding lots, near metropolitan areas, where feed is brought in from a distance, and beef cattle feed lots. The Department of Energy is funding research on gasification of manure in a modified multihearth furnace.[4] Partial predrying is accomplished in a vacuumized externally heated screw conveyor, followed by further drying, pyrolysis and gasification on succeeding hearths of the furnace.

Since manure is high in moisture content, anaerobic digestion to methane appears to be a more attractive alternative. Studies made to date show it extremely important to begin anaerobic digestion while the manure is fresh to obtain best methane yields. This requires so-called "environmental" feedlots with slatted floors, or floors capable of frequent scraping and periodic flushing with water. As a consequence, the volume of liquid to be treated is large and after anaerobic digestion, the amount of effluent liquid is about the same as the input volume. The fertilizer elements remain essentially unchanged in amounts. Disposal of the effluent liquid on land is thus still preferred.

334

MUNICIPAL WASTES AS FUELS

Like manures in metropolitan areas, municipal solid wastes (MSW) have a negative disposal cost. This ranges from about $3.50 to $30.00 per ton, depending on the distance to the landfill site, and the dumping fee at the site. Costs for disposal average about $10/ton for 25–30% moisture content material.

Municipal solid waste contains up to 80% combustible organics. The noncombustibles approximate 8% glass, 8% ferrous metals, 1% nonferrous metals and 5% miscellaneous inorganics. The combustible portion includes:

3.5% plastics	17.5% food scraps
1.5% textiles	31% paper
3.5% wood	Balance: grass-clippings,
2.5% rubber and leather	tree and shrub trimmings

Note the complexity of the waste from the standpoint of type and physical characteristics. To use municipal waste as an energy source efficiently, by any technique other than mass incineration in certain steam generating units, requires shredding to reduce particle size and preferably some separation procedures.

Mass incineration has long been practiced in Europe, in many cases with generation of steam. Incineration of municipal solid waste generally produces stack emissions which need to be treated, because the varying physical characteristics of the waste leads to nonuniform combustion. Complete combustion is favored by higher temperatures, and uniform input feed rates. Europeans led the way to development of water wall boilers, which permit higher temperature than those with refractory walls, and improved input feed and grate systems.

To keep stack emissions within acceptable limits, it is usual to also employ electrostatic precipitators and/or wet scrubbing systems. Many early U.S. incinerators could not be economically modified to meet emission requirements, leading to abandonment of operations. Also, in some cases, no nearby steam customers could be found if new replacement steam generating incinerators were to be installed.

Preparation of Special Fuel
Fractions from MSW

The above limitations of mass incineration of municipal solid waste (MSW) led to much development work on preparation of a "refuse derived fuel" (RDF) consisting of most of the organic, combustible fraction of the municipal solid waste. The RDF can more readily be transported and is more acceptable to a major fuel user than is nonprocessed "whole" municipal waste. Prime outlets for such RDF are coal-burning utility power boilers.

The most common method for preparing refuse derived fuels for power plant use involves dry shredding, air classification to separate the light, mostly paper fraction, from the heavier glass, metals, leather and dense food particles. In the period 1972–1974, such dry mechanical systems were selected for a number of locations, based on the pioneering work at the Bureau of Mines in College Park, Maryland, and a demonstration plant at Union Electric Company in St. Louis. The 300-ton-per-day St. Louis demonstration plant was partially funded by the EPA.

Similar preliminary steps are involved in preparing a fuel fraction for subsequent combustion as pellets in stoker-fired furnaces, for combustion in suspension in patented "Vortex" burners or for flash pyrolysis to produce liquid fuel.

Wet processes can also be used to prepare a fuel fraction. The pioneering system of this type was the Black–Clawson facility at Franklin, Ohio. It involved reduction of particle size by powerful, high-speed dispersing blades operating in water, similar to a household food blender. Heavy objects are removed by density separation. The organics are dewatered to 50% moisture content and can be combusted under conditions like those used for burning wet bark and/or sawdust in combination with coal or oil, or in fluidized bed combustors. Full-scale systems of this type are being constructed in Hempstead, New York and in Dade County, Florida.

The dry process refuse-derived fuel fraction was shown in early trials to be well suited to co-combustion with coal in utility power boilers. In such boilers, the pulverized coal burns in suspension after introduction through nozzles located in the corners of the boilers. RDF is injected into the furnace through separate nozzles at levels of about 10% to 20% of the Btu needs of the boiler. The RDF has a Btu value per pound about half that of coal and, of course, is lower density. Energy efficiency losses obtained and projected are in the 1.5% and 2.6% range for 10% and 20% RDF loadings, respectively.

To be suitable for suspension firing with coal, RDF must be shredded to about 1¼ inches in maximum dimension and should have a minimum of adhering inorganics. A second shredding after air classification is usually employed. The RDF fraction, which is mostly paper, commands a price of $9.00 to $10.00 per ton, or more depending on the price of coal, which it replaces.

Non-Fuel Energy Recovery from
Municipal Solid Waste

Ferrous metals are readily separated from shredded municipal waste by magnetic means.[5] It might be argued that recovery of ferrous metals from municipal solid waste is not "energy recovery." However, energy is saved. It requires about 10 million Btu's per ton less to recover and reuse steel than to smelt it from ore, a reduction of 59%.

Aluminum recovery from municipal solid waste is approaching the stage of proven technology. Aluminum is separated by eddy current metal repulsion. Other nonferrous metals are similarly separated. Aluminum contents in waste approximate ½%, yielding about $2.00 per input ton. Other nonferrous metals which can be recovered are worth about $.95 per ton additional.

Use of recovered glass from waste in glass manufacture saves energy over that required from making it from raw materials. However, glass recovery techniques are less completely developed. It is difficult to separate stones and ceramic pieces from glass and they produce defects in glass bottle manufacture. Furthermore, for best use of recovered glass in bottle manufacture, the glass should be separated for color. Color separation techniques developed at the Franklin, Ohio facility will be employed at the Hempstead, New York facility.

Problems in Use of Municipal
Waste Refuse-Derived Fuel

While refuse-derived fuel from dry separation procedures is mostly paper, it also includes other light materials such as part of the yardwaste, heavier materials which are shaped so as to be lifted by air, textiles, and glass shards adhering to the moist paper fraction.

The textile fraction may clog pneumatic transportation pipes and boiler or burner feed nozzles. The presence of the moisture can cause undesirable compaction of the RDF during transportation and temporary storage before use, leading to difficulties in achieving uniform feed rates into the combustion chamber.

The shards of glass adhering to the waste contribute to the ash removal load and may strain or exceed the capacity of the ash removal system. It is also desirable, of course, to minimize the amount of glass because of the cost of loading, transporting, storage and feeding of non-fuel ingredients. Glass may also be troublesome in boilers with grates because of its relatively low fusion point compared to usual ash ingredients, thus leading to grate blockage.

A number of techniques have been developed for minimizing the amount of glass in the RDF fraction. Drying the fraction, followed by screening with agitation, can remove a large part of the glass. Screening without drying to remove the very fine fraction of the waste also removes much of the glass. However, such screening techniques also remove a portion of the combustibles which would otherwise contribute to the energy yield.

The current direction, therefore, is to remove as much of the glass as possible as early in the separation process as possible. This may be done by "trommeling" prior to primary shredding. Trommeling is a process of passing material through an inclined rotary drum, the barrel of which consists of sections having 4- to 5-inch holes comprising about 35% of the drum surface. The light, generally large, pieces of paper are carried to the end, becoming part of the "overflow;" glass, stones and ceramics pass through the holes and are discharged with the "underflow." The technique was pioneered by the National Center for Resource Recovery in Washington, D.C.[6]

As discussed earlier, most resource recovery plants now in operation, where a fuel fraction is prepared and metals are separated, employ shredding. Shredding produces more uniform particle sizes and liberates entrapped materials which could not otherwise be readily separated and recovered. However, shredders have high operating and maintenance costs. Trommeling before shredding reduces these costs, and several of the new municipal solid waste resource recovery plants will employ trommels as the first step.

Growth of Municipal Resource Recovery
from Solid Waste Systems

It is obvious from the foregoing brief review that energy and resource recovery from municipal solid waste requires somewhat complicated procedures. Both direct combustion systems and separation of recyclables and an RDF for burning off-site are capital intensive. Operating problems have not been fully resolved. Nevertheless, the high cost and waste of resources involved in the alternate of sanitary landfills has resulted in increasing numbers of decisions by officials of metropolitan areas that energy and recyclable resource recovery should be adopted.

There are presently 16 municipal solid waste energy and/or recyclable resource recovery systems in commercial operation, or in the late stages of test runs for acceptance. It is projected that by the end of 1979 35 systems with an annual capacity of 6 million tons of municipal solid waste will be in operation. It will be recalled that about 140 million tons of municipal solid waste are now generated.

The trend in recent planning appears to be toward removal of recyclable metals, followed by mass burning in steam generating water-wall incinerators, rather than preparation of a refuse-derived fuel for co-combustion with coal or oil in utility power boilers. The pioneering water-wall incinerator in this country was that of RESCO in Saugus, Massachusetts, where the steam is sold to the General Electric Company. The system used proven European technology (the Saugus facility was the 54th of this type). It is quite completely described in a recent EPA publication, "Engineering and Economic Analysis of Waste to Energy Systems."[7] A typical RDF-producing facility, that of the City of Chicago, is also described in detail in the same publication.

ADVANCED THERMAL PROCESSES
FOR UTILIZATION OF BIOMASS

Advanced thermal processes, pyrolysis, gasification and lique-faction, continue to be studied intensively at the bench and pilot scale. Several processes have reached the demonstration stage.

As previously noted, almost all of these processes require some preliminary preparation steps, such as shredding, air classification and recyclable metals removal. In the development of many of the bench-scale level processes, prepared synthetic refuse, consisting mostly of paper or of wood, have been used because they provide uniformity and reproducibility. The experience in the development of such processes leads to the conclusion that for the near term, wood wastes are likely to be the source of fuel for these advanced thermal processes rather than municipal wastes, in spite of the nega-tive value of municipal waste and the fact that it is already collected. Even where municipal solid waste might advantageously be used as a fuel source, it is likely that wood wastes will be used to supplement it. In many municipal areas large amounts of wood waste are gener-ated as a result of removal and trimming of trees on city-owned property, frequently involving conversion of limbs and major por-tions of the trunks to chips.

Chips are a preferred fuel for gasification in low cost, low throughput systems. Gasification of wood waste has many advantages for the fuel user, principally the ability to use the gas in existing gas-fired or oil-fired package boilers. However, clean chips, free from fines which are also generated in their manufacture, are expensive and appear not to be an economical raw material for low cost, small-scale gasification of wood, unless there is also a use for the concurrently generated fines.

Gasification of wood may most simply be done in air blown "close coupled" gasifiers in which the condensible oils are burned along with the noncombustible gases without cooling or scrubbing. However, even in these systems some char fines produced in the pyrolysis which precedes gasification tend to be carried out with the generated gases and these with the condensible fractions may coat the cooler parts of the gasifier and the connecting gas piping between the gasifier and combustor. Lesser amounts of such fines are generated when operating with only relatively large pieces of wood as feed. Use of large particles of input feed reduces the chances of the gases forming channels, and produces char which is more readily gasified in the final step.

However, the problem remains as to how best to utilize the wood fines generated in preparation of chips. They, of course, can be dried and pelletized and pellets make excellent feed for moving bed cylindrical updraft or downdraft gasifiers, but pelletization costs from $5 to about $16 per ton.

An alternate is to pyrolyze the wood fines with the production of char, condensible oils and noncondensible gases. The Georgia Institute of Technology developed a system of this sort, aimed particularly at economical pyrolysis of wood sawdust and bark fines. The unit is of the directly fired, countercurrent flow, moving-packed-bed shaft furnace reactor type. The work began about 10 years ago, and continued with a series of pilot units. In 1971, rights to the technology were purchased by Tech-Air Corporation, which subsequently became a wholly-owned subsidiary of American Can Company. The development of the technology is summarized in a recent paper presented at the American Chemical Society's March 15–16, 1978 Symposium on Advanced Thermal Processes for Utilization of

341

Biomass.[9] A char having a controlled uniform high level of volatiles can be produced, which is well suited for manufacture of charcoal barbeque briquettes. The pyrolysis unit can also be operated to produce a lesser yield of lower volatile char, and greater amounts of oils and gas. The char yield is affected primarily by the maximum bed temperature, which in turn is controlled by the air to feed ratio. The distribution of energy among the char, pyrolysis oil and gas correlated against char yield is shown in Figure 13-1.

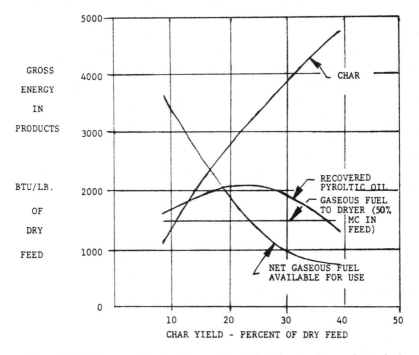

Figure 13-1. Energy Distribution vs. Char Yield for Products of Pyrolysis from Blend of Pine Sawdust and Bark

The gas stream from the reactor contains noncondensible gases, condensible oil vapors, water vapor and entrained particulates. The gases are cooled under controlled conditions to condense the pyrolysis oil, and limit the amount of water vapor condensed. The Btu value of the condensed oils varies with the moisture content, but average about 100,000 per gallon. The oil can be burned blended with No. 6 fuel oil, or in a separate burner by itself.

The pyrolysis gases leaving the oil condenser and demister contain water vapor, some low boiling point liquid fractions in addition to the noncondensible gases, and the mixture is nearly saturated. It is desirable, therefore, to use the gases in a close-coupled combustor. Experience has shown that a desirable use is to generate heat to dry the input waste.

If the input feed to a moving packed bed, partial combustion pyrolysis unit is not predried, some of the feed must be combusted to drive off the moisture before pyrolysis can begin. The extra moisture in the off-gas stream, which is driven from the feed, plus the water and carbon dioxide arising from combustion, dilutes the pyrolysis gas and reduces the Btu value. If combustion air is used to drive the pryolysis reaction, the nitrogen accompanying the oxygen used to combust part of the feed to dry the remainder, also further dilutes the pyrolysis gases.

It is similarly advantageous to use predried feed when air blown gasifiers are used. When the moisture content of the feed is above 30%, the off-gases may be so low in Btu value, and high in moisture, carbon dioxide, and nitrogen that difficulty is incurred in maintaining combustion without a sustaining flame fed by fossil fuel. A 20% moisture content is acceptable, 10–14% is preferred. While rotary drum dryers are satisfactory for drying sawdust and hogged bark, experience with such drying for drying wood chips and large particle size bark have been disappointing.[10] More development work on means of drying wood chips of the preferred size for moving packed bed gasifiers is needed.

An advantage of pyrolysis of wood wastes is that it produces two storable, transportable fuels, the char and the oil. The EPA has funded work on design and preliminary development work on a "portable" pryolysis reactor, which might be moved periodically from one forest area to another, converting wood wastes to char and oil. Alternately, it could be used ro process troublesome crop residues which are generated only part of the year.[11] The crop or wood residues would be shipped to points of fuel use as pyrolysis oil or char, or a blend of the two. Tests have shown good combustion performance of char blended with fuel oil.[12]

An alternate to pyrolysis of the fines generated in the production of chips, is the use of fluidized bed combustors on the whole mass of wood residue. Combustors of this type can operate on green wet wood, but their use would require transport of the wet wood to the fuel use site.

Like gasifiers, fluidized bed combustors operate more efficiently on dry wood, producing a lesser volume of combustion products, thus requiring smaller boilers, when wood free of excess moisture is the fuel.

Fluidized bed pyrolysis is also under study at several locations. Again, dry feed material is preferred and preparation of a uniform sized feed is advantageous. A double-fluidized bed using the char as a recirculating heat carrier, designed by Dr. Richard C. Bailie at West Virginia University, minimizes the dilution of the pyrolysis gases by nitrogen, so use of dry feed is less necessary. It is to be noted that fluidized bed combustors and fluidized bed pyrolysis units require a supplementary system for removing ash from the sand used as the circulating heat carrier.

CONCLUSIONS

The costs of collection, transportation and storage, and special preparation steps will control the degree to which biomass is utilized as fuel in the near future. The physical form and the moisture content of the biomass as produced have major influence on the economics of utilization.

It seems unlikely that large amounts of crop waste will be utilized as fuel in the near future, except for such specialized situations as that of burning sugar cane bagasse.

Animal manures are presently being converted by anaerobic digestion to substitute natural gas on a commercial basis, and the installation of many more systems seems likely.

The use of municipal solid wastes as fuel is attractive because they are collected and have a negative disposal value. However, systems for preparation of the wastes for use as fuel are capital intensive, and need to be large to be economical. Growth in use of mass burning in steam generating water-wall incinerators is anticipated.

The use of wood wastes as fuel seems most likely to increase rapidly. Very large amounts of forest residues are available. Advanced thermal processes such as gasification and pyrolysis, which produce fuel products which can be burned in existing equipment for combustion of natural gas and fuel oils, is expected to grow rapidly. Most of these processes benefit by use of predried wood waste as feed.

REFERENCES

1. Alich, J. A. Jr., and R. E. Inman, "An Evaluation of the Use of Agricultural Residues as an Energy Feedstock," July 1976, Grant AER74–18615A03, NSF/RANN/SE/GI/18615/FR/76/3.

2. Tillman, David A., "Energy from Wastes: An Overview," paper presented at the meeting of the Division of Fuel Chemistry, American Chemical Society, April 6, 1976.

3. "The Feasibility of Using Forest Residues for Energy and Chemicals," prepared for the National Science Foundation by USDA Forest Service, report NSF-RA-760013, March 1976.

4. Garrett, Donald E. and R. D. Mikesell, "A Thermal Process for Energy Recovery from Agricultural Residues," paper presented at American Chemical Society, Symposium on Advanced Thermal Processes for Conversion of Solid Wastes and Residues, March 14–16, 1978.

5. National Center for Resource Recovery, Research Monograph: "The Recovery of Magnetic Metals from Municipal Solid Waste."

6. Bernheisel, J. F., P. M. Bagalman and W. S. Parker: "Trommel Processing of Municipal Solid Waste Prior to Shredding," paper presented at IITRI-Bureau of Mines Symposium, May 2, 1978.

7. Wilson, E. Milton, et al., The Ralph M. Parsons Company, "Engineering and Economic Evaluation of Waste to Energy Systems," EPA–600/7–78–086 (May 1978).

8. Jones, J. L., "Converting Solid Wastes and Residues to Fuel," *Chemical Engineering*, 85 (1), (Jan. 2, 1978).

9. Bowen, M. D., E. D. Smyly, J. A. Knight and K. R. Purdy, "A Vertical Bed Pyrolysis System," paper presented at American Chemical Society Meeting, Anaheim, CA, March 15, 1978.

10. Private communication from Tech-Air, Subsidiary of American Can Co.

11. Tatom, J. W., A. R. Colcord, M. W. Williams and K. R. Purdy, "Prototype Mobile System for Pyrolysis of Agricultural and/or Silvicultural Wastes," EPA report on Grant No. R803430-01, Program Element No. EHE–624, June 1977.

12. Demeter, J., C. R. McCann, et al., "Combustion of Char from Pyrolyzed Wood Waste," Pittsburgh Energy Research Center, PERC/RI–77/9, July 1977.

13. E. Epstein, et al., Energy Resources Company, Cambridge, MA, "Potential Energy Production in Rural Communities from Biomass and Wastes, Using a Fluidized Bed Pyrolysis System," paper presented at Biomass and Wastes Symposium, August 14–18, 1978.

14

Synfuels Technology*

The currently reported recoverable supplies of U.S. fossil fuel liquids are shown in Table 14-1. To meet present demand levels from these resources, the supplies would last just five years for petroleum, and about eleven years for natural gas. (This worst-case assumption is based upon no newly discovered wells or fields.) As Table 14-2 shows, however, recoverable resources in shale oil and tar sands are greater than the total of fossil fuel fluids, and recoverable coal resources are more than tenfold greater yet. It seems evident, therefore, that until we enter the post-fossil-fuel era, our only really abundant domestic energy resource is coal.

Table 14-1. Currently Recoverable U.S. Fossil Fuel Fluids

		Quads
crude oil	29.5×10^9 bbls	171
natural gas liquids	6.0×10^9 bbls	24
natural gas	209×10^{12} scf	213
		408

Source: *C&EN,* 8/79

---

*Source: "Synfuels Technology Update," by William G. Lloyd. Presented at Synthetic Fuel Symposium, Association of Energy Engineers.

Table 14-2. Currently Recoverable U.S. Fossil Fuels

		Quads
fossil fuel fluids		408
shale, tar sands	76.5 × 10^9 bbls	444
coal	218 × 10^9 tons	4796
		5648

Source: *C&EN*, 8/79

Coal is an energy-rich material. One ton of "average" coal is the energy equivalent of 4.5 bbls of oil,

25,000 scf of natural gas,

7,300 kw-hrs of electric power,

208 gallons of 100-octane fuel.

Yet coal has obvious limitations as an all-purpose fuel:

1. It is not a fluid. Hence transportation and distribution pose special problems, and fuel feed to the combustion zone is for some applications (e.g., vehicle fuel) prohibitively difficult.

2. Coal is the dirtiest fuel in common usage. Emissions of SO_2, NO_x and fine particulates from uncontrolled combustion are severe in comparison with emissions associated with liquid fuels.

3. Disposal of ash residue, which may run 5–15% of the feed coal, in an environmentally acceptable way is a significant problem for utilities and other large-scale users, and may be a critical problem for users in highly built-up metropolitan areas.

To address these problems, our national R&D efforts are increasingly concerned with converting coal to gaseous or liquid synthetic fuels. Table 14-3 lists the major classes of coal-conversion technologies. Before looking at specific processes, however, it may be useful to make two points about the nature of coal. First, coal comes in many ranks, and coals differ substantially from one another in

348

Table 14-3. The Strategies of Coal Conversion

Gasification
Low-Btu, Medium-Btu (Syngas)
Synthetic Natural Gas
Liquefaction
Direct Hydroliquefaction
Indirect Liquefaction

chemical composition, physical characteristics, and heating value. Table 14-4 presents some selected average values for American coals of various ranks. In addition, variations of ash, sulfur content and rank within a given seam can be appreciable. Therefore any process must be viewed in terms of a *range* of properties of the feed coal. Second, the ratio of hydrogen to carbon (H/C), commonly 0.6–0.8 in coals, is much lower than that associated with crude and fuel oils (typically 1.4–1.8—see Table 14-5), gasoline (2.0–2.2) or natural gas (3.8–4). Consequently, any coal-liquefaction process must entail the transfer of significant amounts of hydrogen to the feed coal.

Table 14-4. Coal Compositions Vary with Coal Rank

Rank*	H/C Ratio	O/C Ratio	Btu/lb**
an, sa	0.43	0.03	12,600
lvb	0.64	0.03	13,540
mvb	0.63	0.03	14,110
hvAb	0.82	0.07	13,340
hvBb	0.81	0.09	12,060
hvCb	0.91	0.13	12,100
subA	0.84	0.15	12,360
subB	0.90	0.20	10,990
subC	0.87	0.22	11,090
lig	0.84	0.23	10,840

*Rank abbreviations: an—anthracite, sa—semianthracite, lvb—low-volatile bituminous, mvb—medium-volatile bituminous, hvb—high-volatile bituminous, sub—subbituminous, lig—lignite. A, B and C denote subgroups within the hvb and sub ranks.
**Dry, ash—included averages of 196 American coals. Individual coals may typically vary from their rank average by as much as 10%.
Source: Lloyd and Francis (1980).

Table 14-5. H/C Ratios and Heating Values of Several Fuels

	H/C Ratio	Btu/lb
No. 2 fuel oil	1.86	19,600
No. 4 fuel oil	1.51	18,720
Australian shale oil	1.63	18,790
French shale oil	1.61	18,560
H-Coal liquid, ASO	1.45	18,280
H-Coal liquid, ASB	1.37	18,340
Coal (typical hvBb)	0.82	14,320

Source: LLoyd and Davenport, *J. Chem. Educ.,* 57,56 (1980)

In the conversion of coal char to gaseous products, a relatively small number of reactions are available to us. Most of these are shown in Table 14-6. Reactions 1, 2 and 5 are oxidations, exothermic and hence useful in attaining and maintaining operating temperatures. Reaction 3 is the key reaction, the char-steam reaction which is highly endothermic and therefore requires balancing exothermic processes. Reaction 4, a second glasification reaction, is significant but generally slower and less important than Reaction 3. Reaction 6, the "water-gas shift," is nearly thermoneutral, and is used as a separate process when it is desired to adjust the H_2 :CO ratio. Reaction 7, the direct, ane, occurs as a secondary reaction at high temperatures; we do not at present know how to make this reaction proceed at useful rates and moderate temperatures.

Table 14-6. Reactions of Char Carbon

1.	$C* + 1/2\,O_2$	$\rightarrow CO$
2.	$C* + O_2$	$\rightarrow CO_2$
3.	$C* + H_2O$	$\rightarrow CO + H_2$
4.	$C* + CO_2$	$\rightarrow 2CO$
5.	$CO + 1/2\,O_2$	$\rightarrow CO_2$
6.	$CO + H_2O$	$\rightarrow CO_2 + H_2$
7.	$C* + 2H_2$	$\rightarrow CH_4$

The commonly encountered primary reactions in gasification are shown in Table 14-7. The char-steam reaction is essential to every gasification process. The shift reaction occurs in the absence of catalyst, and may be specifically catalyzed. The combustion of char to form CO_2 is required, one way or another, for all gasifications, not for the CO_2 formed but in order that the large exotherm formed can "balance" the endothermic char-steam reaction.

Table 14-7. Coal Gasification

$$C^* + H_2O \;\rightarrow\; CO + H_2$$

optional:

$$CO + H_2O \;\rightarrow\; CO_2 + H_2$$

required:

$$C^* + O_2 \;\rightarrow\; CO_2$$

In the production of synthetic natural gas (SNG, which is essentially methane), the key ractions (omitting char oxidation) are shown in Table 14-8. First, is the familiar char-steam reaction. Second, is the catalyzed shift reaction, carried out to adjust the H_2 :CO ratio to 3:1. Third, the CO-H_2 gas stream (purified to protect the easily poisoned catalyst) is converted in the methanation step to produce methane (CH_4) and CO_2. The sum of these three reactions shows the overall conversion: two pound-moles of char carbon react with steam to produce one pound-mole of methane and an equivalent amount of CO_2. The purified methane is chemically and physically equivalent to high-quality natural gas.

The last equation in Table 14-8 indicates a direct one-step conversion to methane which, as noted above, we do not now know how to carry out selectively.

351

Table 14-8. Synthetic Natural Gas

present technology:

$$2C^* + 2H_2O \rightarrow 2CO + 2H_2$$

$$CO + H_2O \rightarrow CO_2 + H_2$$

$$CO + 3H_2 \rightarrow CH_4 + H_2O$$

$$\overline{2C^* + 2H_2O \rightarrow CH_4 + CO_2}$$

future technology (?)

$$C^* + 2H_2 \rightarrow CH_4$$

The kinds of gases produced by these processes are listed in Table 14-9. Air-blown gasifiers produce low-Btu gas, typically containing 45–65% inert components (mostly N_2 and CO_2). Oxygen-blown gasifiers produce a medium-Btu gas, essentially free of N_2 and typically containing less than 35% inert components. The multistep SNG plants produce substantially pure methane.

Table 14-9. Typical Heating Values for Coal-Derived Gases

Process	Product Gas	Btu/scf
Steam + Air	"Low Btu"	120–180
Steam + Oxygen	"Medium Btu"	280–380
Methanation	"High Btu," "SNG"	900–1050

GASIFICATION

Gasifier technology can be organized in various ways. In this brief overview they are grouped for specific processes in terms of the coal beds: fixed, fluidized, or entrained.

The oldest and most commercialized gasifiers use fixed-bed technology. Most can be either air-blown to produce low-Btu gas or oxygen-blown to produce medium-Btu gas; most have some restrictions or modifications required for use with caking coals. Three examples are noted below.

352

The Wellman–Galusha gasifier has been in use for about 40 years, with over 150 commercial units installed worldwide. Indeed, Wellman has been building coal gasifiers since the 19th century. This unit operates at near-atmosphere pressure, with bed temperatures reaching 2400°F, with all noncaking coals. Its thermal efficiency is 75% (cold gas basis). The *agitated* gasifier (Figure 14-1) can also be used with caking coals. Typical composition of the product gas is shown in Table 14-10.

The Woodall–Duckham gasifier operates at 2200°F and typically at 20–30 atm. Over 100 commercial W–D units have been built in the past few decades, for both air-blown and oxygen-blown operation. These units work well with noncaking coals and with coals of FSI less than 3, but not with highly swelling coals. Thermal efficiency is 77%. In the W–D gasifier (Figure 14-2) the feed coal is heated in falling countercurrent to the product gas stream; consequently this process produces by-product pyrolysis tars and oils. Typical product gas compositions for air- and oxygen-blown operations are shown in Table 14-11.

Table 14-10. Wellman–Galusha Products

(Air-Blown)	
CO	28.6
H_2	15.0
CH_4	2.7
N_2	50.3
CO_2	3.4
HHV (dry) = 168 Btu/scf	

Source: *Handbook of Gasifiers and Gas Treatment Systems* (1976).

353

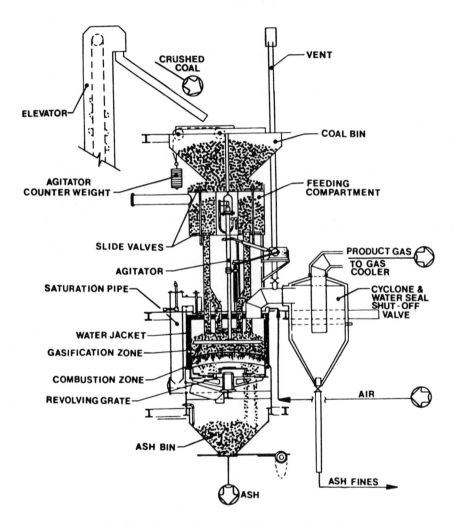

Figure 14-1. Wellman–Galusha Agitated Gasifier

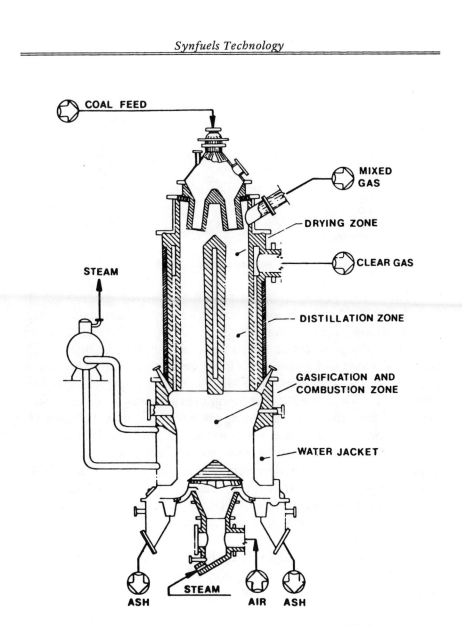

COAL FEED

MIXED GAS

DRYING ZONE

CLEAR GAS

STEAM

DISTILLATION ZONE

GASIFICATION AND COMBUSTION ZONE

WATER JACKET

ASH   STEAM   AIR   ASH

Figure 14-2. Woodall–Duckham Gas Integral Gasifier

Table 14-11. Woodall–Duckham Gasifier Products

	O_2-Blown	Air-Blown
CO	37.5	28.3
H_2	38.4	17.0
CH_4	3.5	2.7
N_2	2.2	47.2
O_2		
CO_2	18.0	4.5
$H_2S + COS$	0.4	0.3
HHV (dry)	280	175

Source: *Handbook of Gasifiers and Gas Treatment Systems* (1976).

The Lurgi gasifier is an improved version of the pre-WW II German unit. There are now a couple of dozen commercial Lurgi gasifier systems, some with multiple gasifier units. Like the W–D, the Lurgi typically operates at 20–30 atm, and can be designed for air-blown or oxygen-blown modes. Lurgi bed temperatures can reach 2500°F; steam is injected into the ash layer, to prevent slagging. Thermal efficiency is 63%. The Lurgi uses a moving spreader to feed fresh coal uniformly, and a moving grate for ash removal (Figure 14-3). The standard Lurgi does not operate efficiently with highly caking coals. For operation with such coals a "slagging Lurgi" version (Figure 14-4) has been develoepd. This operates at considerably higher temperatures, so that the residual ash is melted (slagged) and collected as a granular, low-surface-area material. Tests conducted earlier this year with caking coals by the British Gas Corp. are reported to be successful.

The conventional Lurgi is best known in recent years for its role as the gasifier technology for South Africa's SASOL complex. A typical product composition for oxygen-blown operation is shown in Table 14-12.

Fluid-bed gasifiers use relatively finely crushed feed coal, and operate by passing a steam-gas stream upwards through the coal bed at velocities sufficient to expand and fluidize the bed. Four such gasifiers are described briefly below.

356

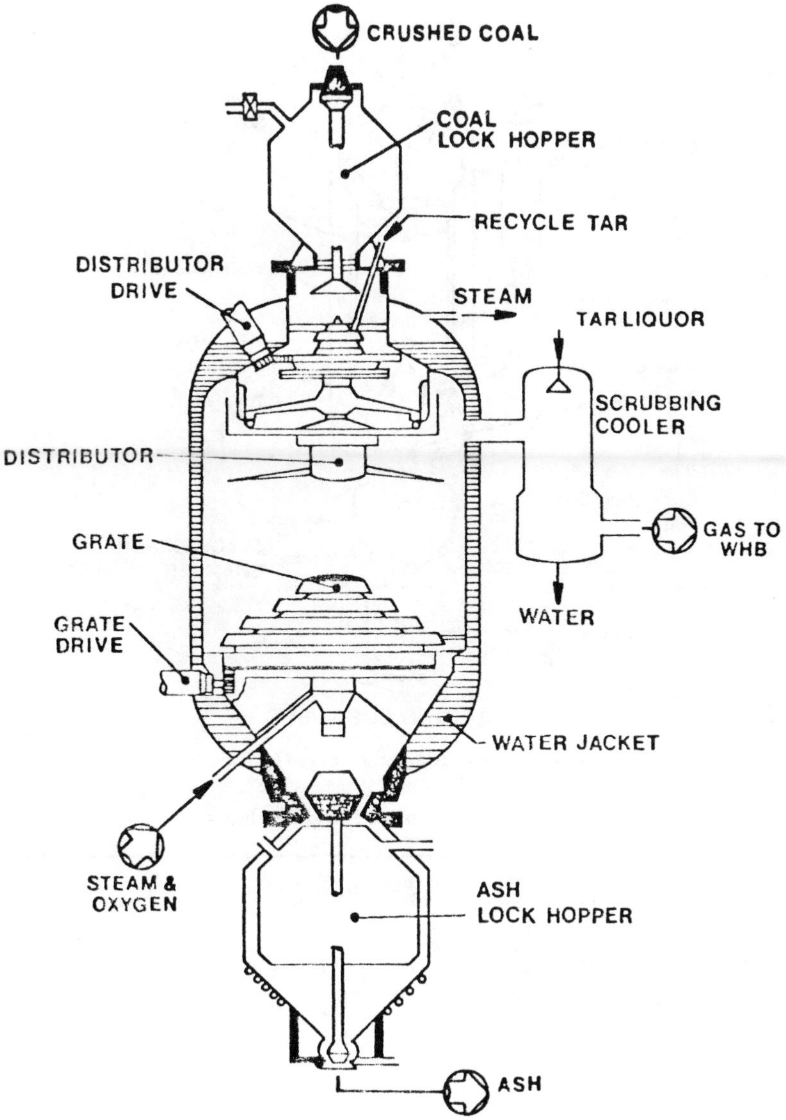

CRUSHED COAL

COAL
LOCK HOPPER

RECYCLE TAR

DISTRIBUTOR
DRIVE

STEAM

TAR LIQUOR

SCRUBBING
COOLER

DISTRIBUTOR

GAS TO
WHB

GRATE

WATER

GRATE
DRIVE

WATER JACKET

STEAM &
OXYGEN

ASH
LOCK HOPPER

ASH

Figure 14-3. Lurgi Pressure Gasifier

357

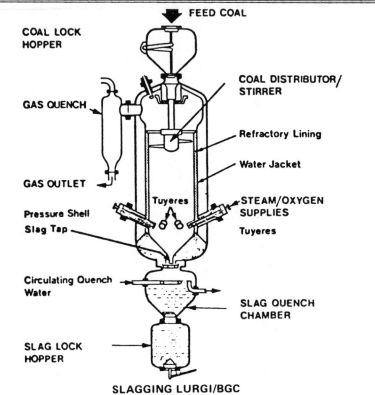

COAL LOCK
HOPPER

FEED COAL

COAL DISTRIBUTOR/
STIRRER

GAS QUENCH

Refractory Lining

Water Jacket

GAS OUTLET

Tuyeres

STEAM/OXYGEN
SUPPLIES

Pressure Shell
Slag Tap

Tuyeres

Circulating Quench
Water

SLAG QUENCH
CHAMBER

SLAG LOCK
HOPPER

SLAGGING LURGI/BGC

Figure 14-4. Slagging Lurgi Gasifier

Table 14-12. Typical Lurgi Products

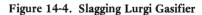

(O$_2$-Blown)			
CO	16.9	CO$_2$	31.5
H$_2$	39.4	H$_2$S + COS	0.8
CH$_4$	9.0	N$_2$ + Ar	1.6
C$_2$H$_6$	0.7		
C$_2$H$_4$	0.1		
HHV = 285 Btu/scf			

Source: *Handbook of Gasifiers and Gas Treatment Systems* (1976).

The Winkler gasifier is the only commercialized fluid-bed unit, with well over a dozen commercial plants. This is a low-pressure (or atmospheric pressure) system, oxygen- or air-blown, and works with noncaking and moderately caking coals. Coals with FSI values above 4 may be used after pretreatment (heating in the presence of a little oxygen to destroy the plastic properties). Bed temperature is low (about 1800°F) and thermal efficiency is 72%. The Winkler (Figure 14-5) may produce ash with as much as 15–30% unconverted carbon.

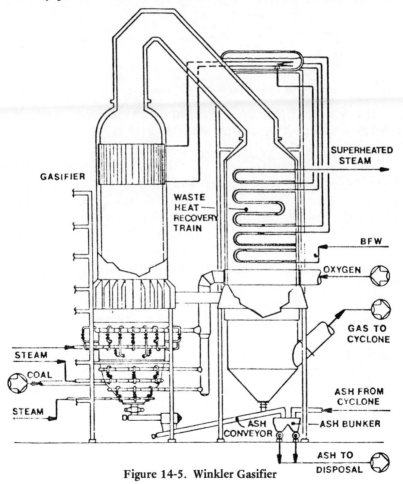

Figure 14-5. Winkler Gasifier

Typical product compositions for air- and oxygen-blown modes are shown in Table 14-13.

Table 14-13. Typical Winkler Products

	O_2-Blown	Air-Blown
CO	48.2	22.0
H_2	35.3	14.0
CH_4	1.8	1.0
CO_2	13.8	7.0
N_2 + Ar	0.9	56.0
HHV	288	126

Source: *Handbook of Gasifiers and Gas Treatment Systems* (1976).

The Synthane process, developed by the U.S. Bureau of Mines at Pittsburgh, is a high-pressure, fluid-bed gasifier, typically operating at 70 atm and 1800°F. With a hot pretreater stage this process can handle all coals. The feed coal falls through the rising product gas stream, devolatizing in the process (Figure 14-6). Char is removed continuously from the bottom. Like the Winkler, the Synthane process can be air-blown or oxygen-blown. From hvb coal this process typically produces about 100 pounds of tar and 600 pounds of char per ton of coal converted. Representative product gas compositions for air- and oxygen-blown operation are shown in Table 14-14.

IGT's Hygas gasifier is another high-pressure, fluid-bed process, operating typically at 80 atm. The feed coal slurry falls through four successive fluid beds, the top a drying bed, the next a first-stage hydrogasifier at about 1300°F, the next a second-stage hydrogasifier at about 1800°F, and the last a steam-oxygen gasifier (Figure 14-7). Overall thermal efficiency is 68%. For use with agglomerating coals an 800°F oxidizing pretreatment stage is needed. This process uses less oxygen than many others, and produces a significant amount of methane (Table 14-15), thereby reducing the amount of shift and methanation required for SNG.

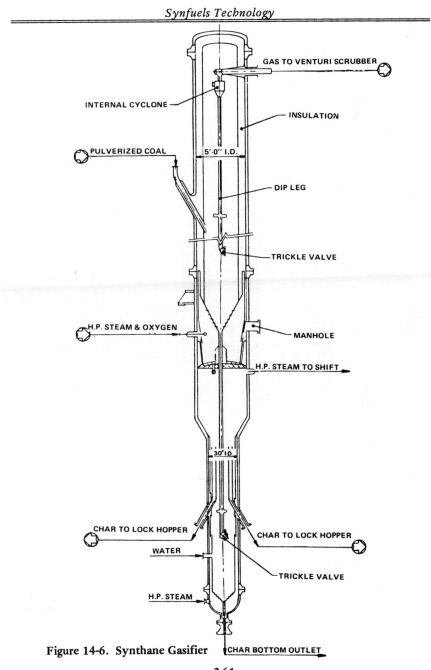

GAS TO VENTURI SCRUBBER

INTERNAL CYCLONE

INSULATION

PULVERIZED COAL

5'-0'' I.D.

DIP LEG

TRICKLE VALVE

H.P. STEAM & OXYGEN

MANHOLE

H.P. STEAM TO SHIFT

30''I.D.

CHAR TO LOCK HOPPER

CHAR TO LOCK HOPPER

WATER

TRICKLE VALVE

H.P. STEAM

Figure 14-6. Synthane Gasifier

CHAR BOTTOM OUTLET

361

Table 14-14. Typical Synthane Products

	O_2-Blown	Air-Blown
CO	13.2	10.1
H_2	32.3	21.5
CH_4	15.0	5.6
C_2H_6	1.6	0.7
$H_2S + COS$	1.6	0.7
$N_2 + Ar$	~0.	43.5
HHV	355	165

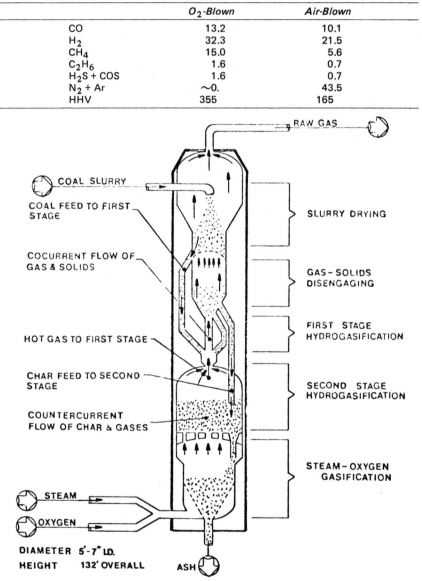

Figure 14-7. Hygas Gasifier (with Steam-Oxygen Gasification)

362

Table 14-15. Hygas Products

(From Pgh No. 8 Seam)			
CO	23.8		
H_2	30.2	N_2 + Ar	0.1
CH_4	18.6	NH_3	0.5
C_2H_6	0.7	H_2S + COS	1.2
BTX	0.4		
HHV (dry) = 375 Btu/scf			

Source: *Handbook of Gasifiers and Gas Treatment Systems* (1976).

Conoco's CO_2 Acceptor gasifier is based upon an entirely different idea. It accepts all coals. In the fluid-bed gasifier, which operates at moderate pressure and temperature (19 atm and 1500°F), the pulverized coal is mixed with calcium oxide (CaO) and is fluidized by steam and inert gas. The endothermic char-steam reaction is slightly counteracted by the moderately exothermic water-gas shift:

$$CO + H_2O \rightarrow CO_2 + H_2$$

and the CO_2 produced by the shift reaction combines with the calcium oxide in a strongly exothermic reaction to form calcium carbonate:

$$CO_2 + CaO \rightarrow CaCO_3$$

This latter reaction provides the bulk of the heat needed for the endothermic char-steam gasification.

The gasifier reactor is shown schematically in Figure 14-8. The bed is continuously bled off, removing carbonated acceptor and char residue. In a second fluid-bed reactor (not shown) the carbonated acceptor and char are contacted with air, so that the char combustion brings the system to 1850–1900°F, where the carbonated acceptor breaks down endothermically to regenerate CaO and release CO_2. The regenerated CaO is returned to the gasification reactor.

The CO_2-Acceptor Process has an overall efficiency of 77%. Although air-blown, its two-chamber process produces a high-quality medium-Btu gas (Table 14-16), with a significant methane content.

363

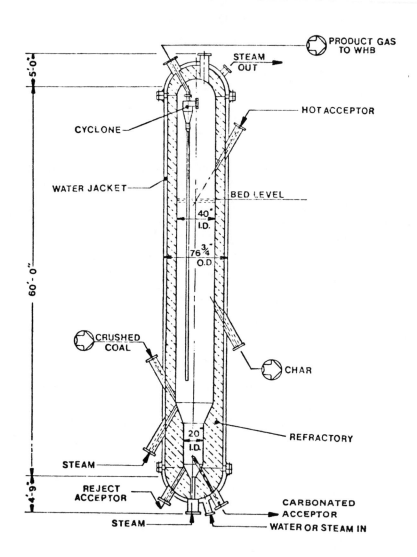

Figure 14-8. CO$_2$ Acceptor Gasifier

364

Table 14-16. CO_2 Acceptor Products

(Air-Blown Regenerator)			
CO	15.5	CO_2	9.1
H_2	58.8	$H_2S + COS$	0.0
CH_4	13.7	$N_2 + Ar$	2.9
HHV = 380 Btu/scf			

Source: *Handbook of Gasifiers and Gas Treatment Systems* (1976).

Entrained-bed gasifiers use pulverized coals at much higher temperatures and much shorter residence times (these latter of the order of 1 sec.). These accept all types of coals and produce gaseous products free of tars and oils. Operating temperatures are such that chars are normally completely reacted and ashes are completely slagged. Two representative processes are noted below.

The Koppers–Totzek process has been commercialized in Europe, Asia, and Africa. Some 40 commercial plants have been commissioned since 1952. These are entrained-bed, oxygen-steam gasifiers, with flame-zone temperatures attaining 3500°F. This system (Figure 14-9) operates at near-atmospheric pressures, with the reaction zone typically 2700–3500°F. Thermal efficiency is about 75%. Product is a medium-Btu gas (Table 14-17).

The Texaco process, developmental for coal, has been commercially proven for hydrocarbon feedstocks (Figure 14-10). Texaco operates at medium to high pressures (20–170 atm) and at temperatures well above ash-slagging ranges. Process details have not been published. Typical product gas composition is given in Table 14-18.

Table 14-17. Koppers–Totzek Products

(O_2-Blown)			
CO	52.5	CO_2	10.0
H_2	36.0	$H_2S + COS$	0.4
		$N_2 + Ar$	1.1
HHV = 286 Btu/scf			

Source: *Handbook of Gasifiers and Gas Treatment Systems* (1976)

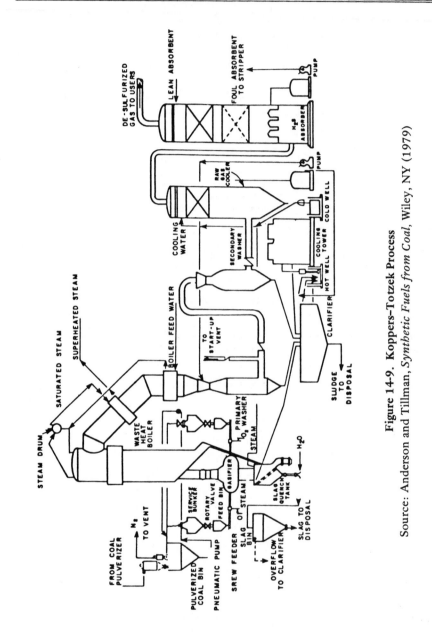

Figure 14-9. Koppers–Totzek Process

Source: Anderson and Tillman, *Synthetic Fuels from Coal*, Wiley, NY (1979)

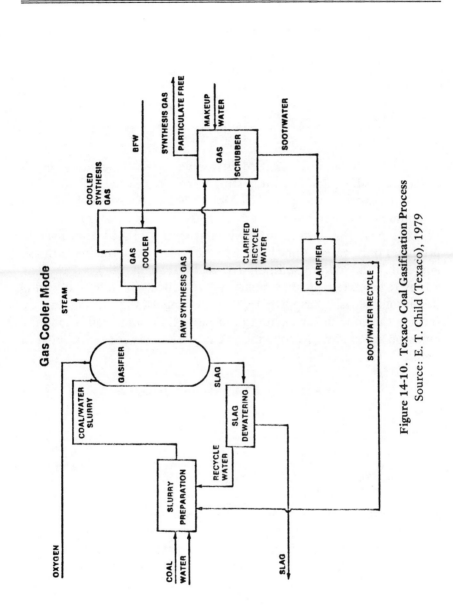

Figure 14-10. Texaco Coal Gasification Process
Source: E. T. Child (Texaco), 1979

Table 14-18. Typical Texaco Products

	(O$_2$-Blown)		
CO	37.6	CO$_2$	20.8
H$_2$	39.0	H$_2$S	1.5
CH$_4$	0.5	H$_2$ + Ar	0.6
HHV = 253 Btu/scf			

Coal may also be gasified in molten media, including metal-oxide slags and salts. An interesting example, piloted although not yet commercial, is the Rockwell Molten Salt Process. Rockwell uses a molten sodium-carbonate bath at 1800°F (Figure 14-11) with air or oxygen at 1–20 atm. This process accepts all kinds of coals, and produces only gaseous products. The carbonate bath retains both ash and sulfur (as sodium sulfide). Thermal efficiency (air) is 78%. The salt is regenerated by quenching and dissolving in water, collecting the insoluble ash, stripping H$_2$S, then evaporating, drying and returning the sodium carbonate to the gasifier. Commercialization may hinge on regeneration economics. A typical product gas is shown in Table 14-19.

Table 14-19. Rockwell Molten Salt Gasifier

	(air blown)		
CO	29.7%	N$_2$	48.0%
H$_2$	13.2%	O$_2$	1.4%
CH$_4$	1.5%	CO$_2$	3.5%
H$_2$S, NH$_3$	5 ppm		
HHV (dry) = 158 Btu/scf			

Source: *Handbook of Gasifiers and Gas Treatment Systems* (1976)

People continue to seek new and better ways to gasify coal char. A recent patent claims methane as the major product of a one-step steam-char gasification in a mixed-oxide melt. Composition ranges of the melt are given in Table 14-20.

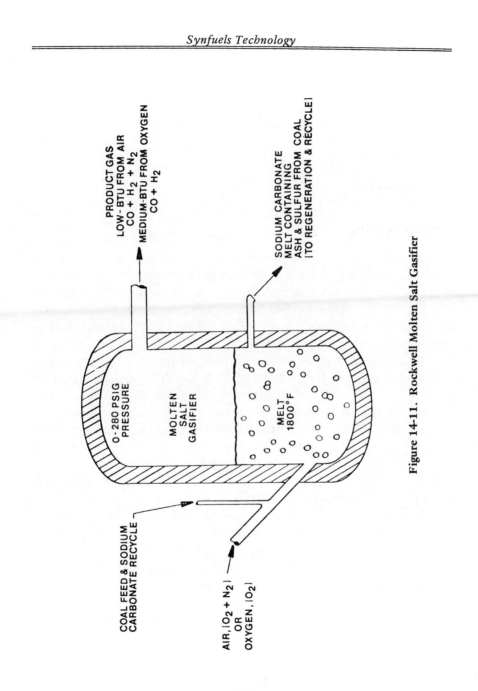

PRODUCT GAS
LOW- BTU FROM AIR
$CO + H_2 + N_2$
MEDIUM-BTU FROM OXYGEN
$CO + H_2$

SODIUM CARBONATE
MELT CONTAINING
ASH & SULFUR FROM COAL
[TO REGENERATION & RECYCLE]

0 - 280 PSIG
PRESSURE

MOLTEN
SALT
GASIFIER

MELT
1800° F

COAL FEED & SODIUM
CARBONATE RECYCLE

AIR, [$O_2 + N_2$]
OR
OXYGEN, [O_2]

Figure 14-11. Rockwell Molten Salt Gasifier

Table 14-20. SNG in One Step

2200°F	
C* + H₂O →mainly methane + oil, char	
ZnO	60--70%
SiO₂	10-15%
B₂O₃	20-5%

Source: K. Nogucki, et al., (Kao Oil Co.)
British 2,016,036 (1979)

LIQUEFACTION

Berthelot in 1869 demonstrated that coal can be substantially converted to liquid products by heating under pressure with hydriodic acid. In later years a number of processes have been developed to bring about the conversion of coal to liquid products. The potentially practical processes can be divided into two groups: *thermal hydroliquefaction* and *indirect liquefaction.*

Thermal hydroliquefaction entails heating crushed or pulverized coal with a hydrocarbon recycle oil to 840–870°F with hydrogen at a system pressure of 120–200 atm, with or without catalysis. Conditions for the old Bergius process and four modern U.S. processes are shown in Table 14-21. It is evident that all of these processes have a great deal in common. In this temperature range the more labile chemical bridges holding together the network structure of coal are broken thermally, and the resulting fragments are stabilized by abstracting hydrogen from the hydrogen-rich recycle oil or from molecular hydrogen. The role of the catalyst (when present) is primarily to promote the regeneration of the hydrogen-depleted recycle oil by facilitating reaction with molecular hydrogen.

Coal-derived liquids are commonly classified as oils (soluble in hexane), asphaltenes (insoluble in hexane but soluble in benzene), and preasphaltenes (insoluble in benzene but soluble in pyridine). Oils are generally the most desired product. Asphaltenes, mainly polynuclear aromatics and compounds with polar functional groups,

Table 14-21. Conditions for Coal Liquefaction

		Catalyst	T, °F	P, atm
Bergius	(1930)	FeO	870	200
H-Coal	(Current)	Co/Mo	840	200
SRC-II	(Current)	none	840	140
EDS	(Current)	Co/Mo	840	120
DOW	(Current)	Mo	860	—

are important early products but are progressively converted to oils as hydroliquefaction time is increased. Preasphaltenes may be the principal initial products of hydroliquefaction; the consensus of coal chemists is that these are converted by further hydrotreatment to asphaltenes, and thence eventually to oils. This pattern of changing composition is shown in Figure 14-12.

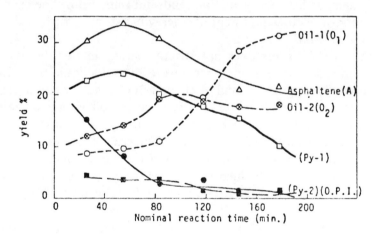

Figure 14-12. The Sequence of Products at 400°C
Source: Y. Mackawa et al., ACS Div. of Fuel Chem., 24, no. 2, 134 (1979)

No hydroliquefaction technology can be described as commercial. Several processes have been brought to the pilot stage, however. A half dozen of these are noted briefly below.

371

The *Synthoil* Process, developed by the Bureau of Mines (now Pittsburgh Energy Technology Center), is an extremely simple and straightforward process (Figure 14-13) in which slurried coal, recycle oil and hydrogen are mixed in a preheater, then passed through a fixed-bed reactor at 840°F and 200 atm. The reactor uses a supported cobalt/molybdenum catalyst, the function of which is to facilitate rehydrogenation of the dehydrogenated recycle oil. Liquefaction occurs when the coal, in its plastic and reactive state, abstracts hydrogen from the hydrogenated recycle oil and from dissolved molecular hydrogen, thereby forming a hydrogen-enriched coal-derived liquid. Typical conversion results are shown in Table 14-22. At the beginning of pilot runs the fresh catalyst produces a liquid containing only about 3% of the sulfur content of the feed coal, and less than 1% of the ash content. By the end of 500 hour runs, however, the data indicate significant deterioration in catalyst activity and the development of high levels of fine solids not removed by the centrifugal separator. The Department of Energy has stopped further development work on this process.

The *SRC-I* Process does not strictly belong in this discussion, since its product (at ambient temperatures) is not a liquid but rather a substantially upgraded solid fuel, or a high-grade reactant for a liquefaction process. Coal slurried in a recycle oil is mixed with a controlled amount of hydrogen, and run through a preheater into a reactor at about 850°F (Figure 14-14). The crude product is filtered hot under pressure, separating unreacted char and mineral matter, and the recycle liquid distilled to leave as the major product a SRC (Solvent Refined Coal) solid which shows little or no hydrogen enrichment, but which has sulfur, oxygen and ash contents sharply reduced (Table 14-23). Gulf is piloting SRC (I and II) at Ft. Lewis, Washington; Air Products and Wheelabrator–Frye are piloting SRC I in Wilsonville, Alabama, and are planning a 6,000 tpd plant to be sited in western Kentucky.

In terms of current U.S. DOE interest, the three front runners for liquids production appear to be the H-Coal, SRC-II and Exxon Donor Solvent (EDS) processes.

372

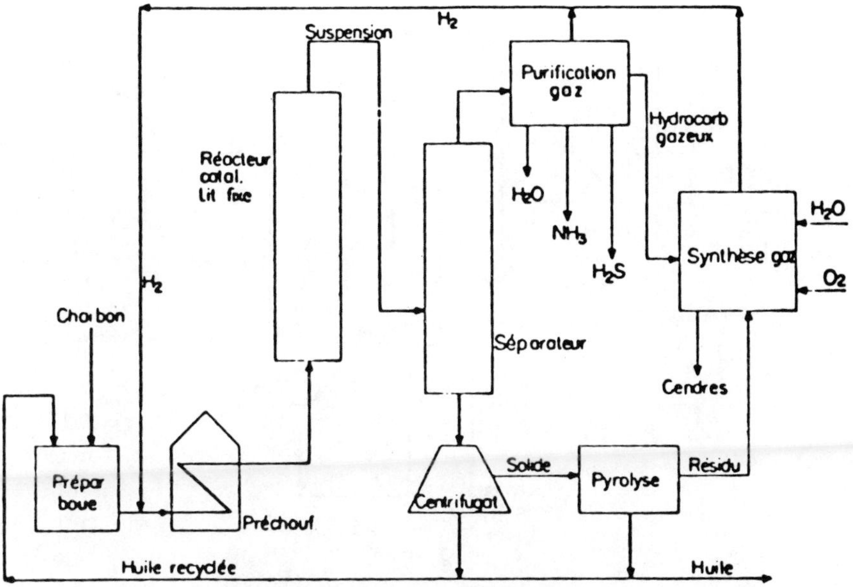

Figure 14-13. Synthoil Process
Source: Aune and Dou, *Rev. Inst. Francais Petrole,* 34, 593 (1979)

Table 14-22. Conversion of Kentucky Coal by Synthoil Process

	Coal	Oil (280 bar)	Oil (140 bar)
sulfur	5.5	0.15–0.50*	0.20–0.70*
ash	16.5	0.10–1.10	0.16–2.90
yield (gal./t)		132	124

*At beginning and end of 500-hr run at 840°F.
Source: Aune and Dou, *Rev. Inst. Francais du Petrole,* 34, 593 (1979)

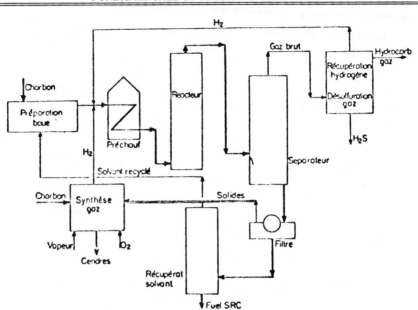

Figure 14-14. Solvent Refined Coal (SRC-I) Process
Source: Aune and Dou, *Rev. Inst. Francais Petrole,* 34, 593 (1979)

Table 14-23. SRC-I Product from Kentucky No. 11 Coal

	Coal	*Product*
Ash	6.9	0.14
C	71.3	88.2
H	5.3	5.0
N	0.94	1.3
S	3.3	0.95
O	12.3	4.4
Btu/lb	14,040	16,020

Source: Aune and Dou, *Rev. Inst. Francais du Petrole,* 34, 593 (1979)

The *H-Coal* Process also feeds slurried coal, recycle oil and hydrogen gas into a preheater and reactor (Figure 14-15) at about 850°F and 200 atm. The distinguishing characteristic of H-Coal is that the reactor is an ebullated bed of cobalt/molybdenum or nickel/molybdenum catalyst, with the upward feed-flow rate adjusted to obtain the desired bed expansion (Figure 14-16). The process can be operated in a "synfuel" mode, providing mainly gasoline and medium distillate oil, or in a "fuel oil" mode, using less hydrogen and making a heavier distillate. The heavy bottoms and unreacted char are gasified to produce process hydrogen. The developer, HRI, and a number of partners are piloting a 600-tpd unit in Catlettsburg, Kentucky, under the leadership of Ashland Oil. Ashland and Airco Energy Co. are planning a full-scale plant in western Kentucky. The expected product slate is shown in Table 14-24. Specific petrochemical volumes are given in Table 14-25.

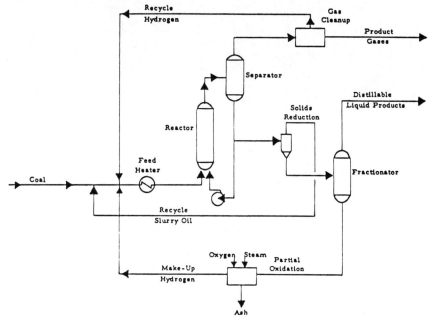

Figure 14-15. H-Coal Process Schematic
Source: Hoertz and Swan, in Pelofsky (ed), *Coal Conversion Technology* (1979)

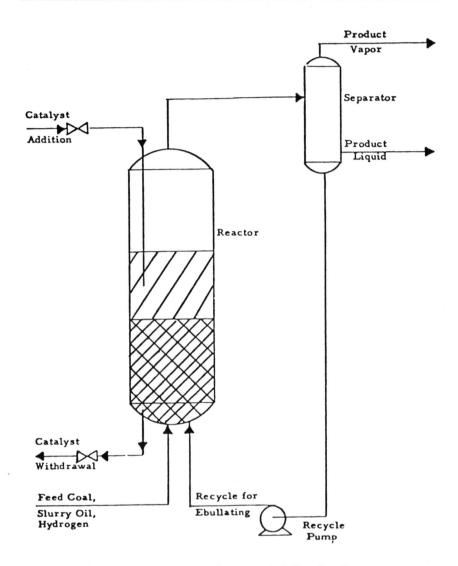

Figure 14-16. H-Coal Reactor (Ebullated Bed)
Source: Hoertz and Swan, in Pelofsky (ed), *Coal Conversion Technology* (1979)

376

Table 14-24. H-Coal Commercial Products

	Commercial Plant Product Slate	
Raw Coal		20,000 TPD
Products		
Reformate		15,300 BPD
Distillate (400--950°F)		27,900 BPD
Butane		3,300 BPD
Propane		3,500 BPD
	Total	50,000 BPD
By-Products		
Sulfur		570 LT/D
Ammonia		170 ST/D
SNG		31.7 MMSCFD

Source: Hoertz and Swan, in Pelofsky (ed), *Coal Conversion Technology* (1979)

Table 14-25. Petrochemicals from a 20,000 TPD H-Coal Plant

Petrochemical Potential	
Benzene	13.6 MM GPY
Toulene	24.5 MM GPY
Xylenes	32.2 MM GPY
Phenols and Cresols	170 MM lb/year
Butanes	222 MM lb/year
Propane	205 MM lb/year
Ethane	222 MM lb/year

Source: Hoertz and Swan, in Pelofsky (ed), *Coal Conversion Technology* (1979)

The *SRC-II* Process is related by history and technology to SRC-I, but is a distinctly different process; it is being developed by Gulf Oil and its subsidiaries. Like SRC-I, this process uses no added catalyst. The SRC-II process (Figure 14-17) uses more severe conditions and more hydrogen than SRC-I, and yields a panel of gaseous and distillable liquid products (Table 14-26) not markedly different from those obtained by H-Coal. A 6,000-tpd demonstration plant is projected at a site near Morgantown, West Virginia, with a completion target date of 1984.

377

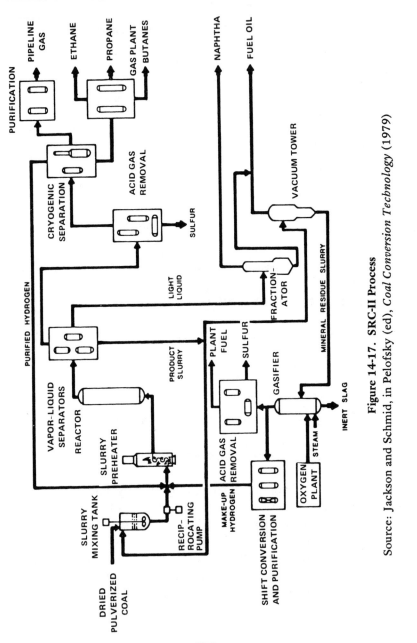

Figure 14-17. SRC-II Process

Source: Jackson and Schmid, in Pelofsky (ed), *Coal Conversion Technology* (1979)

Table 14-26. Products from SRC-II Process

Products from Typical Commercial Plant 33,500 T/SD--High Sulfur Bituminous Coal	
Methane	120 MMSCF/D
Ethane	1,100 T/D
Propane	12,000 B/D
Butanes	8,000 B/D
Naphtha ($C_5 - 350°F$)	13,200 B/D
Fuel Oil (350–900°F)	57,500 B/D
Sulfur	800 T/D
Ammonia	150 T/D
Phenols	35 T/D

Source: Jackson and Schmid, in Pelofsky (ed), *Coal Conversion Technology*, (1979)

The *EDS* (Exxon Donor Solvent) Process uses a coal slurry with hydrogen under similar conditions of temperature and pressure to those of the foregoing processes. Like H-Coal, EDS makes use of a cobalt-molybdenum recycle oil rehydrogenation catalyst, but unlike H-Coal the catalyst is in a fixed bed, separated from coal solids and contacting only liquid distillates (Figure 14-18). Exxon believes this separation is important, to optimize each reaction and to minimize catalyst fouling by coal minerals. Figure 14-19 shows liquid yields for feed coals of various ranks. ("Flexicoking" is an Exxon refining technique converting still bottoms to usable fuel liquids.) The first sizeable pilot operation of this process is a 250-tpd unit at Exxon's Baytown facility, expected to be running by this summer.

One of the new hydroliquefaction processes, unveiled in 1978, is the *Dow* process. Operating under generally similar conditions of temperature and pressure as all of the above (cf. Table 14-21), Dow's general process scheme is similar (Figure 14-20), but it is operationally distinguished by use of a finely divided "emulsion catalyst." Molybdenum compounds in a dispersed slurry are added continuously to the coal slurry feed at about a 100 ppm level; catalyst is partially recycled in actual PDU operation. After hydrocloning, Dow uses a solvent deasphalting step. The product slate from a Midland PDU

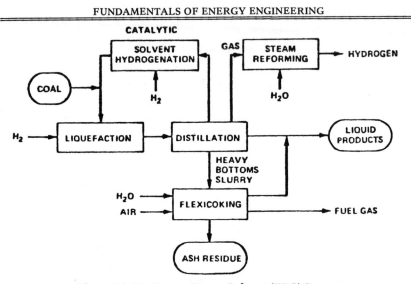

Figure 14-18. Exxon Donor Solvent (EDS) Process
Source: Epperly and Taunton, in Pelofsky (ed), *Coal Conversion Technology*, (1979)

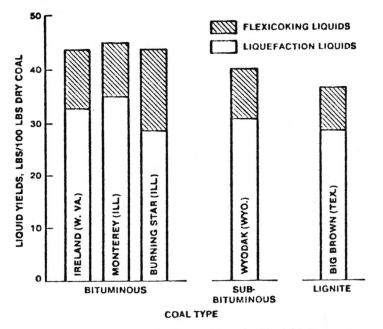

Figure 14-19. Preferred Liquefaction/Coking Liquid Yields in EDS Process
Source: Epperly and Taunton, in Pelofsky (ed), *Coal Conversion Technology*, (1979)

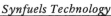

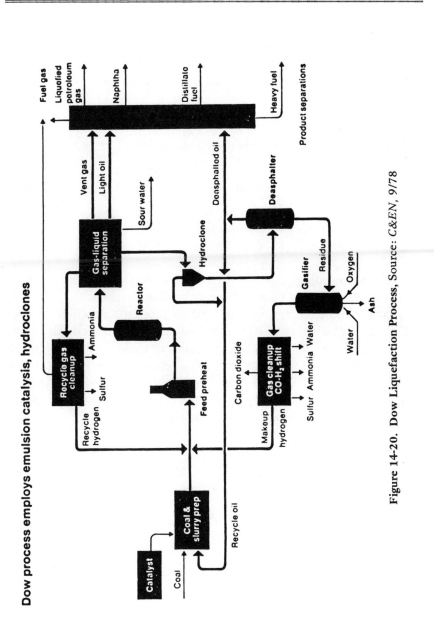

Dow process employs emulsion catalysis, hydroclones

Figure 14-20. Dow Liquefaction Process, Source: *C&EN, 9/78*

run is shown in Table 14-27. These are attractive enough yields to promote interest in larger-scale piloting.

Table 14-27. Dow Liquefaction Process Products

	100 kg coal yields 70.2 liquids
	21.8 kg residue
Liquids:	7.2 kg Methane
	13.1 kg LPG
	13.0 kg Naphtha ($<400^\circ$F)
	33.6 kg Distillate (400–977°F)
	4.3 kg Bottoms ($>977^\circ$F)

Source: Aune and Dou, *Rev. Inst. Francais du Petrole,* 34, 593 (1979)

If the preceding processes share many operating details in common, Conoco's *Zinc Chloride* Process is different. Slurried coal and hydrogen are introduced into a bath of molten zinc chloride (mp 504°F, bp 1350°F) at 800°F and 170 atm. The molten salt serves both as catalyst and heat sink; in addition, sulfur and nitrogen are substantially removed by the salt bath. Products are largely in the gasoline fraction (Table 14-28). The only experience to date has been with a 1-tpd Conoco PDU. Further development may hinge on a showing of the efficiency and economics of the zinc-chlordie regeneration process.

Table 14-28. Zinc Chloride Hydrocracking Products

(2 hrs @ 770–806°F, 2500 psig)	
$C_1 - C_4$	15.9%
$C_5 - 392^\circ$F	53.3%
392–887°F	17.7%
$>887^\circ$F	3.4%

Source: Pell, Maskew, Pasek and Struck, I.E.C.E. Conf. Proceedings, Boston, 1979

All of the above liquefaction processes are examples of thermal hydroliquefaction: in all cases coal is heated in the presence of hydrogen to 800–870°F to bring about the desired change. The alternate route, and indeed the only coal liquefaction route that has been commercially demonstrated, is *indirect* liquefaction, in which the coal is first converted to a carbon monoxide-hydrogen product mixture, and this syngas (synthesis gas) is then catalytically condensed to produce the desired liquid product(s).

Table 14-29 summarizes three routes to liquid products. The Fischer–Tropsch reaction can be catalyzed by a wide variety of supported and homogeneous catalysts, some reactions occurring at room temperature. However, the demonstrated practical processes usually use first-row transition metal catalysts—such as iron, temperatures in the 400–450°F range, and superatmospheric pressures to produce hydrocarbons. The process can be controlled so that the main products are in the gasoline and fuel-oil fractions. The SASOL plant in South Africa uses a high-temperature, entrained-bed reactor at 635°F (Figure 14-21). SASOL has two large Lurgi gasifiers producing a variety of products of which Fischer–Tropsch liquids are the most important. By the time the third SASOL complex is completed, South Africa (which has no domestic crude oils) expects to be virtually self-sufficient with respect to liquid fuels.

Table 14-29. Indirect Liquefaction: $CO + 1.5\text{--}2$ Parts H_2

Process	T, °F	P, atm	Catalysts	Products
Fischer–Tropsch	430	10	Fe, Co, Ni	methane + paraffins
Methanol	570	10	ZnO, Cu, Cr_2O_3, MnO	methanol
Alkanol	480	100	ZnO, Cu, MnO, Cr_2O_3, Na_2O	alcohols C_1 thru C_4

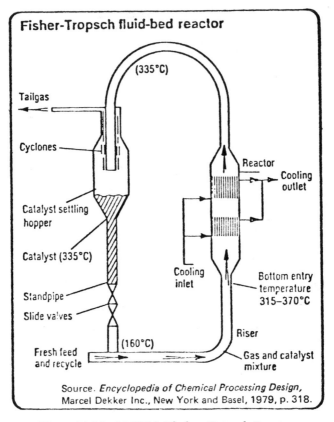

Fisher-Tropsch fluid-bed reactor

(335°C)

Tailgas

Cyclones

Reactor

Cooling outlet

Catalyst settling hopper

Catalyst (335°C)

Cooling inlet

Bottom entry temperature 315–370°C

Standpipe

Slide valves

Riser

Fresh feed and recycle

(160°C)

Gas and catalyst mixture

Source. *Encyclopedia of Chemical Processing Design,*
Marcel Dekker Inc., New York and Basel, 1979, p. 318.

Figure 14-21. SASOL's Fischer–Tropsch Reactor
Source: Hoppe, *E&MU,* December 1979

Another route to liquid fuels starts with methanol, which itself is made from syngas by established chemical technology. Methanol is not a cheap feedstock, as is seen in the economic survey just completed by Kermode and coworkers (Table 14-30). Mobil's new zeolite catalyst converts methanol with syngas to virtually 100% hydrocarbons, treating the vapor mixture at 660–760°F (Figure 14-22). The product slate from a typical PDU run is shown in Table 14-31. Mobil, W. R. Grace, and U.S. DOE are planning a large pilot

Table 14-30. Projected Cost of Producing Methanol from Coal in 1985 (1977 Dollars)

Coal, tons/day	MeOH gal./day	MeOH cost cents/gal.
5,000	1.05 million	72.4
10,000	2.10 million	61.4
20,000	4.19 million	52.7
40,000	8.39 million	45.0

Source: Kermode, Nicholason and Jones, *Chem. Engr.,* 25 Feb. 1980, p.111ff

installation in Baskett, Kentucky. Mobil also has under active investigation at the developmental level a catalytic process which, like Fischer–Tropsch, converts syngas directly to liquid hydrocarbon fuels.

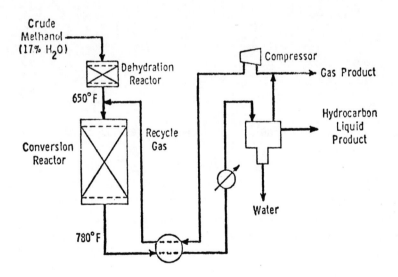

Figure 14-22. Mobil Methanol to Gasoline Process
Source: Mills, I.E.A. Coal Science Conf. (England), 1977

Table 14-31. Yields from Methanol in 4 B/D Fluid-Bed Pilot Unit

Average Bed Temperature, °C	413
Pressure, kPa	275
Space Velocity (WHSV)	1.0
Yields, Wt % of Methanol Charged	
Methanol + Ether	0.2
Hydrocarbons	43.5
Water	56.0
CO, CO_2	0.1
Coke, Other	0.2
	100.0
Hydrocarbon Product, Wt %	
Light Gas	5.6
Propane	5.9
Propylene	5.0
i-Butane	14.5
n-Butane	1.7
Butenes	7.3
C_5 + Gasoline	60.0
	100.0
Gasoline (including Alkylate)	
(96 R + O, RVP = 9.0 psi)	88.0
LPG	6.4
Fuel Gas	5.6
	100.0

Source: Morgan, Warner and Yurchak, Div Pet Chem (ACS) Preprints 25, no. 1 (1980)

Still another option is embodies in a process patent recently issued to the Institut Francais du Petrole (U.S. 4,122,110). Syngas is passed over a mixed metal-oxide catalyst, typically at 60 atm and 480°F with a nominal contact time of about 1 sec. Product compositions for two different gas feed compositions are shown in Table 14-32. This mixed-alcohol product has several advantages over methanol as a fuel component, with regard to volatility, miscibility with petroleum fuels, and heat of combustion. Chem Systems is now evaluating this process for DOE.

Table 14-32. Products from the Alkanol Process

(gas over $Cu/Co/Zn/Cr/K$ at $800^\circ F$)		
	Reformer Gas	*K-T Gas*
Feed		
H	66%	34%
CO	19%	58%
Products		
Methanol	24%	36%
Ethanol	39%	40%
Propanols	20%	11%
Butanols	14%	9%
C_5 and higher	3%	4%

Source: Chem Systems (1979)

The foregoing is a very compressed overview of coal-based synfuels technology, covering the major commercialized processes, as
well as selected examples of processes still in the pilot or laboratory
PDU stages. At least one or two of these latter processes will in all
likelihood become important components of our national energy
production by the early 1990s.

15

Energy Management

A successful energy management program is more than conservation. It is a total program that involves every area of a business. A comprehensive energy management program is not purely technical. It takes into account planning, communication, as well as salesmanship and marketing. It affects the bottom line profits of every business, thus the individual who is assigned the role of "energy manager" has high visibility within the organization.

Energy management includes energy productivity, which is defined as reducing the amount of energy needed to produce one unit of output. Energy management includes energy awareness, which is essential in motivating employees to save energy.

Probably the highest initial rate of return will occur through the establishment of a good maintenance management program.

This chapter reviews the basics of Maintenance Management and Energy Management.

MAINTENANCE MANAGEMENT

There are obvious losses from poor maintenance such as steam and air leaks and uninsulated steam lines. There are also losses from less obvious areas. For example Tables 15-1 and 15-2 illustrate the hidden effect of dirty evaporators and condensors on equipment

Table 15-1. The Effects of Poor Maintenance on the Efficiency of a Reciprocating Compressor, Nominal 15-Ton Capacity

Conditions	(1) °F	(2) °F	(3) Tons	(4) %	(5) HP	(6) HP/T	(7) %
Normal	45	105	17.0	—	15.9	0.93	—
Dirty Condenser	45	115	15.6	8.2	17.5	1.12	20
Dirty Evaporator	35	105	13.8	18.9	15.3	1.10	18
Dirty Condenser and Evaporator	35	115	12.7	25.4	16.4	1.29	39

(1) Suction Temp, °F
(2) Condensing Temp, °F
(3) Tons of refrigerant
(4) Reduction in capacity %
(5) Brake horsepower
(6) Brake horsepower per Ton
(7) Percent increase in compressor bh per/ton

Table 15-2. The Effects of Poor Maintenance on the Efficiency of an Absorption Chiller, 520-Ton Capacity

Condition	Chilled Water °F	Tower Water °F	Tons	Reduction in Capacity %	Steam lb/ton/H	Per Cent
Normal	44	85	520	—	18.7	—
Dirty Condenser	44	90	457	12	19.3	3
Dirty Evaporator	40	85	468	10	19.2	2.5
Dirty Condenser and Evaporator	40	90	396	23.8	20.1	7.5

performance. These losses go undetected and result in decreased capacity of equipment and an increase in energy usage. In Table 15-1, for a reciprocating compressor under the dirty condenser and evaporator conditions capacity is reduced 25.4% and the increase in brake horsepower per ton is 39%. In Table 15-2, for an absorption chiller under similar conditions capacity is reduced 23.8% and power requirements are increased 7.5%.

389

A third major loss is in missing the opportunity to upgrade the facilities within the spare parts program. For example, when a motor burns out it is usually replaced with the same model or sent to a shop for rewinding. If the motor is sent to a rewind shop, it will usually have poorer efficiency and power factor characteristics than before. Thus the energy manager should consider upgrading the replacement with either a high-efficiency motor or requiring a higher specification from the shop rewinding the motor.

The following summarizes key elements of the maintenance management program.

Preventive Maintenance Survey

This survey is made to establish a list of all equipment on the property that requires periodic maintenance and the maintenance that is required. The survey should list all items of equipment according to physical location. The survey sheet should list the following columns:

1. Item
2. Location of Item
3. Frequency of Maintenance
4. Estimated Time Required for Maintenance
5. Time of Day Maintenance Should Be Done
6. Brief Description of Maintenance To Be Done

Preventive Maintenance Schedule

The preventive maintenance schedule is prepared from the information gathered during the survey. Items are to be arranged on schedule sheets according to physical location. The schedule sheet should list the following columns:

1. Item
2. Location of Item
3. Time of Day Maintenance Should Be Done
4. Weekly Schedule with Double Columns for Each Day of the

Week (one column for "scheduled" and one for "completed")

5. Brief Description of Maintenance To Be Done
6. Maintenance Mechanic Assigned To Do the Work

Use of Preventive Maintenance Schedule—At some time before the beginning of the week, the supervisor of maintenance will take a copy of the schedule. The copy the supervisor prepares should be available in a three-ring notebook. He will go over the assignments in person with each mechanic.

After the mechanic has completed the work, he will note this on the schedule by placing a check under the "completed" column for that day and the index card system for cross-checking the PM program.

The supervisor of maintenance or the mechanic will check the schedule daily to determine that all work is being completed according to the plan. At the end of the week, the schedule will be removed from the book and checked to be sure that all work was completed. It will then be filed.

Preventive Maintenance Training

The supervisor of maintenance or mechanic is responsible for assisting department heads in the training of employees in handling, daily care, and the use of equipment. When equipment is mishandled, he must take an active part in correcting this through training.

Spare Parts

All too often equipment is replaced with the exact model as presently installed. Excellent energy conservation opportunities exist in upgrading a plant by installing more efficient replacement parts. Consideration should be given to the following:

- Efflcient line motor to replace standard motors
- Efficient model burners to replace obsolete burners
- Upgrading lighting systems.

391

Leaks—Steam, Water, and Air

The importance of leakage cannot be understated. If a plant has many leaks, this may be indicative of a low standard of operation involving the loss not only of steam, but also water, condensate, compressed air, etc.

If, for example, a valve spindle is worn, or badly packed, giving a clearance of 0.010 inch between the spindle, for a spindle of ¾-inch diameter, the area of leakage will be equal to a 3/32-inch diameter hole. Table 15-3 illustrates fluid loss through small holes:

Table 15-3. Fluid Loss Through Small Holes

Diameter of Hole	Steam—lb/hour		Water—gals/hour		Air SCFM
	100 psig	300 psig	20 psig	100 psig	80 psig
1/16"	14	33	20	45	4
1/8"	56	132	80	180	26
3/16"	126	297	180	405	36
1/4"	224	528	320	720	64

Although the plant may not be in full production for every hour of the entire year (i.e., 8760 hours), the boiler plant water systems and compressed air could be operable. Losses through leakage are usually, therefore, of a continuous nature.

Thermal Insulation

Whatever the pipework system, there is one fundamental: it should be adequately insulated. Table 15-4 gives a guide to the degree of insulation required. Obviously there are a number of types of insulating materials with different properties and at different costs, each one of which will give a variance return on capital. Table 15-4 is based on a good asbestos or magnesia insulation, but most manu-

392

Table 15-4. Pipe Heat Losses

Pipe Dia Inches	Surface Temp °F	Insulation Thickness Inches	Heat Loss (Btu/Ft/Hr)		Insulation Efficiency
			Uninsulated	Insulated	
4	200	1½	300	70	76.7
	300	2	800	120	85.0
	400	2½	1500	150	90.0
6	200	1½	425	95	78.7
	300	2	1300	180	85.8
	400	2½	2000	195	90.25
8	200	1½	550	115	79.1
	300	2	1500	200	86.7
	400	2½	2750	250	91.0

facturers have cataloged data indicating various benefits and savings that can be achieved with their particular product.

Steam Traps

The method of removing condensate is through steam trapping equipment. Most plants will have effective trapping systems. Others may have problems with both the type of traps and the effectiveness of the system.

The problems can vary from the wrong type of trap being installed, to air locking, or steam locking. A well-maintained trap system can be a great steam saver. A bad system can be a notorious steam waster, particularly where traps have to be bypassed or are leaking.

Therefore, the key to efficient trapping of most systems is good installation and maintenance. To facilitate the condensate removal, the pipes should slope in the direction of steam flow. This has two obvious advantages in relationship to the removal of condensate; one is the action of gravity, and the other the pushing action of the steam flow. Under these circumstances the strategic siting of the traps and drainage points is greatly simplified.

393

One common fault that often occurs at the outset is installing the wrong size traps. Traps are very often ordered by the size of the pipe connection. Unfortunately the pipe connection size has nothing whatsoever to do with the capacity of the trap. The discharge capacity of the trap depends upon the area of the valve, the pressure drop across it, and the temperature of the condensate.

It is therefore worth recapping exactly what a steam trap is. It is a device that distinguishes between steam and water and automatically opens a valve to allow the water to pass through but not the steam. There are numerous types of traps with various characteristics. Even within the same category of traps, e.g., ball floats or thermoexpansion traps, there are numerous designs, and the following guide is given for selection purposes:

1. Where a small amount of condensate is to be removed an expansion or thermostatic trap is preferred.
2. Where intermittent discharge is acceptable and air is not a large problem, inverted bucket traps will adequately suffice.
3. Where condensate must be continuously removed at steam temperatures, float traps must be used.
4. When large amounts of condensate have to be removed, relay traps must be used. However, this type of steam trap is unlikely to be required for use in the food industry.

To insure that a steam trap is not stuck open, a weekly inspection should be made and corrective action taken. Steam trap testing can utilize several methods to insure proper operation:

- Install heat sensing tape on trap discharge. The color indicates proper operation.
- Place a screwdriver to the ear lobe with the other end on the trap. If the trap is a bucket-type, listen for the click of the trap operating.
- Use acoustical or infrared instruments to check operation.

ENERGY MANAGEMENT ORGANIZATION

A common problem facing energy managers is that they have too much responsibility and very little authority to get the job done. A second problem is the lack of definition of the job. Probably the biggest problem is the lack of true commitment from top management. As pointed out in the text an overall energy utilization program requires an "investment" in order to get the return desired.

The first phase of the program should start with top management establishing the organization, defining the goals and providing the resources for doing an effective job. Tables 15-5 and 15-6 illustrate a checklist for top management and typical energy conservation goals.

Table 15-5. Checklist for Top Management

A. Inform line supervisors of:
1. The economic reasons for the need to conserve energy
2. Their responsibility for implementing energy saving actions in the areas of their accountability

B. Establish a team having the responsibility for formulating and conducting an energy conservation program and consisting of:
1. Representatives from each department in the plant
2. A coordinator appointed by and reporting to management
   NOTE: In smaller organizations, the manager and his staff may conduct energy conservation activities as part of their management duties.

C. Provide the team with guidelines as to what is expected of them:
1. Plan and participate in energy saving surveys
2. Develop uniform record keeping, reporting, and energy accounting
3. Research and develop ideas on ways to save energy
4. Communicate these ideas and suggestions
5. Suggest tough, but achievable, goals for energy saving
6. Develop ideas and plans for enlisting employee support and participation
7. Plan and conduct a continuing program of activities to stimulate interest in energy conservation efforts

D. Set goals in energy saving:
1. A preliminary goal at the start of the program
2. Later, a revised goal based on savings potential estimates from results of surveys

E. Employ external assistance in surveying the plant and making recommendations, if necessary

F. Communicate periodically to employees regarding management's emphasis on energy conservation action and report on progress

Adapted from *NBS Handbook*, 115.

Table 15-6. Typical Energy Conservation Goals

1. Overall energy reduction goals
    (a) Reduce yearly electrical bills by _____ %.
    (b) Reduce steam usage by _____ %.
    (c) Reduce natural gas usage by _____%.
    (d) Reduce fuel oil usage by _____ %.
    (e) Reduce compressed air usage by _____%.
2. Return on investment goals for individual projects
    (a) Minimum rate of return on investment before taxes is _____
    (b) Minimum payout period is _____
    (c) Minimum ratio of $\dfrac{\text{Btu/year savings}}{\text{capital cost}}$ is _____
    (d) Minimum rate of return on investment after taxes is _____

A key element in any energy management program is "flexibility." For example, in one year natural gas prices may accelerate steeply. Unless an organization reacts quickly it soon may be out of business. Every energy management program should have a contingency or backup plan. Periodically the risk associated with each scenario should be reviewed. Risk assessment can help the energy manager determine the resources required to meet various emergency scenarios. A method of evaluating the cost to business for each scenario is illustrated in Formula 15-1.

$$C = P \times C_1 \qquad\qquad \textit{Formula (15-1)}$$

where

C = cost as a result of emergency where no contingency plans are in effect.

P = the probability or likelihood the emergency will occur.

C_1 = the loss in dollars as a result of an emergency where no plan is in effect.

396

Implementing the Energy
Management Program

The first phase of the energy management program involves the accumulation of data. Table 15-7 illustrates the minimum information required to evaluate the various energy utilization opportunities. The energy manager must also review government regulations which will affect the program.

Table 15-7. Information Required to Set Energy Projects on the Same Base

Fuel	Cost At Present	Estimated Cost Escalation Per Year	Energy Equivalent
1. *Energy equivalents and costs for plant utilities.*			
Natural gas	$_____/1000 ft^3	$_____/1000 ft^3	_____Btu/ft^3
Fuel oil	$_____/gal	$_____/gal	_____Btu/gal
Coal	$_____/ton	$_____/ton	_____Btu/lb
Electric power	$_____/Kwh	$_____/Kwh	_____Btu/Kwh
Steam			
_____psig	$_____/1000 lb	$_____/1000 lb	_____Btu/1000 lb
_____psig	$_____/1000 lb	$_____/1000 lb	_____Btu/1000 lb
_____psig	$_____/1000 lb	$_____/1000 lb	_____Btu/1000 lb
Compressed air	$_____/1000 ft^3	$_____/1000 ft^3	_____Btu/1000 ft^3
Water	$_____/1000 lb	$_____/1000 lb	_____Btu/1000 lb
Boiler make-up water	$_____/1000 lb	$_____/1000 lb	_____Btu/1000 lb

2. *Life Cycle Costing Equivalents*

    A) After tax computations required   _____

       Depreciation method   _____

       Income tax bracket   _____

       Minimum rate of return   _____

       Economic life   _____

       Tax credit   _____

       Method of life cycle costing   _____

       (annual cost method, payout period, etc.)

All too often the energy manager is the last to find out information after the fact. For example, the procurement of motors may be the electrical department's responsibility. The energy manager must get his or her input into the evaluation process prior to purchase. One way is to establish a review process where important documents are initialed by the energy manager before they are issued. Key documents include:

- Bid summaries for major equipment
- One-line diagrams
- Process flow diagrams
- Heat and material balances
- Plot plan
- Piping and instrument diagrams

It is much easier to add local instrumentation to the piping and instrument diagram prior to construction. Another way of insuring that the energy manager gets timely input is to incorporate these activities into the overall planning document, such as the Critical Path or PERT Schedules. Figure 15-1 illustrates a planning schedule incorporating input from the energy manager.

Richard L. Aspenson, Director of Facilities Engineering and Real Estate at 3M summarized* the areas a good energy manager must master as follows:

Technical Expertise

Energy management normally begins with a solid technical background —preferably in mechanical, electrical or plant engineering. Managers will need a good grasp of both design theory and the nuts-and-bolts details of conservation programs. This includes a thorough understanding of the company's processes, products, maintenance procedures and facilities.

---

*"The Skills of the Energy Manager," *Energy Economics, Policy and Management,* Vol. 2, No. 2, 1982.

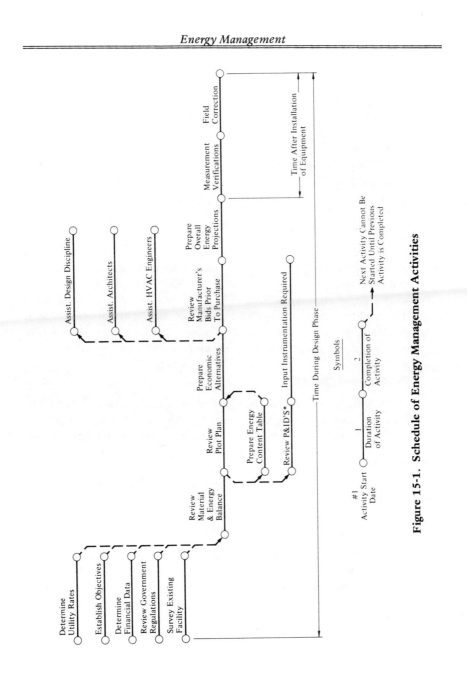

Figure 15-1. Schedule of Energy Management Activities

Communication

In the course of a single week, energy managers might find themselves dealing with lawyers, engineers, accountants, financial planners, public relations specialists, government officials, and even journalists and legislators. A good energy manager has to be able to communicate clearly and persuasively with all of these people—in *their* language. Above all, energy managers must be able to sell the benefits of their programs to top management.

Financial Understanding

To enlist the support of top management, energy managers will have to develop and present their programs as investments, with predictable returns, instead of as unrecoverable costs. They will have to demonstrate what kind of returns—in both energy and cost savings—can be expected from each project, and over what period of time. This means first of all developing some credible way to measure returns—a method that will be understood and accepted by the financial officers of the company.

Planning and Strategy

A basic part of energy management is forecasting future energy supplies and costs with reasonable accuracy. This means coming to grips with the complexities of worldwide supply, market trends, demand projections, and the international political climate. There is no way, of course, to predict all of these things with certainty, but every business—especially energy-intensive ones—will need some kind of reliable forecasting from now on.

Community Relations and Public Policy

Energy managers have some responsibility to go outside their companies to share their ideas and experience with a variety of publics. Trade and professional associations can become clearing houses for new ideas. Legislators and government agencies need, and often welcome, expert help in setting standards and policies. And the public needs help in understanding what is at stake in learning to use limited supplies of energy wisely and efficiently.

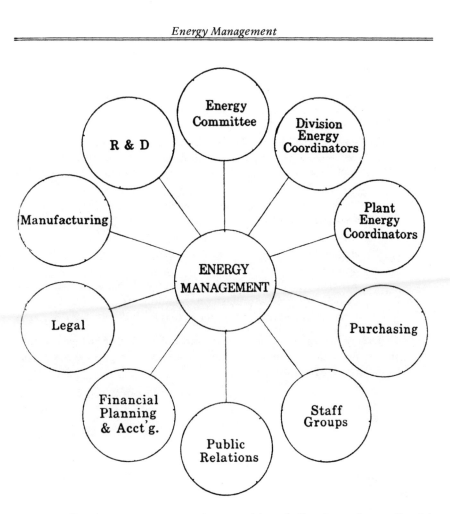

Energy management can be an exciting, challenging and rewarding job if you put your skills to work.

We as Energy Managers are a special breed of people who have unlimited opportunities if we develop the skills and implement the programs that will improve the profitability of our respective businesses. We can and must influence public policy decision by using our technical knowledge in a constructive and objective way.

401

Pointers for the Energy Manager

1. Be aggressive. Learn how to say *no* in a diplomatic way.
2. Energy is not a soft sell. Stick your neck out.
3. Be sure you anticipate questions before you request approval on energy expenditures. Prepare factual data—think ahead.
4. Be creative. Identify needs. Prioritize your action plan tasks.
5. Be a positive person—one with perpetual motion—a catalyst for action.
6. Establish a free-thinking environment. Be a good listener.
7. Develop a five-year plan. Update yearly. Include a strategy for implementation—action plans.
8. Establish credibility through an accurate energy accountability system.
9. Efficient use of energy is an evolutionary process. Be patient but demanding. Follow through on commitments.
10. Give individual recognition for achievements.

Energy managers can turn the problems of a changing energy situation into opportunities. At the same time, we can grow to become better corporate citizens of our society.

16
Appendix

Table 16-1. 10% Interest Factor

Period n	Single-payment compound-amount (SPCA) Future value of $1 $(1 + i)^n$	Single-payment present-worth (SPPW) Present value of $1 $\dfrac{1}{(1 + i)^n}$	Uniform-series compound-amount (USCA) Future value of uniform series of $1 $\dfrac{(1 + i)^n - 1}{i}$	Sinking-fund payment (SFP) Uniform series whose future value is $1 $\dfrac{i}{(1 + i)^n - 1}$	Capital recovery (CR) Uniform series with present value of $1 $\dfrac{i(1 + i)^n}{(1 + i)^n - 1}$	Uniform-series present-worth (USPW) Present value of uniform series of $1 $\dfrac{(1 + i)^n - 1}{i(1 + i)^n}$
1	1.100	0.9091	1.000	1.00000	1.10000	0.909
2	1.210	0.8264	2.100	0.47619	0.57619	1.736
3	1.331	0.7513	3.310	0.30211	0.40211	2.487
4	1.464	0.6830	4.641	0.21547	0.31547	3.170
5	1.611	0.6209	6.105	0.16380	0.26380	3.791
6	1.772	0.5645	7.716	0.12961	0.22961	4.355
7	1.949	0.5132	9.487	0.10541	0.20541	4.868
8	2.144	0.4665	11.436	0.08744	0.18744	5.335
9	2.358	0.4241	13.579	0.07364	0.17364	5.759
10	2.594	0.3855	15.937	0.06275	0.16275	6.144
11	2.853	0.3505	18.531	0.05396	0.15396	6.495
12	3.138	0.3186	21.384	0.04676	0.14676	6.814
13	3.452	0.2897	24.523	0.04078	0.14078	7.103
14	3.797	0.2633	27.975	0.03575	0.13575	7.367
15	4.177	0.2394	31.772	0.03147	0.13147	7.606
16	4.595	0.2176	35.950	0.02782	0.12782	7.824
17	5.054	0.1978	40.545	0.02466	0.12466	8.022
18	5.560	0.1799	45.599	0.02193	0.12193	8.201
19	6.116	0.1635	51.159	0.01955	0.11955	8.365
20	6.727	0.1486	57.275	0.01746	0.11746	8.514
21	7.400	0.1351	64.002	0.01562	0.11562	8.649
22	8.140	0.1228	71.403	0.01401	0.11401	8.772
23	8.954	0.1117	79.543	0.01257	0.11257	8.883
24	9.850	0.1015	88.497	0.01130	0.11130	8.985
25	10.835	0.0923	98.347	0.01017	0.11017	9.077
26	11.918	0.0839	109.182	0.00916	0.10916	9.161
27	13.110	0.0763	121.100	0.00826	0.10826	9.237
28	14.421	0.0693	134.210	0.00745	0.10745	9.307
29	15.863	0.0630	148.631	0.00673	0.10673	9.370
30	17.449	0.0573	164.494	0.00608	0.10608	9.427
35	28.102	0.0356	271.024	0.00369	0.10369	9.644
40	45.259	0.0221	442.593	0.00226	0.10226	9.779
45	72.890	0.0137	718.905	0.00139	0.10139	9.863
50	117.391	0.0085	1163.909	0.00086	0.10086	9.915
55	189.059	0.0053	1880.591	0.00053	0.10053	9.947
60	304.482	0.0033	3034.816	0.00033	0.10033	9.967
65	490.371	0.0020	4893.707	0.00020	0.10020	9.980
70	789.747	0.0013	7887.470	0.00013	0.10013	9.987
75	1271.895	0.0008	12708.954	0.00008	0.10008	9.992
80	2048.400	0.0005	20474.002	0.00005	0.10005	9.995
85	3298.969	0.0003	32979.690	0.00003	0.10003	9.997
90	5313.023	0.0002	53120.226	0.00002	0.10002	9.998
95	8556.676	0.0001	85556.760	0.00001	0.10001	9.999

Table 16-2. 12% Interest Factor

Period *n*	Single-payment compound-amount (SPCA)	Single-payment present-worth (SPPW)	Uniform-series compound-amount (USCA)	Sinking-fund payment (SFP)	Capital recovery (CR)	Uniform-series present-worth (USPW)
	Future value of $1 $(1 + i)^n$	Present value of $1 $\dfrac{1}{(1+i)^n}$	Future value of uniform series of $1 $\dfrac{(1+i)^n - 1}{i}$	Uniform series whose future value is $1 $\dfrac{i}{(1+i)^n - 1}$	Uniform series with present value of $1 $\dfrac{i(1+i)^n}{(1+i)^n - 1}$	Present value of uniform series of $1 $\dfrac{(1+i)^n - 1}{i(1+i)^n}$
1	1.120	0.8929	1.000	1.00000	1.12000	0.893
2	1.254	0.7972	2.120	0.47170	0.59170	1.690
3	1.405	0.7118	3.374	0.29635	0.41635	2.402
4	1.574	0.6355	4.779	0.20923	0.32923	3.037
5	1.762	0.5674	6.353	0.15741	0.27741	3.605
6	1.974	0.5066	8.115	0.12323	0.24323	4.111
7	2.211	0.4523	10.089	0.09912	0.21912	4.564
8	2.476	0.4039	12.300	0.08130	0.20130	4.968
9	2.773	0.3606	14.776	0.06768	0.18768	5.328
10	3.106	0.3220	17.549	0.05698	0.17698	5.650
11	3.479	0.2875	20.655	0.04842	0.16842	5.938
12	3.896	0.2567	24.133	0.04144	0.16144	6.194
13	4.363	0.2292	28.029	0.03568	0.15568	6.424
14	4.887	0.2046	32.393	0.03087	0.15087	6.628
15	5.474	0.1827	37.280	0.02682	0.14682	6.811
16	6.130	0.1631	42.753	0.02339	0.14339	6.974
17	6.866	0.1456	48.884	0.02046	0.14046	7.120
18	7.690	0.1300	55.750	0.01794	0.13794	7.250
19	8.613	0.1161	63.440	0.01576	0.13576	7.366
20	9.646	0.1037	72.052	0.01388	0.13388	7.469
21	10.804	0.0926	81.699	0.01224	0.13224	7.562
22	12.100	0.0826	92.503	0.01081	0.13081	7.645
23	13.552	0.0738	104.603	0.00956	0.12956	7.718
24	15.179	0.0659	118.155	0.00846	0.12846	7.784
25	17.000	0.0588	133.334	0.00750	0.12750	7.843
26	19.040	0.0525	150.334	0.00665	0.12665	7.896
27	21.325	0.0469	169.374	0.00590	0.12590	7.943
28	23.884	0.0419	190.699	0.00524	0.12524	7.984
29	26.750	0.0374	214.583	0.00466	0.12466	8.022
30	29.960	0.0334	241.333	0.00414	0.12414	8.055
35	52.800	0.0189	431.663	0.00232	0.12232	8.176
40	93.051	0.0107	767.091	0.00130	0.12130	8.244
45	163.988	0.0061	1358.230	0.00074	0.12074	8.283
50	289.002	0.0035	2400.018	0.00042	0.12042	8.304
55	509.321	0.0020	4236.005	0.00024	0.12024	8.317
60	897.597	0.0011	7471.641	0.00013	0.12013	8.324
65	1581.872	0.0006	13173.937	0.00008	0.12008	8.328
70	2787.800	0.0004	23223.332	0.00004	0.12004	8.330
75	4913.056	0.0002	40933.799	0.00002	0.12002	8.332
80	8658.483	0.0001	72145.692	0.00001	0.12001	8.332

Table 16-3. 15% Interest Factor

Period n	Single-payment compound-amount (SPCA)	Single-payment present-worth (SPPW)	Uniform-series compound-amount (USCA)	Sinking-fund payment (SFP)	Capital recovery (CR)	Uniform-series present-worth (USPW)
	Future value of \$1 $(1 + i)^n$	Present value of \$1 $\dfrac{1}{(1 + i)^n}$	Future value of uniform series of \$1 $\dfrac{(1 + i)^n - 1}{i}$	Uniform series whose future value is \$1 $\dfrac{i}{(1 + i)^n - 1}$	Uniform series with present value of \$1 $\dfrac{i(1 + i)^n}{(1 + i)^n - 1}$	Present value of uniform series of \$1 $\dfrac{(1 + i)^n - 1}{i(1 + i)^n}$
1	1.200	0.8333	1.000	1.00000	1.20000	0.833
2	1.440	0.6944	2.200	0.45455	0.65455	1.528
3	1.728	0.5787	3.640	0.27473	0.47473	2.106
4	2.074	0.4823	5.368	0.18629	0.38629	2.589
5	2.488	0.4019	7.442	0.13438	0.33438	2.991
6	2.986	0.3349	9.930	0.10071	0.30071	3.326
7	3.583	0.2791	12.916	0.07742	0.27742	3.605
8	4.300	0.2326	16.499	0.06061	0.26061	3.837
9	5.160	0.1938	20.799	0.04808	0.24808	4.031
10	6.192	0.1615	25.959	0.03852	0.23852	4.192
11	7.430	0.1346	32.150	0.03110	0.23110	4.327
12	8.916	0.1122	39.581	0.02526	0.22526	4.439
13	10.699	0.0935	48.497	0.02062	0.22062	4.533
14	12.839	0.0779	59.196	0.01689	0.21689	4.611
15	15.407	0.0649	72.035	0.01388	0.21388	4.675
16	18.488	0.0541	87.442	0.01144	0.21144	4.730
17	22.186	0.0451	105.931	0.00944	0.20944	4.775
18	26.623	0.0376	128.117	0.00781	0.20781	4.812
19	31.948	0.0313	154.740	0.00646	0.20646	4.843
20	38.338	0.0261	186.688	0.00536	0.20536	4.870
21	46.005	0.0217	225.026	0.00444	0.20444	4.891
22	55.206	0.0181	271.031	0.00369	0.20369	4.909
23	66.247	0.0151	326.237	0.00307	0.20307	4.925
24	79.497	0.0126	392.484	0.00255	0.20255	4.937
25	95.396	0.0105	471.981	0.00212	0.20212	4.948
26	114.475	0.0087	567.377	0.00176	0.20176	4.956
27	137.371	0.0073	681.853	0.00147	0.20147	4.964
28	164.845	0.0061	819.223	0.00122	0.20122	4.970
29	197.814	0.0051	984.068	0.00102	0.20102	4.975
30	237.376	0.0042	1181.882	0.00085	0.20085	4.979
35	590.668	0.0017	2948.341	0.00034	0.20034	4.992
40	1469.772	0.0007	7343.858	0.00014	0.20014	4.997
45	3657.262	0.0003	18281.310	0.00005	0.20005	4.999
50	9100.438	0.0001	45497.191	0.00002	0.20002	4.999

Table 16-4. 20% Interest Factor

Period n	Single-payment compound-amount (SPCA)	Single-payment present-worth (SPPW)	Uniform-series compound-amount (USCA)	Sinking-fund payment (SFP)	Capital recovery (CR)	Uniform-series present-worth (USPW)
	Future value of $1 $(1 + i)^n$	Present value of $1 $\dfrac{1}{(1 + i)^n}$	Future value of uniform series of $1 $\dfrac{(1 + i)^n - 1}{i}$	Uniform series whose future value is $1 $\dfrac{i}{(1 + i)^n - 1}$	Uniform series with present value of $1 $\dfrac{i(1 + i)^n}{(1 + i)^n - 1}$	Present value of uniform series of $1 $\dfrac{(1 + i)^n - 1}{i(1 + i)^n}$
1	1.200	0.8333	1.000	1.00000	1.20000	0.833
2	1.440	0.6944	2.200	0.45455	0.65455	1.528
3	1.728	0.5787	3.640	0.27473	0.47473	2.106
4	2.074	0.4823	5.368	0.18629	0.38629	2.589
5	2.488	0.4019	7.442	0.13438	0.33438	2.991
6	2.986	0.3349	9.930	0.10071	0.30071	3.326
7	3.583	0.2791	12.916	0.07742	0.27742	3.605
8	4.300	0.2326	16.499	0.06061	0.26061	3.837
9	5.160	0.1938	20.799	0.04808	0.24808	4.031
10	6.192	0.1615	25.959	0.03852	0.23852	4.192
11	7.430	0.1346	32.150	0.03110	0.23110	4.327
12	8.916	0.1122	39.581	0.02526	0.22526	4.439
13	10.699	0.0935	48.497	0.02062	0.22062	4.533
14	12.839	0.0779	59.196	0.01689	0.21689	4.611
15	15.407	0.0649	72.035	0.01388	0.21388	4.675
16	18.488	0.0541	87.442	0.01144	0.21144	4.730
17	22.186	0.0451	105.931	0.00944	0.20944	4.775
18	26.623	0.0376	128.117	0.00781	0.20781	4.812
19	31.948	0.0313	154.740	0.00646	0.20646	4.843
20	38.338	0.0261	186.688	0.00536	0.20536	4.870
21	46.005	0.0217	225.026	0.00444	0.20444	4.891
22	55.206	0.0181	271.031	0.00369	0.20369	4.909
23	66.247	0.0151	326.237	0.00307	0.20307	4.925
24	79.497	0.0126	392.484	0.00255	0.20255	4.937
25	95.396	0.0105	471.981	0.00212	0.20212	4.948
26	114.475	0.0087	567.377	0.00176	0.20176	4.956
27	137.371	0.0073	681.853	0.00147	0.20147	4.964
28	164.845	0.0061	819.223	0.00122	0.20122	4.970
29	197.814	0.0051	984.068	0.00102	0.20102	4.975
30	237.376	0.0042	1181.882	0.00085	0.20085	4.979
35	590.668	0.0017	2948.341	0.00034	0.20034	4.992
40	1469.772	0.0007	7343.858	0.00014	0.20014	4.997
45	3657.262	0.0003	18281.310	0.00005	0.20005	4.999
50	9100.438	0.0001	45497.191	0.00002	0.20002	4.999

407

Table 16-5. 25% Interest Factor

Period n	Single-payment compound-amount (SPCA) Future value of $1 $(1 + i)^n$	Single-payment present-worth (SPPW) Present value of $1 $\dfrac{1}{(1 + i)^n}$	Uniform-series compound amount (USCA) Future value of uniform series of $1 $\dfrac{(1 + i)^n - 1}{i}$	Sinking-fund payment (SFP) Uniform series whose future value is $1 $\dfrac{i}{(1 + i)^n - 1}$	Capital recovery (CR) Uniform series with present value of $1 $\dfrac{i(1 + i)^n}{(1 + i)^n - 1}$	Uniform-series present-worth (USPW) Present value of uniform series of $1 $\dfrac{(1 + i)^n - 1}{i(1 + i)^n}$
1	1.250	0.8000	1.000	1.00000	1.25000	0.800
2	1.562	0.6400	2.250	0.44444	0.69444	1.440
3	1.953	0.5120	3.812	0.26230	0.51230	1.952
4	2.441	0.4096	5.766	0.17344	0.42344	2.362
5	3.052	0.3277	8.207	0.12185	0.37185	2.689
6	3.815	0.2621	11.259	0.08882	0.33882	2.951
7	4.768	0.2097	15.073	0.06634	0.31634	3.161
8	5.960	0.1678	19.842	0.05040	0.30040	3.329
9	7.451	0.1342	25.802	0.03876	0.28876	3.463
10	9.313	0.1074	33.253	0.03007	0.28007	3.571
11	11.642	0.0859	42.566	0.02349	0.27349	3.656
12	14.552	0.0687	54.208	0.01845	0.26845	3.725
13	18.190	0.0550	68.760	0.01454	0.26454	3.780
14	22.737	0.0440	86.949	0.01150	0.26150	3.824
15	28.422	0.0352	109.687	0.00912	0.25912	3.859
16	35.527	0.0281	138.109	0.00724	0.25724	3.887
17	44.409	0.0225	173.636	0.00576	0.25576	3.910
18	55.511	0.0180	218.045	0.00459	0.25459	3.928
19	69.389	0.0144	273.556	0.00366	0.25366	3.942
20	86.736	0.0115	342.945	0.00292	0.25292	3.954
21	108.420	0.0092	429.681	0.00233	0.25233	3.963
22	135.525	0.0074	538.101	0.00186	0.25186	3.970
23	169.407	0.0059	673.626	0.00148	0.25148	3.976
24	211.758	0.0047	843.033	0.00119	0.25119	3.981
25	264.698	0.0038	1054.791	0.00095	0.25095	3.985
26	330.872	0.0030	1319.489	0.00076	0.25076	3.988
27	413.590	0.0024	1650.361	0.00061	0.25061	3.990
28	516.988	0.0019	2063.952	0.00048	0.25048	3.992
29	646.235	0.0015	2580.939	0.00039	0.25039	3.994
30	807.794	0.0012	3227.174	0.00031	0.25031	3.995
35	2465.190	0.0004	9856.761	0.00010	0.25010	3.998
40	7523.164	0.0001	30088.655	0.00003	0.25003	3.999

408

Table 16-6. 30% Interest Factor

Period n	Single-payment compound-amount (SPCA)	Single-payment present-worth (SPPW)	Uniform-series compound-amount (USCA)	Sinking-fund payment (SFP)	Capital recovery (CR)	Uniform-series present-worth (USPW)
	Future value of \$1 $(1+i)^n$	Present value of \$1 $\dfrac{1}{(1+i)^n}$	Future value of uniform series of \$1 $\dfrac{(1+i)^n-1}{i}$	Uniform series whose future value is \$1 $\dfrac{i}{(1+i)^n-1}$	Uniform series with present value of \$1 $\dfrac{i(1+i)^n}{(1+i)^n-1}$	Present value of uniform series of \$1 $\dfrac{(1+i)^n-1}{i(1+i)^n}$
1	1.300	0.7692	1.000	1.00000	1.30000	0.769
2	1.690	0.5917	2.300	0.43478	0.73478	1.361
3	2.197	0.4552	3.990	0.25063	0.55063	1.816
4	2.856	0.3501	6.187	0.16163	0.46163	2.166
5	3.713	0.2693	9.043	0.11058	0.41058	2.436
6	4.827	0.2072	12.756	0.07839	0.37839	2.643
7	6.275	0.1594	17.583	0.05687	0.35687	2.802
8	8.157	0.1226	23.858	0.04192	0.34192	2.925
9	10.604	0.0943	32.015	0.03124	0.33124	3.019
10	13.786	0.0725	42.619	0.02346	0.32346	3.092
11	17.922	0.0558	56.405	0.01773	0.31773	3.147
12	23.298	0.0429	74.327	0.01345	0.31345	3.190
13	30.288	0.0330	97.625	0.01024	0.31024	3.223
14	39.374	0.0254	127.913	0.00782	0.30782	3.249
15	51.186	0.0195	167.286	0.00598	0.30598	3.268
16	66.542	0.0150	218.472	0.00458	0.30458	3.283
17	86.504	0.0116	285.014	0.00351	0.30351	3.295
18	112.455	0.0089	371.518	0.00269	0.30269	3.304
19	146.192	0.0068	483.973	0.00207	0.30207	3.311
20	190.050	0.0053	630.165	0.00159	0.30159	3.316
21	247.065	0.0040	820.215	0.00122	0.30122	3.320
22	321.184	0.0031	1067.280	0.00094	0.30094	3.323
23	417.539	0.0024	1388.464	0.00072	0.30072	3.325
24	542.801	0.0018	1806.003	0.00055	0.30055	3.327
25	705.641	0.0014	2348.803	0.00043	0.30043	3.329
26	917.333	0.0011	3054.444	0.00033	0.30033	3.330
27	1192.533	0.0008	3971.778	0.00025	0.30025	3.331
28	1550.293	0.0006	5164.311	0.00019	0.30019	3.331
29	2015.381	0.0005	6714.604	0.00015	0.30015	3.332
30	2619.996	0.0004	8729.985	0.00011	0.30011	3.332
35	9727.860	0.0001	32422.868	0.00003	0.30003	3.333

Table 16-7. 40% Interest Factor

Period n	Single-payment compound-amount (SPCA) Future value of $1 $(1 + i)^n$	Single-payment present-worth (SPPW) Present value of $1 $\dfrac{1}{(1 + i)^n}$	Uniform-series compound-amount (USCA) Future value of uniform series of $1 $\dfrac{(1 + i)^n - 1}{i}$	Sinking-fund payment (SFP) Uniform series whose future value is $1 $\dfrac{i}{(1 + i)^n - 1}$	Capital recovery (CR) Uniform series with present value of $1 $\dfrac{i(1 + i)^n}{(1 + i)^n - 1}$	Uniform-series present-worth (USPW) Present value of uniform series of $1 $\dfrac{(1 + i)^n - 1}{i(1 + i)^n}$
1	1.400	0.7143	1.000	1.00000	1.40000	0.714
2	1.960	0.5102	2.400	0.41667	0.81667	1.224
3	2.744	0.3644	4.360	0.22936	0.62936	1.589
4	3.842	0.2603	7.104	0.14077	0.54077	1.849
5	5.378	0.1859	10.946	0.09136	0.49136	2.035
6	7.530	0.1328	16.324	0.06126	0.46126	2.168
7	10.541	0.0949	23.853	0.04192	0.44192	2.263
8	14.758	0.0678	34.395	0.02907	0.42907	2.331
9	20.661	0.0484	49.153	0.02034	0.42034	2.379
10	28.925	0.0346	69.814	0.01432	0.41432	2.414
11	40.496	0.0247	98.739	0.01013	0.41013	2.438
12	56.694	0.0176	139.235	0.00718	0.40718	2.456
13	79.371	0.0126	195.929	0.00510	0.40510	2.469
14	111.120	0.0090	275.300	0.00363	0.40363	2.478
15	155.568	0.0064	386.420	0.00259	0.40259	2.484
16	217.795	0.0046	541.988	0.00185	0.40185	2.489
17	304.913	0.0033	759.784	0.00132	0.40132	2.492
18	426.879	0.0023	1064.697	0.00094	0.40094	2.494
19	597.630	0.0017	1491.576	0.00067	0.40067	2.496
20	836.683	0.0012	2089.206	0.00048	0.40048	2.497
21	1171.356	0.0009	2925.889	0.00034	0.40034	2.498
22	1639.898	0.0006	4097.245	0.00024	0.40024	2.498
23	2295.857	0.0004	5737.142	0.00017	0.40017	2.499
24	3214.200	0.0003	8032.999	0.00012	0.40012	2.499
25	4499.880	0.0002	11247.199	0.00009	0.40009	2.499
26	6299.831	0.0002	15747.079	0.00006	0.40006	2.500
27	8819.764	0.0001	22046.910	0.00005	0.40005	2.500

410

Table 16-8. 50% Interest Factor

Period n	Single-payment compound-amount (SPCA) Future value of \$1 $(1 + i)^n$	Single-payment present-worth (SPPW) Present value of \$1 $\dfrac{1}{(1 + i)^n}$	Uniform-series compound-amount (USCA) Future value of uniform series of \$1 $\dfrac{(1 + i)^n - 1}{i}$	Sinking-fund payment (SFP) Uniform series whose future value is \$1 $\dfrac{i}{(1 + i)^n - 1}$	Capital recovery (CR) Uniform series with present value of \$1 $\dfrac{i(1 + i)^n}{(1 + i)^n - 1}$	Uniform-series present-worth (USPW) Present value of uniform series of \$1 $\dfrac{(1 + i)^n - 1}{i(1 + i)^n}$
1	1.500	0.6667	1.000	1.00000	1.50000	0.667
2	2.250	0.4444	2.500	0.40000	0.90000	1.111
3	3.375	0.2963	4.750	0.21053	0.71053	1.407
4	5.062	0.1975	8.125	0.12308	0.62308	1.605
5	7.594	0.1317	13.188	0.07583	0.57583	1.737
6	11.391	0.0878	20.781	0.04812	0.54812	1.824
7	17.086	0.0585	32.172	0.03108	0.53108	1.883
8	25.629	0.0390	49.258	0.02030	0.52030	1.922
9	38.443	0.0260	74.887	0.01335	0.51335	1.948
10	57.665	0.0173	113.330	0.00882	0.50882	1.965
11	86.498	0.0116	170.995	0.00585	0.50585	1.977
12	129.746	0.0077	257.493	0.00388	0.50388	1.985
13	194.620	0.0051	387.239	0.00258	0.50258	1.990
14	291.929	0.0034	581.859	0.00172	0.50172	1.993
15	437.894	0.0023	873.788	0.00114	0.50114	1.995
16	656.841	0.0015	1311.682	0.00076	0.50076	1.997
17	985.261	0.0010	1968.523	0.00051	0.50051	1.998
18	1477.892	0.0007	2953.784	0.00034	0.50034	1.999
19	2216.838	0.0005	4431.676	0.00023	0.50023	1.999
20	3325.257	0.0003	6648.513	0.00015	0.50015	1.999
21	4987.885	0.0002	9973.770	0.00010	0.50010	2.000
22	7481.828	0.0001	14961.655	0.00007	0.50007	2.000

Table 16-9. Five-Year Escalation Table

Present Worth of a Series of Escalating Payments Compounded Annually
Discount-Escalation Factors for $n = 5$ Years

Discount Rate	Annual Escalation Rate					
	0.10	0.12	0.14	0.16	0.18	0.20
0.10	5.000000	5.279234	5.572605	5.880105	6.202627	6.540569
0.11	4.866862	5.136200	5.420152	5.717603	6.029313	6.355882
0.12	4.738562	5.000000	5.274242	5.561868	5.863289	6.179066
0.13	4.615647	4.869164	5.133876	5.412404	5.704137	6.009541
0.14	4.497670	4.742953	5.000000	5.269208	5.551563	5.847029
0.15	4.384494	4.622149	4.871228	5.131703	5.404955	5.691165
0.16	4.275647	4.505953	4.747390	5.000000	5.264441	5.541511
0.17	4.171042	4.394428	4.628438	4.873699	5.129353	5.397964
0.18	4.070432	4.287089	4.513947	4.751566	5.000000	5.259749
0.19	3.973684	4.183921	4.403996	4.634350	4.875619	5.126925
0.20	3.880510	4.084577	4.298207	4.521178	4.755725	5.000000
0.21	3.790801	3.989001	4.196400	4.413341	4.640260	4.877689
0.22	4.704368	3.896891	4.098287	4.308947	4.529298	4.759649
0.23	3.621094	3.808179	4.003835	4.208479	4.422339	4.645864
0.24	3.540773	3.722628	3.912807	4.111612	4.319417	4.536517
0.25	3.463301	3.640161	3.825008	4.018249	4.220158	4.431144
0.26	3.388553	3.560586	3.740376	3.928286	4.124553	4.329514
0.27	3.316408	3.483803	3.658706	3.841442	4.032275	4.231583
0.28	3.246718	3.409649	3.579870	3.757639	3.943295	4.137057
0.29	3.179393	3.338051	3.503722	3.676771	3.857370	4.045902
0.30	3.114338	3.268861	3.430201	3.598653	3.774459	3.957921
0.31	3.051452	3.201978	3.359143	3.523171	3.694328	3.872901
0.32	2.990618	3.137327	3.290436	3.450224	3.616936	3.790808
0.33	2.931764	3.074780	3.224015	3.379722	3.542100	3.711472
0.34	2.874812	3.014281	3.159770	3.311524	3.469775	3.634758

412

Table 16-10. Ten-Year Escalation Table

Present Worth of a Series of Escalating Payments Compounded Annually
Discount-Escalation Factors for *n* = 10 Years

Discount Rate	Annual Escalation Rate					
	0.10	0.12	0.14	0.16	0.18	0.20
0.10	10.000000	11.056250	12.234870	13.548650	15.013550	16.646080
0.11	9.518405	10.508020	11.613440	12.844310	14.215140	15.741560
0.12	9.068870	10.000000	11.036530	12.190470	13.474590	14.903510
0.13	8.650280	9.526666	10.498990	11.582430	12.786980	14.125780
0.14	8.259741	9.084209	10.000000	11.017130	12.147890	13.403480
0.15	7.895187	8.672058	9.534301	10.490510	11.552670	12.731900
0.16	7.554141	8.286779	9.099380	10.000000	10.998720	12.106600
0.17	7.234974	7.926784	8.693151	9.542653	10.481740	11.524400
0.18	6.935890	7.589595	8.312960	9.113885	10.000000	10.980620
0.19	6.655455	7.273785	7.957330	8.713262	9.549790	10.472990
0.20	6.392080	6.977461	7.624072	8.338518	9.128122	10.000000
0.21	6.144593	6.699373	7.311519	7.987156	8.733109	9.557141
0.22	5.911755	6.437922	7.017915	7.657542	8.363208	9.141752
0.23	5.692557	6.192047	6.742093	7.348193	8.015993	8.752133
0.24	5.485921	5.960481	6.482632	7.057347	7.690163	8.387045
0.25	5.290990	5.742294	6.238276	6.783767	7.383800	8.044173
0.26	5.106956	5.536463	6.008083	6.526298	7.095769	7.721807
0.27	4.933045	5.342146	5.790929	6.283557	6.824442	7.418647
0.28	4.768518	5.158489	5.585917	6.054608	6.568835	7.133100
0.29	4.612762	4.984826	5.392166	5.838531	6.327682	6.864109
0.30	4.465205	4.820429	5.209000	5.634354	6.100129	6.610435
0.31	4.325286	4.664669	5.035615	5.441257	5.885058	6.370867
0.32	4.192478	4.517015	4.871346	5.258512	5.681746	6.144601
0.33	4.066339	4.376884	4.715648	5.085461	5.489304	5.930659
0.34	3.946452	4.243845	4.567942	4.921409	5.307107	5.728189

413

Table 16-11. Fifteen-Year Escalation Table

Present Worth of a Series of Escalating Payments Compounded Annually
Discount-Escalation Factors for n = 15 years

Discount Rate	Annual Escalation Rate					
	0.10	0.12	0.14	0.16	0.18	0.20
0.10	15.000000	17.377880	20.199780	23.549540	27.529640	32.259620
0.11	13.964150	16.126230	18.690120	21.727370	25.328490	29.601330
0.12	13.026090	15.000000	17.332040	20.090360	23.355070	27.221890
0.13	12.177030	13.981710	16.105770	18.616160	21.581750	25.087260
0.14	11.406510	13.057790	15.000000	17.287320	19.985530	23.169060
0.15	10.706220	12.220570	13.998120	16.086500	18.545150	21.442230
0.16	10.068030	11.459170	13.088900	15.000000	17.244580	19.884420
0.17	9.485654	10.766180	12.262790	14.015480	16.066830	18.477610
0.18	8.953083	10.133630	11.510270	13.118840	15.000000	17.203010
0.19	8.465335	9.555676	10.824310	12.303300	14.030830	16.047480
0.20	8.017635	9.026333	10.197550	11.560150	13.148090	15.000000
0.21	7.606115	8.540965	9.623969	10.881130	12.343120	14.046400
0.22	7.227109	8.094845	9.097863	10.259820	11.608480	13.176250
0.23	6.877548	7.684317	8.614813	9.690559	10.936240	12.381480
0.24	6.554501	7.305762	8.170423	9.167798	10.320590	11.655310
0.25	6.255518	6.956243	7.760848	8.687104	9.755424	10.990130
0.26	5.978393	6.632936	7.382943	8.244519	9.236152	10.379760
0.27	5.721101	6.333429	7.033547	7.836080	8.757889	9.819020
0.28	5.481814	6.055485	6.710042	7.458700	8.316982	9.302823
0.29	5.258970	5.797236	6.410005	7.109541	7.909701	8.827153
0.30	5.051153	5.556882	6.131433	6.785917	7.533113	8.388091
0.31	4.857052	5.332839	5.872303	6.485500	7.184156	7.982019
0.32	4.675478	5.123753	5.630905	6.206250	6.860492	7.606122
0.33	4.505413	4.928297	5.405771	5.946343	6.559743	7.257569
0.34	4.345926	4.745399	5.195502	5.704048	6.280019	6.933897

Table 16-12. Twenty-Year Escalation Table

Present Worth of a Series of Escalating Payments Compounded Annually
Discount-Escalation Factors for *n* = 20 Years

Discount Rate	Annual Escalation Rate					
	0.10	0.12	0.14	0.16	0.18	0.20
0.10	20.000000	24.295450	29.722090	36.592170	45.308970	56.383330
0.11	18.213210	22.002090	26.776150	32.799710	40.417480	50.067940
0.12	16.642370	20.000000	24.210030	29.505400	36.181240	44.614710
0.13	15.259850	18.243100	21.964990	26.634490	32.502270	39.891400
0.14	14.038630	16.694830	20.000000	24.127100	29.298170	35.789680
0.15	12.957040	15.329770	18.271200	21.929940	26.498510	32.218060
0.16	11.995640	14.121040	16.746150	20.000000	24.047720	29.098950
0.17	11.138940	13.048560	15.397670	18.300390	21.894660	26.369210
0.18	10.373120	12.093400	14.201180	16.795710	20.000000	23.970940
0.19	9.686791	11.240870	13.137510	15.463070	18.326720	21.860120
0.20	9.069737	10.477430	12.188860	14.279470	16.844020	20.000000
0.21	8.513605	9.792256	11.340570	13.224610	15.527270	18.353210
0.22	8.010912	9.175267	10.579620	12.282120	14.355520	16.890730
0.23	7.555427	8.618459	9.895583	11.438060	13.309280	15.589300
0.24	7.141531	8.114476	9.278916	10.679810	12.373300	14.429370
0.25	6.764528	7.657278	8.721467	9.997057	11.533310	13.392180
0.26	6.420316	7.241402	8.216490	9.380883	10.778020	12.462340
0.27	6.105252	6.862203	7.757722	8.823063	10.096710	11.626890
0.28	5.816151	6.515563	7.339966	8.316995	9.480940	10.874120
0.29	5.550301	6.198027	6.958601	7.856833	8.922847	10.194520
0.30	5.305312	5.906440	6.609778	7.437339	8.416060	9.579437
0.31	5.079039	5.638064	6.289875	7.054007	7.954518	9.021190
0.32	4.869585	5.390575	5.995840	6.702967	7.533406	8.513612
0.33	4.675331	5.161809	5.725066	6.380829	7.148198	8.050965
0.34	4.494838	4.949990	5.475180	6.084525	6.795200	7.628322

Table 16-13. Saturated Steam: Temperature

Temp Fahr t	Abs Press lb per Sq In. p	Specific Volume Sat Liquid v_f	Specific Volume Evap v_{fg}	Specific Volume Sat Vapor v_g	Enthalpy Sat Liquid h_f	Enthalpy Evap h_{fg}	Enthalpy Sat Vapor h_g	Entropy Sat Liquid s_f	Entropy Evap s_{fg}	Entropy Sat Vapor s_g	Temp Fahr t
32.0*	0.08859	0.016022	3304.7	3304.7	0.0179	1075.5	1075.5	0.0000	2.1873	2.1873	32.0*
34.0	0.09600	0.016021	3061.9	3061.9	1.996	1074.4	1076.4	0.0041	2.1762	2.1802	34.0
36.0	0.10395	0.016020	2839.0	2839.0	4.008	1073.2	1077.2	0.0081	2.1651	2.1732	36.0
38.0	0.11249	0.016019	2634.1	2634.2	6.018	1072.1	1078.1	0.0122	2.1541	2.1663	38.0
40.0	0.12163	0.016019	2445.8	2445.8	8.027	1071.0	1079.0	0.0162	2.1432	2.1594	40.0
42.0	0.13143	0.016019	2272.4	2272.4	10.035	1069.8	1079.9	0.0202	2.1325	2.1527	42.0
44.0	0.14192	0.016019	2112.8	2112.8	12.041	1068.7	1080.7	0.0242	2.1217	2.1459	44.0
46.0	0.15314	0.016020	1965.7	1965.7	14.047	1067.6	1081.6	0.0282	2.1111	2.1393	46.0
48.0	0.16514	0.016021	1830.0	1830.0	16.051	1066.4	1082.5	0.0321	2.1006	2.1327	48.0
50.0	0.17796	0.016023	1704.8	1704.8	18.054	1065.3	1083.4	0.0361	2.0901	2.1262	50.0
52.0	0.19165	0.016024	1589.2	1589.2	20.057	1064.2	1084.2	0.0400	2.0798	2.1197	52.0
54.0	0.20625	0.016026	1482.4	1482.4	22.058	1063.1	1085.1	0.0439	2.0695	2.1134	54.0
56.0	0.22183	0.016028	1383.6	1383.6	24.059	1061.9	1086.0	0.0478	2.0593	2.1070	56.0
58.0	0.23843	0.016031	1292.2	1292.2	26.060	1060.8	1086.9	0.0516	2.0491	2.1008	58.0
60.0	0.25611	0.016033	1207.6	1207.6	28.060	1059.7	1087.7	0.0555	2.0391	2.0946	60.0
62.0	0.27494	0.016036	1129.2	1129.2	30.059	1058.5	1088.6	0.0593	2.0291	2.0885	62.0
64.0	0.29497	0.016039	1056.5	1056.5	32.058	1057.4	1089.5	0.0632	2.0192	2.0824	64.0
66.0	0.31626	0.016043	989.1	989.1	34.056	1056.3	1090.4	0.0670	2.0094	2.0764	66.0
68.0	0.33889	0.016046	926.5	925.5	36.054	1055.2	1091.2	0.0708	1.9996	2.0704	68.0
70.0	0.36292	0.016050	868.3	868.4	38.052	1054.0	1092.1	0.0745	1.9900	2.0645	70.0
72.0	0.38844	0.016054	814.3	814.3	40.049	1052.9	1093.0	0.0783	1.9804	2.0587	72.0
74.0	0.41550	0.016058	764.1	764.1	42.046	1051.8	1093.8	0.0821	1.9708	2.0529	74.0
76.0	0.44420	0.016063	717.4	717.4	44.043	1050.7	1094.7	0.0858	1.9614	2.0472	76.0
78.0	0.47461	0.016067	673.8	673.9	46.040	1049.5	1095.6	0.0895	1.9520	2.0415	78.0
80.0	0.50683	0.016072	633.3	633.3	48.037	1048.4	1096.4	0.0932	1.9426	2.0359	80.0
82.0	0.54093	0.016077	595.5	595.5	50.033	1047.3	1097.3	0.0969	1.9334	2.0303	82.0
84.0	0.57702	0.016082	560.3	560.3	52.029	1046.1	1098.2	0.1006	1.9242	2.0248	84.0
86.0	0.61518	0.016087	527.5	527.5	54.026	1045.0	1099.0	0.1043	1.9151	2.0193	86.0
88.0	0.65551	0.016093	496.8	496.8	56.022	1043.9	1099.9	0.1079	1.9060	2.0139	88.0
90.0	0.69813	0.016099	468.1	468.1	58.018	1042.7	1100.8	0.1115	1.8970	2.0086	90.0
92.0	0.74313	0.016105	441.3	441.3	60.014	1041.6	1101.6	0.1152	1.8881	2.0033	92.0
94.0	0.79062	0.016111	416.3	416.3	62.010	1040.5	1102.5	0.1188	1.8792	1.9980	94.0
96.0	0.84072	0.016117	392.8	392.9	64.006	1039.3	1103.3	0.1224	1.8704	1.9928	96.0
98.0	0.89356	0.016123	370.9	370.9	66.003	1038.2	1104.2	0.1260	1.8617	1.9876	98.0

Temp.												Temp.
100.0	0.94924	0.016130	350.4	350.4	67.999	1037.1	1105.1	0.1295	1.8530	1.9825	100.0
102.0	1.00789	0.016137	331.1	331.1	69.995	1035.9	1105.9	0.1331	1.8444	1.9775	102.0
104.0	1.06965	0.016144	313.1	313.1	71.992	1034.8	1106.8	0.1366	1.8358	1.9725	104.0
106.0	1.1347	0.016151	296.16	296.18	73.99	1033.6	1107.6	0.1402	1.8273	1.9675	106.0
108.0	1.2030	0.016158	280.28	280.30	75.98	1032.5	1108.5	0.1437	1.8188	1.9626	108.0
110.0	1.2750	0.016165	265.37	265.39	77.98	1031.4	1109.3	0.1472	1.8105	1.9577	110.0
112.0	1.3505	0.016173	251.37	251.38	79.98	1030.2	1110.2	0.1507	1.8021	1.9528	112.0
114.0	1.4299	0.016180	238.21	238.22	81.97	1029.1	1111.0	0.1542	1.7938	1.9480	114.0
116.0	1.5133	0.016188	225.84	225.85	83.97	1027.9	1111.9	0.1577	1.7856	1.9433	116.0
118.0	1.6009	0.016196	214.20	214.21	85.97	1026.8	1112.7	0.1611	1.7774	1.9386	118.0
120.0	1.6927	0.016204	203.25	203.26	87.97	1025.6	1113.6	0.1646	1.7693	1.9339	120.0
122.0	1.7891	0.016213	192.94	192.95	89.96	1024.5	1114.4	0.1680	1.7613	1.9293	122.0
124.0	1.8901	0.016221	183.23	183.24	91.96	1023.3	1115.3	0.1715	1.7533	1.9247	124.0
126.0	1.9959	0.016229	174.08	174.09	93.96	1022.2	1116.1	0.1749	1.7453	1.9202	126.0
128.0	2.1068	0.016238	165.45	165.47	95.96	1021.0	1117.0	0.1783	1.7374	1.9157	128.0
130.0	2.2230	0.016247	157.32	157.33	97.96	1019.8	1117.8	0.1817	1.7295	1.9112	130.0
132.0	2.3445	0.016256	149.64	149.66	99.95	1018.7	1118.6	0.1851	1.7217	1.9068	132.0
134.0	2.4717	0.016265	142.40	142.41	101.95	1017.5	1119.5	0.1884	1.7140	1.9024	134.0
136.0	2.6047	0.016274	135.55	135.57	103.95	1016.4	1120.3	0.1918	1.7063	1.8980	136.0
138.0	2.7438	0.016284	129.09	129.11	105.95	1015.2	1121.1	0.1951	1.6986	1.8937	138.0
140.0	2.8892	0.016293	122.98	123.00	107.95	1014.0	1122.0	0.1985	1.6910	1.8895	140.0
142.0	3.0411	0.016303	117.21	117.22	109.95	1012.9	1122.8	0.2018	1.6834	1.8852	142.0
144.0	3.1997	0.016312	111.74	111.76	111.95	1011.7	1123.6	0.2051	1.6759	1.8810	144.0
146.0	3.3653	0.016322	106.58	106.59	113.95	1010.5	1124.5	0.2084	1.6684	1.8769	146.0
148.0	3.5381	0.016332	101.68	101.70	115.95	1009.3	1125.3	0.2117	1.6610	1.8727	148.0
150.0	3.7184	0.016343	97.05	97.07	117.95	1008.2	1126.1	0.2150	1.6536	1.8686	150.0
152.0	3.9065	0.016353	92.66	92.68	119.95	1007.0	1126.9	0.2183	1.6463	1.8646	152.0
154.0	4.1025	0.016363	88.50	88.52	121.95	1005.8	1127.7	0.2216	1.6390	1.8606	154.0
156.0	4.3068	0.016374	84.56	84.57	123.95	1004.6	1128.6	0.2248	1.6318	1.8566	156.0
158.0	4.5197	0.016384	80.82	80.83	125.96	1003.4	1129.4	0.2281	1.6245	1.8526	158.0
160.0	4.7414	0.016395	77.27	77.29	127.96	1002.2	1130.2	0.2313	1.6174	1.8487	160.0
162.0	4.9722	0.016406	73.90	73.92	129.96	1001.0	1131.0	0.2345	1.6103	1.8448	162.0
164.0	5.2124	0.016417	70.70	70.72	131.96	999.8	1131.8	0.2377	1.6032	1.8409	164.0
166.0	5.4623	0.016428	67.67	67.68	133.97	998.6	1132.6	0.2409	1.5961	1.8371	166.0
168.0	5.7223	0.016440	64.78	64.80	135.97	997.4	1133.4	0.2441	1.5892	1.8333	168.0
170.0	5.9926	0.016451	62.04	62.06	137.97	996.2	1134.2	0.2473	1.5822	1.8295	170.0
172.0	6.2736	0.016463	59.43	59.45	139.98	995.0	1135.0	0.2505	1.5753	1.8258	172.0
174.0	6.5656	0.016474	56.95	56.97	141.98	993.8	1135.8	0.2537	1.5684	1.8221	174.0
176.0	6.8690	0.016486	54.59	54.61	143.99	992.6	1136.6	0.2568	1.5616	1.8184	176.0
178.0	7.1840	0.016498	52.35	52.36	145.99	991.4	1137.4	0.2600	1.5548	1.8147	178.0

[a] The states shown are mestable

417

Table 16-13. (Continued)

Temp Fahr t	Abs Press. Lb per Sq In. p	Specific Volume			Enthalpy			Entropy			Temp Fahr t
		Sat. Liquid v_f	Evap v_{fg}	Sat. Vapor v_g	Sat. Liquid h_f	Evap h_{fg}	Sat. Vapor h_g	Sat. Liquid s_f	Evap s_{fg}	Sat. Vapor s_g	
180.0	7.5110	0.016510	50.21	50.22	148.00	990.2	1138.2	0.2631	1.5480	1.8111	180.0
182.0	7.850	0.016522	48.172	48.189	150.01	989.0	1139.0	0.2662	1.5413	1.8075	182.0
184.0	8.203	0.016534	46.232	46.249	152.01	987.8	1139.8	0.2694	1.5346	1.8040	184.0
186.0	8.568	0.016547	44.383	44.400	154.02	986.5	1140.5	0.2725	1.5279	1.8004	186.0
188.0	8.947	0.016559	42.621	42.638	156.03	985.3	1141.3	0.2756	1.5213	1.7969	188.0
190.0	9.340	0.016572	40.941	40.957	158.04	984.1	1142.1	0.2787	1.5148	1.7934	190.0
192.0	9.747	0.016585	39.337	39.354	160.05	982.8	1142.9	0.2818	1.5082	1.7900	192.0
194.0	10.168	0.016598	37.808	37.824	162.05	981.6	1143.7	0.2848	1.5017	1.7865	194.0
196.0	10.605	0.016611	36.348	36.364	164.06	980.4	1144.4	0.2879	1.4952	1.7831	196.0
198.0	11.058	0.016624	34.954	34.970	166.08	979.1	1145.2	0.2910	1.4888	1.7798	198.0
200.0	11.526	0.016637	33.622	33.639	168.09	977.9	1146.0	0.2940	1.4824	1.7764	200.0
204.0	12.512	0.016664	31.135	31.151	172.11	975.4	1147.5	0.3001	1.4697	1.7698	204.0
208.0	13.568	0.016691	28.862	28.878	176.14	972.8	1149.0	0.3061	1.4571	1.7632	208.0
212.0	14.696	0.016719	26.782	26.799	180.17	970.3	1150.5	0.3121	1.4447	1.7568	212.0
216.0	15.901	0.016747	24.878	24.894	184.20	967.8	1152.0	0.3181	1.4323	1.7505	216.0
220.0	17.186	0.016775	23.131	23.148	188.23	965.2	1153.4	0.3241	1.4201	1.7442	220.0
224.0	18.556	0.016805	21.529	21.545	192.27	962.6	1154.9	0.3300	1.4081	1.7380	224.0
228.0	20.015	0.016834	20.056	20.073	196.31	960.0	1156.3	0.3359	1.3961	1.7320	228.0
232.0	21.567	0.016864	18.701	18.718	200.35	957.4	1157.8	0.3417	1.3842	1.7260	232.0
236.0	23.216	0.016895	17.454	17.471	204.40	954.8	1159.2	0.3476	1.3725	1.7201	236.0
240.0	24.968	0.016926	16.304	16.321	208.45	952.1	1160.6	0.3533	1.3609	1.7142	240.0
244.0	26.826	0.016958	15.243	15.260	212.50	949.5	1162.0	0.3591	1.3494	1.7085	244.0
248.0	28.796	0.016990	14.264	14.281	216.56	946.8	1163.4	0.3649	1.3379	1.7028	248.0
252.0	30.883	0.017022	13.358	13.375	220.62	944.1	1164.7	0.3706	1.3266	1.6972	252.0
256.0	33.091	0.017055	12.520	12.538	224.69	941.4	1166.1	0.3763	1.3154	1.6917	256.0
260.0	35.427	0.017089	11.745	11.762	228.76	938.6	1167.4	0.3819	1.3043	1.6862	260.0
264.0	37.894	0.017123	11.025	11.042	232.83	935.9	1168.7	0.3876	1.2933	1.6808	264.0
268.0	40.500	0.017157	10.358	10.375	236.91	933.1	1170.0	0.3932	1.2823	1.6755	268.0
272.0	43.249	0.017193	9.738	9.755	240.99	930.3	1171.3	0.3987	1.2715	1.6702	272.0
276.0	46.147	0.017228	9.162	9.180	245.08	927.5	1172.5	0.4043	1.2607	1.6650	276.0
280.0	49.200	0.017264	8.627	8.644	249.17	924.6	1173.8	0.4098	1.2501	1.6599	280.0
284.0	52.414	0.01730	8.1280	8.1453	253.3	921.7	1175.0	0.4154	1.2395	1.6548	284.0
288.0	55.795	0.01734	7.6634	7.6807	257.4	918.8	1176.2	0.4208	1.2290	1.6498	288.0
292.0	59.350	0.01738	7.2301	7.2475	261.5	915.9	1177.4	0.4263	1.2186	1.6449	292.0
296.0	63.084	0.01741	6.8259	6.8433	265.6	913.0	1178.6	0.4317	1.2082	1.6400	296.0

Temp.											Temp.
300.0	67.005	0.01745	6.4483	6.4658	269.7	910.0	1179.7	0.4372	1.1979	1.6351	300.0
304.0	71.119	0.01749	6.0955	6.1130	273.8	907.0	1180.9	0.4426	1.1877	1.6303	304.0
308.0	75.433	0.01753	5.7655	5.7830	278.0	904.0	1182.0	0.4479	1.1776	1.6256	308.0
312.0	79.953	0.01757	5.4566	5.4742	282.1	901.0	1183.1	0.4533	1.1676	1.6209	312.0
316.0	84.688	0.01761	5.1673	5.1849	286.3	897.9	1184.1	0.4586	1.1576	1.6162	316.0
320.0	89.643	0.01766	4.8961	4.9138	290.4	894.8	1185.2	0.4640	1.1477	1.6116	320.0
324.0	94.826	0.01770	4.6418	4.6595	294.6	891.6	1186.2	0.4692	1.1378	1.6071	324.0
328.0	100.245	0.01774	4.4030	4.4208	298.7	888.5	1187.2	0.4745	1.1280	1.6025	328.0
332.0	105.907	0.01779	4.1788	4.1966	302.9	885.3	1188.2	0.4798	1.1183	1.5981	332.0
336.0	111.820	0.01783	3.9681	3.9859	307.1	882.1	1189.1	0.4850	1.1086	1.5936	336.0
340.0	117.992	0.01787	3.7699	3.7878	311.3	878.8	1190.1	0.4902	1.0990	1.5892	340.0
344.0	124.430	0.01792	3.5834	3.6013	315.5	875.5	1191.0	0.4954	1.0894	1.5849	344.0
348.0	131.142	0.01797	3.4078	3.4258	319.7	872.2	1191.9	0.5006	1.0799	1.5806	348.0
352.0	138.138	0.01801	3.2423	3.2603	323.9	868.9	1192.7	0.5058	1.0705	1.5763	352.0
356.0	145.424	0.01806	3.0863	3.1044	328.1	865.5	1193.6	0.5110	1.0611	1.5721	356.0
360.0	153.010	0.01811	2.9392	2.9573	332.3	862.1	1194.4	0.5161	1.0517	1.5678	360.0
364.0	160.903	0.01816	2.8002	2.8184	336.5	858.6	1195.2	0.5212	1.0424	1.5637	364.0
368.0	169.113	0.01821	2.6691	2.6873	340.8	855.1	1195.9	0.5263	1.0332	1.5595	368.0
372.0	177.648	0.01826	2.5451	2.5633	345.0	851.6	1196.7	0.5314	1.0240	1.5554	372.0
376.0	186.517	0.01831	2.4279	2.4462	349.3	848.1	1197.4	0.5365	1.0148	1.5513	376.0
380.0	195.729	0.01836	2.3170	2.3353	353.6	844.5	1198.0	0.5416	1.0057	1.5473	380.0
384.0	205.294	0.01842	2.2120	2.2304	357.9	840.8	1198.7	0.5466	0.9966	1.5432	384.0
388.0	215.220	0.01847	2.1126	2.1311	362.2	837.2	1199.3	0.5516	0.9876	1.5392	388.0
392.0	225.516	0.01853	2.0184	2.0369	366.5	833.4	1199.9	0.5567	0.9786	1.5352	392.0
396.0	236.193	0.01858	1.9291	1.9477	370.8	829.7	1200.4	0.5617	0.9696	1.5313	396.0
400.0	247.259	0.01864	1.8444	1.8630	375.1	825.9	1201.0	0.5667	0.9607	1.5274	400.0
404.0	258.725	0.01870	1.7640	1.7827	379.4	822.0	1201.5	0.5717	0.9518	1.5234	404.0
408.0	270.600	0.01875	1.6877	1.7064	383.8	818.2	1202.0	0.5766	0.9429	1.5195	408.0
412.0	282.894	0.01881	1.6152	1.6340	388.1	814.2	1202.4	0.5816	0.9341	1.5157	412.0
416.0	295.617	0.01887	1.5463	1.5651	392.5	810.2	1202.8	0.5866	0.9253	1.5118	416.0
420.0	308.780	0.01894	1.4808	1.4997	396.9	806.2	1203.1	0.5915	0.9165	1.5080	420.0
424.0	322.391	0.01900	1.4184	1.4374	401.3	802.2	1203.5	0.5964	0.9077	1.5042	424.0
428.0	336.463	0.01906	1.3591	1.3782	405.7	798.0	1203.7	0.6014	0.8990	1.5004	428.0
432.0	351.00	0.01913	1.30266	1.32179	410.1	793.9	1204.0	0.6063	0.8903	1.4966	432.0
436.0	366.03	0.01919	1.24887	1.26806	414.6	789.7	1204.2	0.6112	0.8816	1.4928	436.0
440.0	381.54	0.01926	1.19761	1.21687	419.0	785.4	1204.4	0.6161	0.8729	1.4890	440.0
444.0	397.56	0.01933	1.14874	1.16806	423.5	781.1	1204.6	0.6210	0.8643	1.4853	444.0
448.0	414.09	0.01940	1.10212	1.12152	428.0	776.7	1204.7	0.6259	0.8557	1.4815	448.0
452.0	431.14	0.01947	1.05764	1.07711	432.5	772.3	1204.8	0.6308	0.8471	1.4778	452.0
456.0	448.73	0.01954	1.01518	1.03472	437.0	767.8	1204.8	0.6356	0.8385	1.4741	456.0

419

Table 16-13. (Continued)

Temp Fahr t	Abs Press. Lb per Sq In p	Specific Volume Sat. Liquid v_f	Specific Volume Evap v_{fg}	Specific Volume Sat Vapor v_g	Enthalpy Sat. Liquid h_f	Enthalpy Evap h_{fg}	Enthalpy Sat Vapor h_g	Entropy Sat Liquid s_f	Entropy Evap s_{fg}	Entropy Sat Vapor s_g	Temp Fahr t
460.0	466.87	0.01961	0.97463	0.99424	441.5	763.2	1204.8	0.6405	0.8299	1.4704	460.0
464.0	485.56	0.01969	0.93588	0.95557	446.1	758.6	1204.7	0.6454	0.8213	1.4667	464.0
468.0	504.83	0.01976	0.89885	0.91862	450.7	754.0	1204.6	0.6502	0.8127	1.4629	468.0
472.0	524.67	0.01984	0.86345	0.88329	455.2	749.3	1204.5	0.6551	0.8042	1.4592	472.0
476.0	545.11	0.01992	0.82958	0.84950	459.9	744.5	1204.3	0.6599	0.7956	1.4555	476.0
480.0	566.15	0.02000	0.79716	0.81717	464.5	739.6	1204.1	0.6648	0.7871	1.4518	480.0
484.0	587.81	0.02009	0.76613	0.78622	469.1	734.7	1203.8	0.6696	0.7785	1.4481	484.0
488.0	610.10	0.02017	0.73641	0.75658	473.8	729.7	1203.5	0.6745	0.7700	1.4444	488.0
492.0	633.03	0.02026	0.70794	0.72820	478.5	724.6	1203.1	0.6793	0.7614	1.4407	492.0
496.0	656.61	0.02034	0.68065	0.70100	483.2	719.5	1202.7	0.6842	0.7528	1.4370	496.0
500.0	680.86	0.02043	0.65448	0.67492	487.9	714.3	1202.2	0.6890	0.7443	1.4333	500.0
504.0	705.78	0.02053	0.62938	0.64991	492.7	709.0	1201.7	0.6939	0.7357	1.4296	504.0
508.0	731.40	0.02062	0.60530	0.62592	497.5	703.7	1201.1	0.6987	0.7271	1.4258	508.0
512.0	757.72	0.02072	0.58218	0.60289	502.3	698.2	1200.5	0.7036	0.7185	1.4221	512.0
516.0	784.76	0.02081	0.55997	0.58079	507.1	692.7	1199.8	0.7085	0.7099	1.4183	516.0
520.0	812.53	0.02091	0.53864	0.55956	512.0	687.0	1199.0	0.7133	0.7013	1.4146	520.0
524.0	841.04	0.02102	0.51814	0.53916	516.9	681.3	1198.2	0.7182	0.6926	1.4108	524.0
528.0	870.31	0.02112	0.49843	0.51955	521.8	675.5	1197.3	0.7231	0.6839	1.4070	528.0
532.0	900.34	0.02123	0.47947	0.50070	526.8	669.6	1196.4	0.7280	0.6752	1.4032	532.0
536.0	931.17	0.02134	0.46123	0.48257	531.7	663.6	1195.4	0.7329	0.6665	1.3993	536.0
540.0	962.79	0.02146	0.44367	0.46513	536.8	657.5	1194.3	0.7378	0.6577	1.3954	540.0
544.0	995.22	0.02157	0.42677	0.44834	541.8	651.3	1193.1	0.7427	0.6489	1.3915	544.0
548.0	1028.49	0.02169	0.41048	0.43217	546.9	645.0	1191.9	0.7476	0.6400	1.3876	548.0
552.0	1062.59	0.02182	0.39479	0.41660	552.0	638.5	1190.6	0.7525	0.6311	1.3837	552.0
556.0	1097.55	0.02194	0.37966	0.40160	557.2	632.0	1189.2	0.7575	0.6222	1.3797	556.0
560.0	1133.38	0.02207	0.36507	0.38714	562.4	625.3	1187.7	0.7625	0.6132	1.3757	560.0
564.0	1170.10	0.02221	0.35099	0.37320	567.6	618.5	1186.1	0.7674	0.6041	1.3716	564.0
568.0	1207.72	0.02235	0.33741	0.35975	572.9	611.5	1184.5	0.7725	0.5950	1.3675	568.0
572.0	1246.26	0.02249	0.32429	0.34678	578.3	604.5	1182.7	0.7775	0.5859	1.3634	572.0
576.0	1285.74	0.02264	0.31162	0.33426	583.7	597.2	1180.9	0.7825	0.5766	1.3592	576.0
580.0	1326.17	0.02279	0.29937	0.32216	589.1	589.9	1179.0	0.7876	0.5673	1.3550	580.0
584.0	1367.7	0.02295	0.28753	0.31048	594.6	582.4	1176.9	0.7927	0.5580	1.3507	584.0
588.0	1410.0	0.02311	0.27608	0.29919	600.1	574.7	1174.8	0.7978	0.5485	1.3464	588.0
592.0	1453.3	0.02328	0.26499	0.28827	605.7	566.8	1172.6	0.8030	0.5390	1.3420	592.0
596.0	1497.8	0.02345	0.25425	0.27770	611.4	558.8	1170.2	0.8082	0.5293	1.3375	596.0

600.0	1543.2	0.02364	0.24384	0.26747	617.1	550.6	1167.7	0.8134	0.5196	1.3330	**600.0**
604.0	1589.7	0.02382	0.23374	0.25757	622.9	542.2	1165.1	0.8187	0.5097	1.3284	**604.0**
608.0	1637.3	0.02402	0.22394	0.24796	628.8	533.6	1162.4	0.8240	0.4997	1.3238	**608.0**
612.0	1686.1	0.02422	0.21442	0.23865	634.8	524.7	1159.5	0.8294	0.4896	1.3190	**612.0**
616.6	1735.9	0.02444	0.20516	0.22960	640.8	515.6	1156.4	0.8348	0.4794	1.3141	**616.0**
620.0	1786.9	0.02466	0.19615	0.22081	646.9	506.3	1153.2	0.8403	0.4689	1.3092	**620.0**
624.0	1839.0	0.02489	0.18737	0.21226	653.1	496.6	1149.8	0.8458	0.4583	1.3041	**624.0**
628.0	1892.4	0.02514	0.17880	0.20394	659.5	486.7	1146.1	0.8514	0.4474	1.2988	**628.0**
632.0	1947.0	0.02539	0.17044	0.19583	665.9	476.4	1142.2	0.8571	0.4364	1.2934	**632.0**
636.0	2002.8	0.02566	0.16226	0.18792	672.4	465.7	1138.1	0.8628	0.4251	1.2879	**636.0**
640.0	2059.9	0.02595	0.15427	0.18021	679.1	454.6	1133.7	0.8686	0.4134	1.2821	**640.0**
644.0	2118.3	0.02625	0.14644	0.17269	685.9	443.1	1129.0	0.8746	0.4015	1.2761	**644.0**
648.0	2178.1	0.02657	0.13876	0.16534	692.9	431.1	1124.0	0.8806	0.3893	1.2699	**648.0**
652.0	2239.2	0.02691	0.13124	0.15816	700.0	418.7	1118.7	0.8868	0.3767	1.2634	**652.0**
656.0	2301.7	0.02728	0.12387	0.15115	707.4	405.7	1113.1	0.8931	0.3637	1.2567	**656.0**
660.0	2365.7	0.02768	0.11663	0.14431	714.9	392.1	1107.0	0.8995	0.3502	1.2498	**660.0**
664.0	2431.1	0.02811	0.10947	0.13757	722.9	377.7	1100.6	0.9064	0.3361	1.2425	**664.0**
668.0	2498.1	0.02858	0.10229	0.13087	731.5	362.1	1093.5	0.9137	0.3210	1.2347	**668.0**
672.0	2566.6	0.02911	0.09514	0.12424	740.2	345.7	1085.9	0.9212	0.3054	1.2266	**672.0**
676.0	2636.8	0.02970	0.08799	0.11769	749.2	328.5	1077.6	0.9287	0.2892	1.2179	**676.0**
680.0	2708.6	0.03037	0.08080	0.11117	758.5	310.1	1068.5	0.9365	0.2720	1.2086	**680.0**
684.0	2782.1	0.03114	0.07349	0.10463	768.2	290.2	1058.4	0.9447	0.2537	1.1984	**684.0**
688.0	2857.4	0.03204	0.06595	0.09799	778.8	268.2	1047.0	0.9535	0.2337	1.1872	**688.0**
692.0	2934.5	0.03313	0.05797	0.09110	790.5	243.1	1033.6	0.9634	0.2110	1.1744	**692.0**
696.0	3013.4	0.03455	0.04916	0.08371	804.4	212.8	1017.2	0.9749	0.1841	1.1591	**696.0**
700.0	3094.3	0.03662	0.03857	0.07519	822.4	172.7	995.2	0.9901	0.1490	1.1390	**700.0**
702.0	3135.5	0.03824	0.03173	0.06997	835.0	144.7	979.7	1.0006	0.1246	1.1252	**702.0**
704.0	3177.2	0.04108	0.02192	0.06300	854.2	102.0	956.2	1.0169	0.0876	1.1046	**704.0**
705.0	3198.3	0.04427	0.01304	0.05730	873.0	61.4	934.4	1.0329	0.0527	1.0856	**705.0**
705.47*	3208.2	0.05078	0.00000	0.05078	906.0	0.0	906.0	1.0612	0.0000	1.0612	**705.47***

*Critical temperature

421

Table 16-14. Saturated Steam: Pressure Table

Abs Press. Lb/Sq In. p	Temp Fahr t	Specific Volume			Enthalpy			Entropy			Abs Press. Lb/Sq In. p
		Sat. Liquid v_f	Evap v_fg	Sat. Vapor v_g	Sat. Liquid h_f	Evap h_fg	Sat. Vapor h_g	Sat. Liquid s_f	Evap s_fg	Sat. Vapor s_g	
0.08865	32.018	0.016022	3302.4	3302.4	0.0003	1075.5	1075.5	0.0000	2.1872	2.1872	0.08865
0.25	59.323	0.016032	1235.5	1235.5	27.382	1060.1	1087.4	0.0542	2.0425	2.0967	0.25
0.50	79.586	0.016071	641.5	641.5	47.623	1048.6	1096.3	0.0925	1.9446	2.0370	0.50
1.0	101.74	0.016136	333.59	333.60	69.73	1036.1	1105.8	0.1326	1.8455	1.9781	1.0
5.0	162.24	0.016407	73.515	73.532	130.20	1000.9	1131.1	0.2349	1.6094	1.8443	5.0
10.0	193.21	0.016592	38.404	38.420	161.26	982.1	1143.3	0.2836	1.5043	1.7879	10.0
14.696	212.00	0.016719	26.782	26.799	180.17	970.3	1150.5	0.3121	1.4447	1.7568	14.696
15.0	213.03	0.016726	26.274	26.290	181.21	969.7	1150.9	0.3137	1.4415	1.7552	15.0
20.0	227.96	0.016834	20.070	20.087	196.27	960.1	1156.3	0.3358	1.3962	1.7320	20.0
30.0	250.34	0.017009	13.7266	13.7436	218.9	945.2	1164.1	0.3682	1.3313	1.6995	30.0
40.0	267.25	0.017151	10.4794	10.4965	236.1	933.6	1169.8	0.3921	1.2844	1.6765	40.0
50.0	281.02	0.017274	8.4967	8.5140	250.2	923.9	1174.1	0.4112	1.2474	1.6586	50.0
60.0	292.71	0.017383	7.1562	7.1736	262.2	915.4	1177.6	0.4273	1.2167	1.6440	60.0
70.0	302.93	0.017482	6.1875	6.2050	272.7	907.8	1180.6	0.4411	1.1905	1.6316	70.0
80.0	312.04	0.017573	5.4536	5.4711	282.1	900.9	1183.1	0.4534	1.1675	1.6208	80.0
90.0	320.28	0.017659	4.8779	4.8953	290.7	894.6	1185.3	0.4643	1.1470	1.6113	90.0
100.0	327.82	0.017740	4.4133	4.4310	298.5	888.6	1187.2	0.4743	1.1284	1.6027	100.0
110.0	334.79	0.017782	4.0306	4.0484	305.8	883.1	1188.9	0.4834	1.1115	1.5950	110.0
120.0	341.27	0.017789	3.7097	3.7275	312.6	877.8	1190.4	0.4919	1.0960	1.5879	120.0
130.0	347.33	0.017796	3.4364	3.4544	319.0	872.8	1191.7	0.4998	1.0815	1.5813	130.0
140.0	353.04	0.017803	3.2010	3.2190	325.0	868.0	1193.0	0.5071	1.0681	1.5752	140.0
150.0	358.43	0.017809	2.9958	3.0139	330.6	863.4	1194.1	0.5141	1.0554	1.5695	150.0
160.0	363.55	0.017815	2.8155	2.8336	336.1	859.0	1195.1	0.5206	1.0435	1.5641	160.0
170.0	368.42	0.017821	2.6556	2.6738	341.2	854.8	1196.0	0.5269	1.0322	1.5591	170.0
180.0	373.08	0.017827	2.5129	2.5312	346.2	850.7	1196.9	0.5328	1.0215	1.5543	180.0
190.0	377.53	0.017833	2.3847	2.4030	350.9	846.7	1197.6	0.5384	1.0113	1.5498	190.0
200.0	381.80	0.017839	2.2689	2.2873	355.5	842.8	1198.3	0.5438	1.0016	1.5454	200.0
210.0	385.91	0.017844	2.16373	2.18217	359.9	839.1	1199.0	0.5490	0.9923	1.5413	210.0
220.0	389.88	0.017850	2.06779	2.08629	364.2	835.4	1199.6	0.5540	0.9834	1.5374	220.0
230.0	393.70	0.017855	1.97991	1.99846	368.3	831.8	1200.1	0.5588	0.9748	1.5336	230.0
240.0	397.39	0.017860	1.89909	1.91769	372.3	828.4	1200.6	0.5634	0.9665	1.5299	240.0
250.0	400.97	0.017865	1.82452	1.84317	376.1	825.0	1201.1	0.5679	0.9585	1.5264	250.0
260.0	404.44	0.017870	1.75548	1.77418	379.9	821.6	1201.5	0.5722	0.9508	1.5230	260.0
270.0	407.80	0.017875	1.69137	1.71013	383.6	818.3	1201.9	0.5764	0.9433	1.5197	270.0
280.0	411.07	0.017880	1.63169	1.65049	387.1	815.1	1202.3	0.5805	0.9361	1.5166	280.0
290.0	414.25	0.017885	1.57597	1.59482	390.6	812.0	1202.6	0.5844	0.9291	1.5135	290.0
300.0	417.35	0.01889	1.52384	1.54274	394.0	808.9	1202.9	0.5882	0.9223	1.5105	300.0
350.0	431.73	0.01912	1.30642	1.32554	409.8	794.2	1204.0	0.6059	0.8909	1.4968	350.0
400.0	444.60	0.01934	1.14162	1.16095	424.2	780.4	1204.6	0.6217	0.8630	1.4847	400.0

450.0	456.28	0.01954	1.01224	1.03179	437.3	767.5	1204.8	0.6360	0.8378	1.4738	450.0
500.0	467.01	0.01975	0.90787	0.92762	449.5	755.1	1204.7	0.6490	0.8148	1.4639	500.0
550.0	476.94	0.01994	0.82183	0.84177	460.9	743.3	1204.3	0.6611	0.7936	1.4547	550.0
600.0	486.20	0.02013	0.74962	0.76975	471.7	732.0	1203.7	0.6723	0.7738	1.4461	600.0
650.0	494.89	0.02032	0.68811	0.70843	481.9	720.9	1202.8	0.6828	0.7552	1.4381	650.0
700.0	503.08	0.02050	0.63505	0.65556	491.6	710.2	1201.8	0.6928	0.7377	1.4304	700.0
750.0	510.84	0.02069	0.58880	0.60949	500.9	699.8	1200.7	0.7022	0.7210	1.4232	750.0
800.0	518.21	0.02087	0.54809	0.56896	509.8	689.6	1199.4	0.7111	0.7051	1.4163	800.0
850.0	525.24	0.02105	0.51197	0.53302	518.4	679.5	1198.0	0.7197	0.6899	1.4096	850.0
900.0	531.95	0.02123	0.47968	0.50091	526.7	669.7	1196.4	0.7279	0.6753	1.4032	900.0
950.0	538.39	0.02141	0.45064	0.47205	534.7	660.0	1194.7	0.7358	0.6612	1.3970	950.0
1000.0	544.58	0.02159	0.42436	0.44596	542.6	650.4	1192.9	0.7434	0.6476	1.3910	1000.0
1050.0	550.53	0.02177	0.40047	0.42224	550.1	640.9	1191.0	0.7507	0.6344	1.3851	1050.0
1100.0	556.28	0.02195	0.37863	0.40058	557.5	631.5	1189.1	0.7578	0.6216	1.3794	1100.0
1150.0	561.82	0.02214	0.35859	0.38073	564.8	622.2	1187.0	0.7647	0.6091	1.3738	1150.0
1200.0	567.19	0.02232	0.34013	0.36245	571.9	613.0	1184.8	0.7714	0.5969	1.3683	1200.0
1250.0	572.38	0.02250	0.32306	0.34556	578.8	603.8	1182.6	0.7780	0.5850	1.3630	1250.0
1300.0	577.42	0.02269	0.30722	0.32991	585.6	594.6	1180.2	0.7843	0.5733	1.3577	1300.0
1350.0	582.32	0.02288	0.29250	0.31537	592.3	585.4	1177.8	0.7906	0.5620	1.3525	1350.0
1400.0	587.07	0.02307	0.27871	0.30178	598.8	576.5	1175.3	0.7966	0.5507	1.3474	1400.0
1450.0	591.70	0.02327	0.26584	0.28911	605.3	567.4	1172.8	0.8026	0.5397	1.3423	1450.0
1500.0	596.20	0.02346	0.25372	0.27719	611.7	558.4	1170.1	0.8085	0.5288	1.3373	1500.0
1550.0	600.59	0.02366	0.24235	0.26601	618.0	549.4	1167.4	0.8142	0.5182	1.3324	1550.0
1600.0	604.87	0.02387	0.23159	0.25545	624.2	540.3	1164.5	0.8199	0.5076	1.3274	1600.0
1650.0	609.05	0.02407	0.22143	0.24551	630.4	531.3	1161.6	0.8254	0.4971	1.3225	1650.0
1700.0	613.13	0.02428	0.21178	0.23607	636.5	522.2	1158.6	0.8309	0.4867	1.3176	1700.0
1750.0	617.12	0.02450	0.20263	0.22713	642.5	513.1	1155.6	0.8363	0.4765	1.3128	1750.0
1800.0	621.02	0.02472	0.19390	0.21861	648.5	503.8	1152.3	0.8417	0.4662	1.3079	1800.0
1850.0	624.83	0.02495	0.18558	0.21052	654.5	494.6	1149.0	0.8470	0.4561	1.3030	1850.0
1900.0	628.56	0.02517	0.17761	0.20278	660.4	485.2	1145.6	0.8522	0.4459	1.2981	1900.0
1950.0	632.22	0.02541	0.16999	0.19540	666.3	475.8	1142.0	0.8574	0.4358	1.2931	1950.0
2000.0	635.80	0.02565	0.16266	0.18831	672.1	466.2	1138.3	0.8625	0.4256	1.2881	2000.0
2100.0	642.76	0.02615	0.14885	0.17501	683.8	446.7	1130.5	0.8727	0.4053	1.2780	2100.0
2200.0	649.45	0.02669	0.13603	0.16272	695.5	426.7	1122.2	0.8828	0.3848	1.2676	2200.0
2300.0	655.89	0.02727	0.12406	0.15133	707.2	406.0	1113.2	0.8929	0.3640	1.2569	2300.0
2400.0	662.11	0.02790	0.11287	0.14076	719.0	384.8	1103.7	0.9031	0.3430	1.2460	2400.0
2500.0	668.11	0.02859	0.10209	0.13068	731.7	361.6	1093.3	0.9139	0.3206	1.2345	2500.0
2600.0	673.91	0.02938	0.09172	0.12110	744.5	337.6	1082.0	0.9247	0.2977	1.2225	2600.0
2700.0	679.53	0.03029	0.08165	0.11194	757.3	312.3	1069.7	0.9356	0.2741	1.2097	2700.0
2800.0	684.96	0.03134	0.07171	0.10305	770.7	285.1	1055.8	0.9468	0.2491	1.1958	2800.0
2900.0	690.22	0.03262	0.06158	0.09420	785.1	254.7	1039.8	0.9588	0.2215	1.1803	2900.0
3000.0	695.33	0.03428	0.05073	0.08500	801.8	218.4	1020.3	0.9728	0.1891	1.1619	3000.0
3100.0	700.28	0.03681	0.03771	0.07452	824.0	169.3	993.3	0.9914	0.1460	1.1373	3100.0
3200.0	705.08	0.04472	0.01191	0.05663	875.5	56.1	931.6	1.0351	0.0482	1.0832	3200.0
3208.2*	705.47	0.05078	0.00000	0.05078	906.0	0.0	906.0	1.0612	0.0000	1.0612	3208.2*

*Critical pressure

423

Table 16-15. Superheated Steam

Abs Press. Lb/Sq In. (Sat Temp)		Sat. Water	Sat. Steam	Temperature — Degrees Fahrenheit													
				200	250	300	350	400	450	500	600	700	800	900	1000	1100	1200
1 (101.74)	Sh	0.01614	333.6	98.26	148.26	198.26	248.26	298.26	348.26	398.26	498.26	598.26	698.26	798.26	898.26	998.26	1098.26
	v	69.73	1105.8	392.5	422.4	452.3	482.1	511.9	541.7	571.5	631.1	690.7	750.3	809.8	869.4	929.0	988.6
	h	0.1326	1.9781	1150.2	1172.9	1195.7	1218.7	1241.8	1265.1	1288.6	1336.1	1384.5	1433.7	1483.8	1534.9	1586.8	1639.7
	s			2.0509	2.0841	2.1152	2.1445	2.1722	2.1985	2.2237	2.2708	2.3144	2.3551	2.3934	2.4296	2.4640	2.4969
5 (162.24)	Sh	0.01641	73.53	37.76	87.76	137.76	187.76	237.76	287.76	337.76	437.76	537.76	637.76	737.76	837.76	937.76	1037.76
	v	130.20	1131.1	78.14	84.21	90.24	96.25	102.23	108.23	114.21	126.15	138.08	150.01	161.94	173.86	185.78	197.70
	h	0.2349	1.8443	1148.6	1171.7	1194.8	1218.0	1241.3	1264.7	1288.2	1335.9	1384.3	1433.6	1483.7	1534.7	1586.7	1639.6
	s			1.8716	1.9054	1.9369	1.9664	1.9943	2.0208	2.0460	2.0932	2.1369	2.1776	2.2159	2.2521	2.2866	2.3194
10 (193.21)	Sh	0.01659	38.42	6.79	56.79	106.79	156.79	206.79	256.79	306.79	406.79	506.79	606.79	706.79	806.79	906.79	1006.79
	v	161.26	1143.3	38.84	41.93	44.98	48.02	51.03	54.04	57.04	63.03	69.00	74.98	80.94	86.91	92.87	98.84
	h	0.2836	1.7879	1146.6	1170.2	1193.7	1217.1	1240.6	1264.1	1287.8	1335.5	1384.0	1433.4	1483.5	1534.6	1586.6	1639.5
	s			1.7928	1.8273	1.8593	1.8892	1.9173	1.9439	1.9692	2.0166	2.0603	2.1011	2.1394	2.1757	2.2101	2.2430
14.696 (212.00)	Sh	0.167	26.799		38.00	88.00	138.00	188.00	238.00	288.00	388.00	488.00	588.00	688.00	788.00	888.00	988.00
	v	180.17	1150.5		28.42	30.52	32.60	34.67	36.72	38.77	42.86	46.93	51.00	55.06	59.13	63.19	67.25
	h	0.3121	1.7568		1168.8	1192.6	1216.3	1239.9	1263.6	1287.4	1335.2	1383.8	1433.2	1483.4	1534.5	1586.5	1639.4
	s				1.7833	1.8158	1.8459	1.8743	1.9010	1.9265	1.9739	2.0177	2.0585	2.0969	2.1332	2.1676	2.2005
15 (213.03)	Sh	0.01673	26.290		36.97	86.97	136.97	186.97	236.97	286.97	386.97	486.97	586.97	686.97	786.97	886.97	986.97
	v	181.21	1150.9		27.837	29.899	31.939	33.963	35.977	37.985	41.986	45.978	49.964	53.946	57.926	61.905	65.882
	h	0.3137	1.7552		1168.7	1192.5	1216.2	1239.9	1263.6	1287.3	1335.2	1383.8	1433.4	1483.4	1534.5	1586.5	1639.4
	s				1.7809	1.8134	1.8437	1.8720	1.8988	1.9242	1.9717	2.0155	2.0563	2.0946	2.1309	2.1653	2.1982
20 (227.96)	Sh	0.01683	20.087		22.04	72.04	122.04	172.04	222.04	272.04	372.04	472.04	572.04	672.04	772.04	872.04	972.04
	v	196.27	1156.3		20.788	22.356	23.900	25.428	26.946	28.457	31.466	34.465	37.458	40.447	43.435	46.420	49.405
	h	0.3358	1.7320		1167.1	1191.4	1215.4	1239.2	1263.0	1286.9	1334.9	1383.5	1432.9	1483.2	1534.3	1586.3	1639.3
	s				1.7475	1.7805	1.8111	1.8397	1.8666	1.8921	1.9397	1.9836	2.0244	2.0628	2.0991	2.1336	2.1665
25 (240.07)	Sh	0.01693	16.301		9.93	59.93	109.93	159.93	209.93	259.93	359.93	459.93	559.93	659.93	759.93	859.93	959.93
	v	208.52	1160.6		16.558	17.829	19.076	20.307	21.527	22.740	25.153	27.557	29.954	32.348	34.740	37.130	39.518
	h	0.3535	1.7141		1165.6	1190.2	1214.5	1238.5	1262.5	1286.4	1334.6	1383.3	1432.7	1483.0	1534.2	1586.2	1639.2
	s				1.7212	1.7547	1.7856	1.8145	1.8415	1.8672	1.9149	1.9588	1.9997	2.0381	2.0744	2.1089	2.1418
30 (250.34)	Sh	0.01701	13.744			49.66	99.66	149.66	199.66	249.66	349.66	449.66	549.66	649.66	749.66	849.66	949.66
	v	218.93	1164.1			14.810	15.859	16.892	17.914	18.929	20.945	22.951	24.952	26.949	28.943	30.936	32.927
	h	0.3682	1.6995			1189.0	1213.6	1237.8	1261.9	1286.0	1334.2	1383.0	1432.5	1482.8	1534.0	1586.1	1639.0
	s					1.7334	1.7647	1.7937	1.8210	1.8467	1.8946	1.9386	1.9795	2.0179	2.0543	2.0888	2.1217

Abs Press, psia (Sat Temp, F)		Sat. Liquid	Sat. Vapor	300	350	400	450	500	600	700	800	900	1000	1100	1200
35 (259.29)	Sh			40.71	90.71	140.71	190.71	240.71	340.71	440.71	540.71	640.71	740.71	840.71	940.71
	v	0.01708	11.896	12.654	13.562	14.453	15.334	16.207	17.939	19.662	21.379	23.092	24.803	26.512	28.220
	h	228.03	1167.1	1187.8	1212.7	1237.1	1261.3	1285.5	1333.9	1382.8	1432.3	1482.7	1533.9	1586.0	1638.9
	s	0.3809	1.6872	1.7152	1.7468	1.7761	1.8035	1.8294	1.8774	1.9214	1.9624	2.0009	2.0372	2.0717	2.1046
40 (267.25)	Sh			32.75	82.75	132.75	182.75	232.75	332.75	432.75	532.75	632.75	732.75	832.75	932.75
	v	0.01715	10.497	11.036	11.838	12.624	13.398	14.165	15.685	17.195	18.699	20.199	21.697	23.194	24.689
	h	236.14	1169.8	1186.6	1211.7	1236.4	1260.8	1285.0	1333.6	1382.5	1432.1	1482.5	1533.7	1585.8	1638.8
	s	0.3921	1.6765	1.6992	1.7312	1.7608	1.7883	1.8143	1.8624	1.9065	1.9476	1.9860	2.0224	2.0569	2.0899
45 (274.44)	Sh			25.56	75.56	125.56	175.56	225.56	325.56	425.56	525.56	625.56	725.56	825.56	925.56
	v	0.01721	9.399	9.777	10.497	11.201	11.892	12.577	13.932	15.276	16.614	17.950	19.282	20.613	21.943
	h	243.49	1172.1	1185.4	1210.4	1235.7	1260.2	1284.6	1333.3	1382.3	1431.9	1482.3	1533.6	1585.7	1638.7
	s	0.4021	1.6671	1.6849	1.7173	1.7471	1.7748	1.8010	1.8492	1.8934	1.9345	1.9730	2.0093	2.0439	2.0768
50 (281.02)	Sh			18.98	68.98	118.98	168.98	218.98	318.98	418.98	518.98	618.98	718.98	818.98	918.98
	v	0.01727	8.514	8.769	9.424	10.062	10.688	11.306	12.529	13.741	14.947	16.150	17.350	18.549	19.746
	h	250.21	1174.1	1184.1	1209.9	1234.9	1259.6	1284.1	1332.9	1382.0	1431.7	1482.2	1533.4	1585.6	1638.6
	s	0.4112	1.6586	1.6720	1.7048	1.7349	1.7628	1.7890	1.8374	1.8816	1.9227	1.9613	1.9977	2.0322	2.0652
55 (287.07)	Sh			12.93	62.93	112.93	162.93	212.93	312.93	412.93	512.93	612.93	712.93	812.93	912.93
	v	0.01733		7.945	8.546	9.130	9.702	10.267	11.381	12.485	13.583	14.677	15.769	16.859	17.948
	h	256.43		1182.9	1208.9	1234.2	1259.1	1283.6	1332.6	1381.8	1431.5	1482.0	1533.3	1585.5	1638.5
	s	0.4196		1.6601	1.6933	1.7237	1.7518	1.7781	1.8266	1.8710	1.9121	1.9507	1.987	2.022	2.055
60 (292.71)	Sh			7.29	57.29	107.29	157.29	207.29	307.29	407.29	507.29	607.29	707.29	807.29	907.29
	v	0.01738	7.174	7.257	7.815	8.354	8.881	9.400	10.425	11.438	12.446	13.450	14.452	15.452	16.450
	h	262.21	1177.6	1181.6	1208.0	1233.5	1258.5	1283.2	1332.3	1381.5	1431.3	1481.8	1533.2	1585.3	1638.4
	s	0.4273	1.6440	1.6492	1.6834	1.7134	1.7417	1.7681	1.8168	1.8612	1.9024	1.9410	1.9774	2.0120	2.0450
65 (297.98)	Sh			2.02	52.02	102.02	152.02	202.02	302.02	402.02	502.02	602.02	702.02	802.02	902.02
	v	0.01743	6.653	6.675	7.195	7.697	8.186	8.667	9.615	10.552	11.484	12.412	13.337	14.261	15.183
	h	267.63	1179.1	1180.3	1207.0	1232.7	1257.9	1282.7	1331.9	1381.3	1431.1	1481.6	1533.0	1585.2	1638.3
	s	0.4344	1.6375	1.6390	1.6731	1.7040	1.7324	1.7590	1.8077	1.8522	1.8935	1.9321	1.9685	2.0031	2.0361
70 (302.93)	Sh				47.07	97.07	147.07	197.07	297.07	397.07	497.07	597.07	697.07	797.07	897.07
	v	0.01748	6.205		6.664	7.133	7.590	8.039	8.922	9.793	10.659	11.522	12.382	13.240	14.097
	h	272.74	1180.6		1206.0	1232.0	1257.3	1282.2	1331.6	1381.0	1430.9	1481.5	1532.9	1585.1	1638.2
	s	0.4411	1.6316		1.6640	1.6951	1.7237	1.7504	1.7993	1.8439	1.8852	1.9238	1.9603	1.9949	2.0279
75 (307.61)	Sh				42.39	92.39	142.39	192.39	292.39	392.39	492.39	592.39	692.39	792.39	892.39
	v	0.01753	5.814		6.204	6.645	7.074	7.494	8.320	9.135	9.945	10.750	11.553	12.355	13.155
	h	277.56	1181.9		1205.0	1231.2	1256.7	1281.7	1331.3	1380.7	1430.7	1481.3	1532.7	1585.0	1638.1
	s	0.4474	1.6260		1.6554	1.6868	1.7156	1.7424	1.7915	1.8361	1.8774	1.9161	1.9526	1.9872	2.0202

Sh = superheat, F

v = specific volume, cu ft per lb

h = enthalpy, Btu per lb

s = entropy, Btu per F per lb

425

Table 16-15. (Continued)

Abs Press. Lb/Sq In (Sat Temp)		Sat. Water	Sat. Steam	350	400	450	500	550	600	700	800	900	1000	1100	1200	1300	1400
80 (312.04)	Sh			37.96	87.96	137.96	187.96	237.96	287.96	387.96	487.96	587.96	687.96	787.96	887.96	987.96	1087.96
	v	0.01757	5.471	5.801	6.218	6.622	7.018	7.408	7.794	8.560	9.319	10.075	10.829	11.581	12.331	13.081	13.829
	h	282.15	1183.1	1204.0	1230.5	1256.1	1281.3	1306.2	1330.9	1380.5	1430.5	1481.1	1532.6	1584.9	1638.0	1692.0	1746.8
	s	0.4534	1.6208	1.6473	1.6790	1.7080	1.7349	1.7602	1.7842	1.8289	1.8702	1.9089	1.9454	1.9800	2.0131	2.0446	2.0750
85 (316.26)	Sh			33.74	83.74	133.74	183.74	233.74	283.74	383.74	483.74	583.74	683.74	783.74	883.74	983.74	1083.74
	v	0.01762	5.167	5.445	5.840	6.223	6.597	6.966	7.330	8.052	8.768	9.480	10.190	10.898	11.604	12.310	13.014
	h	286.52	1184.2	1203.0	1229.7	1255.5	1280.8	1305.8	1330.6	1380.2	1430.3	1481.0	1532.4	1584.7	1637.9	1691.9	1746.8
	s	0.4590	1.6159	1.6396	1.6716	1.7008	1.7279	1.7532	1.7772	1.8220	1.8634	1.9021	1.9386	1.9733	2.0063	2.0379	2.0682
90 (320.28)	Sh			29.72	79.72	129.72	179.72	229.72	279.72	379.72	479.72	579.72	679.72	779.72	879.72	979.72	1079.72
	v	0.01766	4.895	5.128	5.505	5.869	6.223	6.572	6.917	7.600	8.277	8.950	9.621	10.290	10.958	11.625	12.290
	h	290.69	1185.3	1202.0	1228.9	1254.9	1280.3	1305.4	1330.2	1380.0	1430.1	1480.8	1532.3	1584.6	1637.8	1691.8	1746.7
	s	0.4643	1.6113	1.6323	1.6646	1.6940	1.7212	1.7467	1.7707	1.8156	1.8570	1.8957	1.9323	1.9669	2.0000	2.0316	2.0619
95 (324.13)	Sh			25.87	75.87	125.87	175.87	225.87	275.87	375.87	475.87	575.87	675.87	775.87	875.87	975.87	1075.87
	v	0.01770	4.651	4.845	5.205	5.551	5.889	6.221	6.548	7.196	7.838	8.477	9.113	9.747	10.380	11.012	11.643
	h	294.70	1186.2	1200.9	1228.1	1254.3	1279.8	1305.0	1329.9	1379.7	1429.9	1480.6	1532.1	1584.5	1637.7	1691.7	1746.6
	s	0.4694	1.6069	1.6253	1.6580	1.6876	1.7149	1.7404	1.7645	1.8094	1.8509	1.8897	1.9262	1.9609	1.9940	2.0256	2.0559
100 (327.82)	Sh			22.18	72.18	122.18	172.18	222.18	272.18	372.18	472.18	572.18	672.18	772.18	872.18	972.18	1072.18
	v	0.01774	4.431	4.590	4.935	5.266	5.588	5.904	6.216	6.833	7.443	8.050	8.655	9.258	9.860	10.460	11.060
	h	298.54	1187.2	1199.9	1227.4	1253.7	1279.3	1304.6	1329.6	1379.5	1429.7	1480.4	1532.0	1584.4	1637.6	1691.6	1746.5
	s	0.4743	1.6027	1.6187	1.6516	1.6814	1.7088	1.7344	1.7586	1.8036	1.8451	1.8839	1.9205	1.9552	1.9883	2.0199	2.0502
105 (331.37)	Sh			18.63	68.63	118.63	168.63	218.63	268.63	368.63	468.63	568.63	668.63	768.63	868.63	968.63	1068.63
	v	0.01778	4.231	4.359	4.690	5.007	5.315	5.617	5.915	6.504	7.086	7.665	8.241	8.816	9.389	9.961	10.532
	h	302.24	1188.0	1198.8	1226.6	1253.1	1278.8	1304.2	1329.2	1379.2	1429.4	1480.3	1531.8	1584.2	1637.5	1691.5	1746.4
	s	0.4790	1.5988	1.6122	1.6455	1.6755	1.7031	1.7288	1.7530	1.7981	1.8396	1.8785	1.9151	1.9498	1.9828	2.0145	2.0448
110 (334.79)	Sh			15.21	65.21	115.21	165.21	215.21	265.21	365.21	465.21	565.21	665.21	765.21	865.21	965.21	1065.21
	v	0.01782	4.048	4.149	4.468	4.772	5.068	5.357	5.642	6.205	6.761	7.314	7.865	8.413	8.961	9.507	10.053
	h	305.80	1188.9	1197.7	1225.8	1252.5	1278.3	1303.8	1328.9	1379.0	1429.2	1480.1	1531.7	1584.1	1637.4	1691.4	1746.4
	s	0.4834	1.5950	1.6061	1.6396	1.6698	1.6975	1.7233	1.7476	1.7928	1.8344	1.8732	1.9099	1.9446	1.9777	2.0093	2.0397
115 (338.08)	Sh			11.92	61.92	111.92	161.92	211.92	261.92	361.92	461.92	561.92	661.92	761.92	861.92	961.92	1061.92
	v	0.01785	3.881	3.957	4.265	4.558	4.841	5.119	5.392	5.932	6.465	6.994	7.521	8.046	8.570	9.093	9.615
	h	309.25	1189.6	1196.1	1225.0	1251.8	1277.9	1303.3	1328.6	1378.7	1429.0	1479.9	1531.6	1584.0	1637.2	1691.4	1746.3
	s	0.4877	1.5913	1.6001	1.6340	1.6644	1.6922	1.7181	1.7425	1.7877	1.8294	1.8682	1.9049	1.9396	1.9727	2.0044	2.0347

Superheated steam properties. The header value in each block is the superheat, Sh (°F); the three property rows are v, h, and s.

120 (341.27)

	Sat. liq.	Sat. vap.	8.73	58.73	108.73	158.73	208.73	258.73	358.73	458.73	558.73	658.73	758.73	858.73	958.73	1058.73
v	0.01789	3.7275	3.7815	4.0786	4.3610	4.6341	4.9009	5.1637	5.6813	6.1928	6.7006	7.2060	7.7096	8.2219	8.7130	9.2134
h	312.58	1190.4	1195.6	1224.1	1251.2	1277.4	1302.9	1328.2	1378.4	1428.8	1479.8	1531.4	1583.9	1637.1	1691.3	1746.2
s	0.4919	1.5879	1.5943	1.6286	1.6592	1.6872	1.7132	1.7376	1.7829	1.8246	1.8635	1.9001	1.9349	1.9680	1.9996	2.0300

130 (347.33)

	Sat. liq.	Sat. vap.	2.67	52.67	102.67	152.67	202.67	252.67	352.67	452.67	552.67	652.67	752.67	852.67	952.67	1052.67
v	0.01796	3.4544	3.4699	3.7489	4.0129	4.2672	4.5151	4.7589	5.2384	5.7118	6.1814	6.6486	7.1140	7.5781	8.0411	8.5033
h	318.95	1191.7	1193.4	1222.5	1249.9	1276.4	1302.1	1327.5	1377.9	1428.4	1479.4	1531.1	1583.6	1636.9	1691.1	1746.1
s	0.4998	1.5813	1.5833	1.6182	1.6493	1.6775	1.7037	1.7283	1.7737	1.8155	1.8545	1.8911	1.9259	1.9591	1.9907	2.0211

140 (353.04)

	Sat. liq.	Sat. vap.		46.96	96.96	146.96	196.96	246.96	346.96	446.96	546.96	646.96	746.96	846.96	946.96	1046.96
v	0.01803	3.2190		3.4661	3.7143	3.9526	4.1844	4.4119	4.8588	5.2995	5.7364	6.1709	6.6036	7.0049	7.4652	7.8946
h	324.96	1193.0		1220.8	1248.7	1275.3	1301.3	1326.8	1377.4	1428.0	1479.1	1530.8	1583.4	1636.7	1690.9	1745.9
s	0.5071	1.5752		1.6085	1.6400	1.6686	1.6949	1.7196	1.7652	1.8071	1.8461	1.8828	1.9176	1.9508	1.9825	2.0129

150 (358.43)

	Sat. liq.	Sat. vap.		41.57	91.57	141.57	191.57	241.57	341.57	441.57	541.57	641.57	741.57	841.57	941.57	1041.57
v	0.01809	3.0139		3.2208	3.4555	3.6799	3.8978	4.1112	4.5298	4.9421	5.3507	5.7568	6.1612	6.5642	6.9661	7.3671
h	330.65	1194.1		1219.1	1247.4	1274.3	1300.5	1326.1	1376.9	1427.6	1478.7	1530.5	1583.1	1636.5	1690.7	1745.7
s	0.5141	1.5695		1.5993	1.6313	1.6602	1.6867	1.7115	1.7573	1.7992	1.8383	1.8751	1.9099	1.9431	1.9748	2.0052

160 (363.55)

	Sat. liq.	Sat. vap.		36.45	86.45	136.45	186.45	236.45	336.45	436.45	536.45	636.45	736.45	836.45	936.45	1036.45
v	0.01815	2.8336		3.0060	3.2288	3.4413	3.6469	3.8480	4.2420	4.6295	5.0132	5.3945	5.7741	6.1522	6.5293	6.9055
h	336.07	1195.1		1217.4	1246.0	1273.3	1299.6	1325.4	1376.4	1427.2	1478.4	1530.3	1582.9	1636.3	1690.5	1745.6
s	0.5206	1.5641		1.5906	1.6231	1.6522	1.6790	1.7039	1.7499	1.7919	1.8310	1.8678	1.9027	1.9359	1.9676	1.9980

170 (368.42)

	Sat. liq.	Sat. vap.		31.58	81.58	131.58	181.58	231.58	331.58	431.58	531.58	631.58	731.58	831.58	931.58	1031.58
v	0.01821	2.6738		2.8162	3.0288	3.2306	3.4255	3.6158	3.9879	4.3536	4.7155	5.0749	5.4325	5.7888	6.1440	6.4983
h	341.24	1196.0		1215.6	1244.7	1272.2	1298.8	1324.7	1375.8	1426.8	1478.0	1530.0	1582.6	1636.1	1690.4	1745.4
s	0.5269	1.5591		1.5823	1.6152	1.6447	1.6717	1.6968	1.7428	1.7850	1.8241	1.8610	1.8959	1.9291	1.9608	1.9913

180 (373.08)

	Sat. liq.	Sat. vap.		26.92	76.92	126.92	176.92	226.92	326.92	426.92	526.92	626.92	726.92	826.92	926.92	1026.92
v	0.01827	2.5312		2.6474	2.8508	3.0433	3.2286	3.4093	3.7621	4.1084	4.4508	4.7907	5.1289	5.4657	5.8014	6.1363
h	346.19	1196.9		1213.8	1243.4	1271.2	1297.9	1324.0	1375.3	1426.3	1477.7	1529.7	1582.4	1635.9	1690.2	1745.3
s	0.5328	1.5543		1.5743	1.6078	1.6376	1.6647	1.6900	1.7362	1.7784	1.8176	1.8545	1.8894	1.9227	1.9545	1.9849

190 (377.53)

	Sat. liq.	Sat. vap.		22.47	72.47	122.47	172.47	222.47	322.47	422.47	522.47	622.47	722.47	822.47	922.47	1022.47
v	0.01833	2.4030		2.4961	2.6915	2.8756	3.0525	3.2246	3.5601	3.8889	4.2140	4.5365	4.8572	5.1766	5.4949	5.8124
h	350.94	1197.6		1212.0	1242.0	1270.1	1297.1	1323.3	1374.8	1425.9	1477.4	1529.4	1582.1	1635.7	1690.0	1745.1
s	0.5384	1.5498		1.5667	1.6006	1.6307	1.6581	1.6835	1.7299	1.7722	1.8115	1.8484	1.8834	1.9166	1.9484	1.9789

200 (381.80)

	Sat. liq.	Sat. vap.		18.20	68.20	118.20	168.20	218.20	318.20	418.20	518.20	618.20	718.20	818.20	918.20	1018.20
v	0.01839	2.2873		2.3598	2.5480	2.7247	2.8939	3.0583	3.3783	3.6915	4.0008	4.3077	4.6128	4.9165	5.2191	5.5209
h	355.51	1198.3		1210.1	1240.6	1269.0	1296.2	1322.6	1374.3	1425.5	1477.0	1529.1	1581.9	1635.4	1689.8	1745.0
s	0.5438	1.5454		1.5593	1.5938	1.6242	1.6518	1.6773	1.7239	1.7663	1.8057	1.8426	1.8776	1.9109	1.9427	1.9732

Sh = superheat, F
v = specific volume, cu ft per lb
h = enthalpy, Btu per lb
s = entropy, Btu per F per lb

Table 16-15. (Continued)

Abs Press Lb/Sq In. (Sat. Temp)		Sat Water	Sat Steam	400	450	500	550	600	700	800	900	1000	1100	1200	1300	1400	1500
				Temperature – Degrees Fahrenheit													
210 (385.91)	Sh			14.09	64.09	114.09	164.09	214.09	314.09	414.09	514.09	614.09	714.09	814.09	914.09	1014.09	1114.09
	v	0.01844	2.1822	2.2364	2.4181	2.5880	2.7504	2.9078	3.2137	3.5128	3.8080	4.1007	4.3915	4.6811	4.9695	5.2571	5.5440
	h	359.91	1199.0	1208.00	1239.2	1268.0	1295.3	1321.9	1373.7	1425.1	1476.7	1528.8	1581.6	1635.2	1689.6	1744.8	1800.8
	s	0.5490	1.5413	1.5522	1.5872	1.6180	1.6458	1.6715	1.7182	1.7607	1.8001	1.8371	1.8721	1.9054	1.9372	1.9677	1.9970
220 (389.88)	Sh			10.12	60.12	110.12	160.12	210.12	310.12	410.12	510.12	610.12	710.12	810.12	910.12	1010.12	1110.12
	v	0.01850	2.0863	2.1240	2.2999	2.4638	2.6199	2.7710	3.0642	3.3504	3.6327	3.9125	4.1905	4.4671	4.7426	5.0173	5.2913
	h	364.17	1199.6	1206.3	1237.8	1266.9	1294.5	1321.2	1373.2	1424.7	1476.3	1528.5	1581.4	1635.0	1689.4	1744.7	1800.6
	s	0.5540	1.5374	1.5453	1.5808	1.6120	1.6400	1.6658	1.7128	1.7553	1.7948	1.8318	1.8668	1.9002	1.9320	1.9625	1.9919
230 (393.70)	Sh			6.30	56.30	106.30	156.30	206.30	306.30	406.30	506.30	606.30	706.30	806.30	906.30	1006.30	1106.30
	v	0.01855	1.9985	2.0212	2.1919	2.3503	2.5008	2.6461	2.9276	3.2020	3.4726	3.7406	4.0068	4.2717	4.5355	4.7984	5.0606
	h	368.28	1200.1	1204.4	1236.3	1265.7	1293.6	1320.4	1372.7	1424.2	1476.0	1528.2	1581.1	1634.8	1689.3	1744.5	1800.5
	s	0.5588	1.5336	1.5385	1.5747	1.6062	1.6344	1.6604	1.7075	1.7502	1.7897	1.8268	1.8618	1.8952	1.9270	1.9576	1.9869
240 (397.39)	Sh			2.61	52.61	102.61	152.61	202.61	302.61	402.61	502.61	602.61	702.61	802.61	902.61	1002.61	1102.61
	v	0.01860	1.9177	1.9268	2.0928	2.2462	2.3915	2.5316	2.8024	3.0661	3.3259	3.5831	3.8385	4.0926	4.3456	4.5977	4.8492
	h	372.27	1200.6	1202.4	1234.9	1264.6	1292.7	1319.7	1372.1	1423.8	1475.6	1527.9	1580.9	1634.6	1689.1	1744.3	1800.4
	s	0.5634	1.5299	1.5320	1.5687	1.6006	1.6291	1.6552	1.7025	1.7452	1.7848	1.8219	1.8570	1.8904	1.9223	1.9528	1.9822
250 (400.97)	Sh				49.03	99.03	149.03	199.03	299.03	399.03	499.03	599.03	699.03	799.03	899.03	999.03	1099.03
	v	0.01865	1.8432		2.0016	2.1504	2.2909	2.4262	2.6872	2.9410	3.1909	3.4382	3.6837	3.9278	4.1709	4.4131	4.6546
	h	376.14	1201.1		1233.4	1263.5	1291.8	1319.0	1371.6	1423.4	1475.3	1527.6	1580.6	1634.4	1688.9	1744.2	1800.2
	s	0.5679	1.5264		1.5629	1.5951	1.6239	1.6502	1.6976	1.7405	1.7801	1.8173	1.8524	1.8858	1.9177	1.9482	1.9776
260 (404.44)	Sh				45.56	95.56	145.56	195.56	295.56	395.56	495.56	595.56	695.56	795.56	895.56	995.56	1095.56
	v	0.01870	1.7742		1.9173	2.0619	2.1981	2.3289	2.5808	2.8256	3.0663	3.3044	3.5408	3.7758	4.0097	4.2427	4.4750
	h	379.90	1201.5		1231.9	1262.4	1290.9	1318.2	1371.1	1423.0	1474.9	1527.3	1580.4	1634.2	1688.7	1744.0	1800.1
	s	0.5722	1.5230		1.5573	1.5899	1.6189	1.6453	1.6930	1.7359	1.7756	1.8128	1.8480	1.8814	1.9133	1.9439	1.9732
270 (407.80)	Sh				42.20	92.20	142.20	192.20	292.20	392.20	492.20	592.20	692.20	792.20	892.20	992.20	1092.20
	v	0.01875	1.7101		1.8391	1.9799	2.1121	2.2388	2.4824	2.7186	2.9509	3.1806	3.4084	3.6349	3.8603	4.0849	4.3087
	h	383.56	1201.9		1230.4	1261.2	1290.0	1317.5	1370.5	1422.6	1474.6	1527.1	1580.1	1634.0	1688.5	1743.9	1800.0
	s	0.5764	1.5197		1.5518	1.5848	1.6140	1.6406	1.6885	1.7315	1.7713	1.8085	1.8437	1.8771	1.9090	1.9396	1.9690
280 (411.07)	Sh				38.93	88.93	138.93	188.93	288.93	388.93	488.93	588.93	688.93	788.93	888.93	988.93	1088.93
	v	0.01880	1.6505		1.7665	1.9037	2.0322	2.1551	2.3909	2.6194	2.8437	3.0655	3.2855	3.5042	3.7217	3.9384	4.1543
	h	387.12	1202.3		1228.8	1260.0	1289.1	1316.8	1370.0	1422.1	1474.2	1526.8	1579.9	1633.8	1688.4	1743.7	1799.8
	s	0.5805	1.5166		1.5464	1.5798	1.6093	1.6361	1.6841	1.7273	1.7671	1.8043	1.8395	1.8730	1.9050	1.9356	1.9649

Superheated steam table (pressures 290–380 psia). For each pressure the saturation temperature (°F) is given in parentheses. The saturation‑liquid and saturation‑vapor columns are followed by points of increasing superheat.

290 (414.25)

Sh	v	h	s
(sat. liq.)	0.01885	390.60	0.5844
(sat. vap.)	1.5948	1202.6	1.5135
35.75	1.6988	1227.3	1.5412
85.75	1.8327	1258.9	1.5750
135.75	1.9578	1288.1	1.6048
185.75	2.0772	1316.0	1.6317
285.75	2.3058	1369.5	1.6799
385.75	2.5269	1421.7	1.7232
485.75	2.7440	1473.9	1.7630
585.75	2.9585	1526.5	1.8003
685.75	3.1711	1579.6	1.8356
785.75	3.3824	1633.5	1.8690
885.75	3.5926	1688.2	1.9010
985.75	3.8019	1743.6	1.9316
1085.75	4.0106	1799.7	1.9610

300 (417.35)

Sh	v	h	s
(sat. liq.)	0.01889	393.99	0.5882
(sat. vap.)	1.5427	1202.9	1.5105
32.65	1.6356	1225.7	1.5361
82.65	1.7665	1257.7	1.5703
132.65	1.8883	1287.2	1.6003
182.65	2.0044	1315.2	1.6274
282.65	2.2263	1368.9	1.6758
382.65	2.4407	1421.3	1.7192
482.65	2.6509	1473.6	1.7591
582.65	2.8585	1526.2	1.7964
682.65	3.0643	1579.4	1.8317
782.65	3.2688	1633.3	1.8652
882.65	3.4721	1688.0	1.8972
982.65	3.6746	1743.4	1.9278
1082.65	3.8764	1799.6	1.9572

310 (420.36)

Sh	v	h	s
(sat. liq.)	0.01894	397.30	0.5920
(sat. vap.)	1.4939	1203.2	1.5076
29.64	1.5763	1224.1	1.5311
79.64	1.7044	1256.5	1.5657
129.64	1.8233	1286.3	1.5960
179.64	1.9363	1314.5	1.6233
279.64	2.1520	1368.4	1.6719
379.64	2.3600	1420.9	1.7153
479.64	2.5638	1473.2	1.7553
579.64	2.7650	1525.9	1.7927
679.64	2.9644	1579.2	1.8280
779.64	3.1625	1633.1	1.8615
879.64	3.3594	1687.8	1.8935
979.64	3.5555	1743.3	1.9241
1079.64	3.7509	1799.4	1.9536

320 (423.31)

Sh	v	h	s
(sat. liq.)	0.01899	400.53	0.5956
(sat. vap.)	1.4480	1203.4	1.5048
26.69	1.5207	1222.5	1.5261
76.69	1.6462	1255.2	1.5612
126.69	1.7623	1285.3	1.5918
176.69	1.8725	1313.7	1.6192
276.69	2.0823	1367.8	1.6680
376.69	2.2843	1420.5	1.7116
476.69	2.4821	1472.9	1.7516
576.69	2.6774	1525.6	1.7890
676.69	2.8708	1578.9	1.8243
776.69	3.0628	1632.9	1.8579
876.69	3.2538	1687.6	1.8899
976.69	3.4438	1743.1	1.9206
1076.69	3.6332	1799.3	1.9500

330 (426.18)

Sh	v	h	s
(sat. liq.)	0.01903	403.70	0.5991
(sat. vap.)	1.4048	1203.6	1.5021
23.82	1.4684	1220.9	1.5213
73.82	1.5915	1254.0	1.5568
123.82	1.7050	1284.4	1.5876
173.82	1.8125	1313.0	1.6153
273.82	2.0168	1367.3	1.6643
373.82	2.2132	1420.0	1.7079
473.82	2.4054	1472.5	1.7480
573.82	2.5950	1525.3	1.7855
673.82	2.7828	1578.7	1.8208
773.82	2.9692	1632.7	1.8544
873.82	3.1545	1687.5	1.8864
973.82	3.3389	1742.9	1.9171
1073.82	3.5227	1799.2	1.9466

340 (428.99)

Sh	v	h	s
(sat. liq.)	0.01908	406.80	0.6026
(sat. vap.)	1.3640	1203.8	1.4994
21.01	1.4191	1219.2	1.5165
71.01	1.5399	1252.8	1.5525
121.01	1.6511	1283.4	1.5836
171.01	1.7561	1312.2	1.6114
271.01	1.9552	1366.7	1.6606
371.01	2.1463	1419.6	1.7044
471.01	2.3333	1472.2	1.7445
571.01	2.5175	1525.0	1.7820
671.01	2.7000	1578.4	1.8174
771.01	2.8811	1632.5	1.8510
871.01	3.0611	1687.3	1.8831
971.01	3.2402	1742.8	1.9138
1071.01	3.4186	1799.0	1.9432

350 (431.73)

Sh	v	h	s
(sat. liq.)	0.01912	409.83	0.6059
(sat. vap.)	1.3255	1204.0	1.4968
18.27	1.3725	1217.5	1.5119
68.27	1.4913	1251.5	1.5483
118.27	1.6002	1282.4	1.5797
168.27	1.7028	1311.4	1.6077
268.27	1.8970	1366.2	1.6571
368.27	2.0832	1419.2	1.7009
468.27	2.2652	1471.8	1.7411
568.27	2.4445	1524.7	1.7787
668.27	2.6219	1578.2	1.8141
768.27	2.7980	1632.3	1.8477
868.27	2.9730	1687.1	1.8798
968.27	3.1471	1742.6	1.9105
1068.27	3.3205	1798.9	1.9400

360 (434.41)

Sh	v	h	s
(sat. liq.)	0.01917	412.81	0.6092
(sat. vap.)	1.2891	1204.1	1.4943
15.59	1.3285	1215.8	1.5073
65.59	1.4454	1250.3	1.5441
115.59	1.5521	1281.5	1.5758
165.59	1.6525	1310.6	1.6040
265.59	1.8421	1365.6	1.6536
365.59	2.0237	1418.7	1.6976
465.59	2.2009	1471.5	1.7379
565.59	2.3755	1524.4	1.7754
665.59	2.5482	1577.9	1.8109
765.59	2.7196	1632.1	1.8445
865.59	2.8898	1686.9	1.8766
965.59	3.0592	1742.5	1.9073
1065.59	3.2279	1798.8	1.9368

380 (439.61)

Sh	v	h	s
(sat. liq.)	0.01925	418.59	0.6156
(sat. vap.)	1.2218	1204.4	1.4894
10.39	1.2472	1212.4	1.4982
60.39	1.3606	1247.7	1.5360
110.39	1.4635	1279.5	1.5683
160.39	1.5598	1309.0	1.5969
260.39	1.7410	1364.5	1.6470
360.39	1.9139	1417.9	1.6911
460.39	2.0825	1470.8	1.7315
560.39	2.2484	1523.8	1.7692
660.39	2.4124	1577.4	1.8047
760.39	2.5750	1631.6	1.8384
860.39	2.7366	1686.5	1.8705
960.39	2.8973	1742.2	1.9012
1060.39	3.0572	1798.5	1.9307

Sh = superheat, F
v = specific volume, cu ft per lb
h = enthalpy, Btu per lb
s = entropy, Btu per F per lb

Table 16-15. (Continued)

Abs Press. Lb/Sq In. (Sat. Temp)		Sat. Water	Sat. Steam	Temperature – Degrees Fahrenheit													
				450	500	550	600	650	700	800	900	1000	1100	1200	1300	1400	1500
400 (444.60)	Sh			5.40	55.40	105.40	155.40	205.40	255.40	355.40	455.40	555.40	655.40	755.40	855.40	955.40	1055.40
	v	0.01934	1.1610	1.1738	1.2841	1.3836	1.4763	1.5646	1.6499	1.8151	1.9759	2.1339	2.2901	2.4450	2.5987	2.7515	2.9037
	h	424.17	1204.6	1208.8	1245.1	1277.5	1307.4	1335.9	1363.4	1417.0	1470.1	1523.3	1576.9	1631.2	1686.2	1741.9	1798.2
	s	0.6217	1.4847	1.4894	1.5282	1.5611	1.5901	1.6163	1.6406	1.6850	1.7255	1.7632	1.7988	1.8325	1.8647	1.8955	1.9250
420 (449.40)	Sh			60	50.60	100.60	150.60	200.60	250.60	350.60	450.60	550.60	650.60	750.60	850.60	950.60	1050.60
	v	0.01942	1.1057	1.1071	1.2148	1.3113	1.4007	1.4856	1.5676	1.7258	1.8795	2.0304	2.1795	2.3273	2.4739	2.6196	2.7647
	h	429.56	1204.7	1205.2	1242.4	1275.4	1305.8	1334.5	1362.3	1416.2	1469.4	1522.7	1576.4	1630.8	1685.8	1741.6	1798.0
	s	0.6276	1.4802	1.4808	1.5206	1.5542	1.5835	1.6100	1.6345	1.6791	1.7197	1.7575	1.7932	1.8269	1.8591	1.8899	1.9195
440 (454.03)	Sh				45.97	95.97	145.97	195.97	245.97	345.97	445.97	545.97	645.97	745.97	845.97	945.97	1045.97
	v	0.01950	1.0554		1.1517	1.2454	1.3319	1.4138	1.4926	1.6445	1.7918	1.9363	2.0790	2.2203	2.3605	2.4998	2.6384
	h	434.77	1204.8		1239.7	1273.4	1304.2	1333.2	1361.1	1415.3	1468.7	1522.1	1575.9	1630.4	1685.5	1741.2	1797.7
	s	0.6332	1.4759		1.5132	1.5474	1.5772	1.6040	1.6286	1.6734	1.7142	1.7521	1.7878	1.8216	1.8538	1.8847	1.9143
460 (458.50)	Sh				41.50	91.50	141.50	191.50	241.50	341.50	441.50	541.50	641.50	741.50	841.50	941.50	1041.50
	v	0.01959	1.0092		1.0939	1.1852	1.2691	1.3482	1.4242	1.5703	1.7117	1.8504	1.9872	2.1226	2.2569	2.3903	2.5230
	h	439.83	1204.8		1236.9	1271.3	1302.5	1331.8	1360.0	1414.4	1468.0	1521.5	1575.4	1629.9	1685.1	1740.9	1797.4
	s	0.6387	1.4718		1.5060	1.5409	1.5711	1.5982	1.6230	1.6680	1.7089	1.7469	1.7826	1.8165	1.8488	1.8797	1.9093
480 (462.82)	Sh				37.18	87.18	137.18	187.18	237.18	337.18	437.18	537.18	637.18	737.18	837.18	937.18	1037.18
	v	0.01967	0.9668		1.0409	1.1300	1.2115	1.2889	1.3615	1.5023	1.6384	1.7716	1.9030	2.0330	2.1619	2.2900	2.4173
	h	444.75	1204.8		1234.1	1269.1	1300.8	1330.5	1358.8	1413.6	1467.3	1520.9	1574.9	1629.5	1684.7	1740.6	1797.2
	s	0.6439	1.4677		1.4990	1.5346	1.5652	1.5925	1.6176	1.6628	1.7038	1.7419	1.7777	1.8116	1.8439	1.8748	1.9045
500 (467.01)	Sh				32.99	82.99	132.99	182.99	232.99	332.99	432.99	532.99	632.99	732.99	832.99	932.99	1032.99
	v	0.01975	0.9276		0.9919	1.0791	1.1584	1.2327	1.3037	1.4397	1.5708	1.6992	1.8256	1.9507	2.0746	2.1977	2.3200
	h	449.52	1204.7		1231.2	1267.0	1299.1	1329.1	1357.7	1412.7	1466.6	1520.3	1574.4	1629.1	1684.4	1740.3	1796.9
	s	0.6490	1.4639		1.4921	1.5284	1.5595	1.5871	1.6123	1.6578	1.6990	1.7371	1.7730	1.8069	1.8393	1.8702	1.8998
520 (471.07)	Sh				28.93	78.93	128.93	178.93	228.93	328.93	428.93	528.93	628.93	728.93	828.93	928.93	1028.93
	v	0.01982	0.8914		0.9466	1.0321	1.1094	1.1816	1.2504	1.3819	1.5085	1.6323	1.7542	1.8746	1.9940	2.1125	2.2302
	h	454.18	1204.5		1228.3	1264.8	1297.4	1327.7	1356.5	1411.8	1465.9	1519.7	1573.9	1628.7	1684.0	1740.0	1796.7
	s	0.6540	1.4601		1.4853	1.5223	1.5539	1.5818	1.6072	1.6530	1.6943	1.7325	1.7684	1.8024	1.8348	1.8657	1.8954
540 (475.01)	Sh				24.99	74.99	124.99	174.99	224.99	324.99	424.99	524.99	624.99	724.99	824.99	924.99	1024.99
	v	0.01990	0.8577		0.9045	0.9884	1.0640	1.1342	1.2010	1.3284	1.4508	1.5704	1.6880	1.8042	1.9193	2.0336	2.1471
	h	458.71	1204.4		1225.3	1262.5	1295.7	1326.3	1355.3	1410.9	1465.1	1519.1	1573.4	1628.2	1683.6	1739.7	1796.4
	s	0.6587	1.4565		1.4786	1.5164	1.5485	1.5767	1.6023	1.6483	1.6897	1.7280	1.7640	1.7981	1.8305	1.8615	1.8911

Superheated steam table (abs. pressure, psia, with saturation temperature in °F shown in parentheses). Columns correspond to total temperatures of 500, 550, 600, 650, 700, 800, 900, 1000, 1100, 1200, 1300, 1400 and 1500 °F.

Abs. press. (sat. temp)	Prop.	Sat. liq.	Sat. vap.	500	550	600	650	700	800	900	1000	1100	1200	1300	1400	1500
560 (478.84)	Sh		0	21.16	71.16	121.16	171.16	221.16	321.16	421.16	521.16	621.16	721.16	821.16	921.16	1021.16
	v	0.01998	0.8264	0.8653	0.9479	1.0217	1.0902	1.1552	1.2787	1.3972	1.5129	1.6266	1.7388	1.8500	1.9603	2.0699
	h	463.14	1204.2	1222.2	1260.3	1293.9	1324.9	1354.2	1410.0	1464.4	1518.6	1572.9	1627.8	1683.3	1739.4	1796.1
	s	0.6634	1.4529	1.4720	1.5106	1.5431	1.5717	1.5975	1.6438	1.6853	1.7237	1.7598	1.7939	1.8263	1.8573	1.8870
580 (482.57)	Sh		0	17.43	67.43	117.43	167.43	217.43	317.43	417.43	517.43	617.43	717.43	817.43	917.43	1017.43
	v	0.02006	0.7971	0.8287	0.9100	0.9824	1.0492	1.1125	1.2324	1.3473	1.4593	1.5693	1.6780	1.7855	1.8921	1.9980
	h	467.47	1203.9	1219.1	1258.0	1292.1	1323.4	1353.0	1409.2	1463.7	1518.0	1572.4	1627.4	1682.9	1739.1	1795.9
	s	0.6679	1.4495	1.4654	1.5049	1.5380	1.5668	1.5929	1.6394	1.6811	1.7196	1.7556	1.7898	1.8223	1.8533	1.8831
600 (486.20)	Sh		0	13.80	63.80	113.80	163.80	213.80	313.80	413.80	513.80	613.80	713.80	813.80	913.80	1013.80
	v	0.02013	0.7697	0.7944	0.8746	0.9456	1.0109	1.0726	1.1892	1.3008	1.4093	1.5160	1.6211	1.7252	1.8284	1.9309
	h	471.70	1203.7	1215.9	1255.6	1290.3	1322.0	1351.8	1408.3	1463.0	1517.4	1571.9	1627.0	1682.6	1738.8	1795.6
	s	0.6723	1.4461	1.4590	1.4993	1.5329	1.5621	1.5884	1.6351	1.6769	1.7155	1.7517	1.7859	1.8184	1.8494	1.8792
650 (494.89)	Sh		0	5.11	55.11	105.11	155.11	205.11	305.11	405.11	505.11	605.11	705.11	805.11	905.11	1005.11
	v	0.02032	0.7084	0.7173	0.7954	0.8634	0.9254	0.9835	1.0929	1.1969	1.2979	1.3969	1.4944	1.5909	1.6864	1.7813
	h	481.89	1202.8	1207.6	1249.6	1285.7	1318.3	1348.7	1406.0	1461.2	1515.9	1570.7	1625.9	1681.6	1738.0	1794.9
	s	0.6828	1.4381	1.4430	1.4858	1.5207	1.5507	1.5775	1.6249	1.6671	1.7059	1.7422	1.7765	1.8092	1.8403	1.8701
700 (503.08)	Sh		0		46.92	96.92	146.92	196.92	296.92	396.92	496.92	596.92	696.92	796.92	896.92	996.92
	v	0.02050	0.6556		0.7271	0.7928	0.8520	0.9072	1.0102	1.1078	1.2023	1.2948	1.3858	1.4757	1.5647	1.6530
	h	491.60	1201.8		1243.4	1281.0	1314.6	1345.6	1403.7	1459.4	1514.4	1569.4	1624.8	1680.7	1737.2	1794.3
	s	0.6928	1.4304		1.4726	1.5090	1.5399	1.5673	1.6154	1.6580	1.6970	1.7335	1.7679	1.8006	1.8318	1.8617
750 (510.84)	Sh		0		39.16	89.16	139.16	189.16	289.16	389.16	489.16	589.16	689.16	789.16	889.16	989.16
	v	0.02069	0.6095		0.6676	0.7313	0.7882	0.8409	0.9386	1.0306	1.1195	1.2063	1.2916	1.3759	1.4592	1.5419
	h	500.89	1200.7		1236.9	1276.1	1310.7	1342.5	1401.5	1457.6	1512.9	1568.2	1623.8	1679.8	1736.4	1793.6
	s	0.7022	1.4232		1.4598	1.4977	1.5296	1.5577	1.6065	1.6494	1.6886	1.7252	1.7598	1.7926	1.8239	1.8538
800 (518.21)	Sh		0		31.79	81.79	131.79	181.79	281.79	381.79	481.79	581.79	681.79	781.79	881.79	981.79
	v	0.02087	0.5690		0.6151	0.6774	0.7323	0.7828	0.8759	0.9631	1.0470	1.1289	1.2093	1.2885	1.3669	1.4446
	h	509.81	1199.4		1230.1	1271.1	1306.8	1339.3	1399.1	1455.8	1511.4	1566.9	1622.7	1678.9	1735.7	1792.9
	s	0.7111	1.4163		1.4472	1.4869	1.5198	1.5484	1.5980	1.6413	1.6807	1.7175	1.7522	1.7851	1.8164	1.8464
850 (525.24)	Sh		0		24.76	74.76	124.76	174.76	274.76	374.76	474.76	574.76	674.76	774.76	874.76	974.76
	v	0.02105	0.5330		0.5683	0.6296	0.6829	0.7315	0.8205	0.9034	0.9830	1.0606	1.1366	1.2115	1.2855	1.3588
	h	518.40	1198.0		1223.0	1265.9	1302.8	1336.0	1396.8	1454.0	1510.0	1565.7	1621.6	1678.0	1734.9	1792.3
	s	0.7197	1.4096		1.4347	1.4763	1.5102	1.5396	1.5899	1.6336	1.6733	1.7102	1.7450	1.7780	1.8094	1.8395
900 (531.95)	Sh		0		18.05	68.05	118.05	168.05	268.05	368.05	468.05	568.05	668.05	768.05	868.05	968.05
	v	0.02123	0.5009		0.5263	0.5869	0.6388	0.6858	0.7713	0.8504	0.9262	0.9998	1.0720	1.1430	1.2131	1.2825
	h	526.70	1196.4		1215.5	1260.6	1298.6	1332.7	1394.4	1452.2	1508.5	1564.4	1620.6	1677.1	1734.1	1791.6
	s	0.7279	1.4032		1.4223	1.4659	1.5010	1.5311	1.5822	1.6263	1.6662	1.7033	1.7382	1.7713	1.8028	1.8329

Sh = superheat, F
v = specific volume, cu ft per lb
h = enthalpy, Btu per lb
s = entropy, Btu per F per lb

Table 16-15. (Continued)

Abs Press. Lb/Sq In. (Sat. Temp)		Sat Water	Sat Steam	Temperature – Degrees Fahrenheit													
				550	600	650	700	750	800	850	900	1000	1100	1200	1300	1400	1500
950 (538.39)	Sh			11.61	61.61	111.61	161.61	211.61	261.61	311.61	361.61	461.61	561.61	661.61	761.61	861.61	961.61
	v	0.02141	0.4721	0.4883	0.5485	0.5993	0.6449	0.6871	0.7272	0.7656	0.8030	0.8753	0.9455	1.0142	1.0817	1.1484	1.2143
	h	534.74	1194.7	1207.6	1255.1	1294.4	1329.3	1361.5	1392.0	1421.5	1450.3	1507.0	1563.2	1619.5	1676.2	1733.3	1791.0
	s	0.7358	1.3970	1.4098	1.4557	1.4921	1.5228	1.5500	1.5748	1.5977	1.6193	1.6595	1.6967	1.7317	1.7649	1.7965	1.8267
1000 (544.58)	Sh			5.42	55.42	105.42	155.42	205.42	255.42	305.42	355.42	455.42	555.42	655.42	755.42	855.42	955.42
	v	0.02159	0.4460	0.4535	0.5137	0.5636	0.6080	0.6489	0.6875	0.7245	0.7603	0.8295	0.8966	0.9622	1.0266	1.0901	1.1529
	h	542.55	1192.9	1199.3	1249.3	1290.1	1325.9	1358.7	1389.6	1419.4	1448.5	1505.4	1561.9	1618.4	1675.3	1732.5	1790.3
	s	0.7434	1.3910	1.3973	1.4457	1.4833	1.5149	1.5426	1.5677	1.5908	1.6126	1.6530	1.6905	1.7256	1.7589	1.7905	1.8207
1050 (550.53)	Sh				49.47	99.47	149.47	199.47	249.47	299.47	349.47	449.47	549.47	649.47	749.47	849.47	949.47
	v	0.02177	0.4222		0.4821	0.5312	0.5745	0.6142	0.6515	0.6872	0.7216	0.7881	0.8524	0.9151	0.9767	1.0373	1.0973
	h	550.15	1191.0		1243.4	1285.7	1322.4	1355.4	1387.2	1417.3	1446.6	1503.9	1560.7	1617.4	1674.4	1731.8	1789.6
	s	0.7507	1.3851		1.4358	1.4748	1.5072	1.5354	1.5608	1.5842	1.6062	1.6469	1.6845	1.7197	1.7531	1.7848	1.8151
1100 (556.28)	Sh				43.72	93.72	143.72	193.72	243.72	293.72	343.72	443.72	543.72	643.72	743.72	843.72	943.72
	v	0.02195	0.4006		0.4531	0.5017	0.5440	0.5826	0.6188	0.6533	0.6865	0.7505	0.8121	0.8723	0.9313	0.9894	1.0468
	h	557.55	1189.1		1237.3	1281.2	1318.8	1352.9	1384.7	1415.2	1444.7	1502.4	1559.4	1616.3	1673.5	1731.0	1789.0
	s	0.7578	1.3794		1.4259	1.4664	1.4996	1.5284	1.5542	1.5779	1.6000	1.6410	1.6787	1.7141	1.7475	1.7793	1.8097
1150 (561.82)	Sh				39.18	89.18	139.18	189.18	239.18	289.18	339.18	439.18	539.18	639.18	739.18	839.18	939.18
	v	0.02214	0.3807		0.4263	0.4746	0.5162	0.5538	0.5889	0.6223	0.6544	0.7161	0.7754	0.8332	0.8899	0.9456	1.0007
	h	564.78	1187.0		1230.9	1276.6	1315.2	1349.9	1382.2	1413.0	1442.8	1500.9	1558.1	1615.2	1672.6	1730.2	1788.3
	s	0.7647	1.3738		1.4160	1.4582	1.4923	1.5216	1.5478	1.5717	1.5941	1.6353	1.6732	1.7087	1.7422	1.7741	1.8045
1200 (567.19)	Sh				32.81	82.81	132.81	182.81	232.81	282.81	332.81	432.81	532.81	632.81	732.81	832.81	932.81
	v	0.02232	0.3624		0.4016	0.4497	0.4905	0.5273	0.5615	0.5939	0.6250	0.6845	0.7418	0.7974	0.8519	0.9055	0.9584
	h	571.85	1184.8		1224.2	1271.8	1311.5	1346.9	1379.7	1410.8	1440.9	1499.4	1556.9	1614.2	1671.6	1729.4	1787.6
	s	0.7714	1.3683		1.4061	1.4501	1.4851	1.5150	1.5415	1.5658	1.5883	1.6298	1.6679	1.7035	1.7371	1.7691	1.7996
1300 (577.42)	Sh				22.58	72.58	122.58	172.58	222.58	272.58	322.58	422.58	522.58	622.58	722.58	822.58	922.58
	v	0.02269	0.3299		0.3570	0.4052	0.4451	0.4804	0.5129	0.5436	0.5729	0.6287	0.6822	0.7341	0.7847	0.8345	0.8836
	h	585.58	1180.2		1209.9	1261.9	1303.9	1340.8	1374.6	1406.4	1437.1	1496.3	1554.3	1612.0	1669.8	1727.9	1786.3
	s	0.7843	1.3577		1.3860	1.4340	1.4711	1.5022	1.5296	1.5544	1.5773	1.6194	1.6578	1.6937	1.7275	1.7596	1.7902
1400 (587.07)	Sh				12.93	62.93	112.93	162.93	212.93	262.93	312.93	412.93	512.93	612.93	712.93	812.93	912.93
	v	0.02307	0.3018		0.3176	0.3667	0.4059	0.4400	0.4712	0.5004	0.5282	0.5809	0.6311	0.6798	0.7272	0.7737	0.8195
	h	598.83	1175.3		1194.1	1251.4	1296.1	1334.5	1369.3	1402.0	1433.2	1493.2	1551.8	1609.9	1668.0	1726.3	1785.0
	s	0.7966	1.3474		1.3652	1.4181	1.4575	1.4900	1.5182	1.5436	1.5670	1.6096	1.6484	1.6845	1.7185	1.7508	1.7815

Column headings (°F superheat at top of each temperature column, from the 1500 psia row): 3.80, 53.80, 103.80, 153.80, 203.80, 253.80, 303.80, 403.80, 503.80, 603.80, 703.80, 803.80, 903.80

Sh = superheat, F
v = specific volume, cu ft per lb
h = enthalpy, Btu per lb
s = entropy, Btu per F per lb

P (Tsat)		Sat. liq	Sat. vap	3.80	53.80	103.80	153.80	203.80	253.80	303.80	403.80	503.80	603.80	703.80	803.80	903.80
1500 (596.20)	Sh				53.80	103.80	153.80	203.80	253.80	303.80	403.80	503.80	603.80	703.80	803.80	903.80
	v	0.02346	0.2772	0.2820	0.3328	0.3717	0.4049	0.4350	0.4629	0.4894	0.5394	0.5869	0.6327	0.6773	0.7210	0.7639
	h	611.68	1170.1	1176.3	1240.2	1287.9	1328.0	1364.0	1397.4	1429.2	1490.1	1549.2	1607.7	1666.2	1724.8	1783.7
	s	0.8085	1.3373	1.3431	1.4022	1.4443	1.4782	1.5073	1.5333	1.5572	1.6004	1.6395	1.6759	1.7101	1.7425	1.7734
1600 (604.87)	Sh				45.13	95.13	145.13	195.13	245.13	295.13	395.13	495.13	595.13	695.13	795.13	895.13
	v	0.02387	0.2555		0.3026	0.3415	0.3741	0.4032	0.4301	0.4555	0.5031	0.5482	0.5915	0.6336	0.6748	0.7153
	h	624.20	1164.5		1228.3	1279.4	1321.4	1358.5	1392.8	1425.2	1486.9	1546.6	1605.6	1664.3	1723.2	1782.3
	s	0.8199	1.3274		1.3861	1.4312	1.4667	1.4968	1.5235	1.5478	1.5916	1.6312	1.6678	1.7022	1.7347	1.7657
1700 (613.13)	Sh				36.87	86.87	136.87	186.87	236.87	286.87	386.87	486.87	586.87	686.87	786.87	886.87
	v	0.02428	0.2361		0.2754	0.3147	0.3468	0.3751	0.4011	0.4255	0.4711	0.5140	0.5552	0.5951	0.6341	0.6724
	h	636.45	1158.6		1215.3	1270.5	1314.5	1352.9	1388.1	1421.2	1483.8	1544.0	1603.4	1662.5	1721.7	1781.0
	s	0.8309	1.3176		1.3697	1.4183	1.4555	1.4867	1.5140	1.5388	1.5833	1.6232	1.6601	1.6947	1.7274	1.7585
1800 (621.02)	Sh				28.98	78.98	128.98	178.98	228.98	278.98	378.98	478.98	578.98	678.98	778.98	878.98
	v	0.02472	0.2186		0.2505	0.2906	0.3223	0.3500	0.3752	0.3988	0.4426	0.4836	0.5229	0.5609	0.5980	0.6343
	h	648.49	1152.3		1201.2	1261.1	1307.4	1347.2	1383.3	1417.1	1480.6	1541.4	1601.2	1660.7	1720.1	1779.7
	s	0.8417	1.3079		1.3526	1.4054	1.4446	1.4768	1.5049	1.5302	1.5753	1.6156	1.6528	1.6876	1.7204	1.7516
1900 (628.56)	Sh				21.44	71.44	121.44	171.44	221.44	271.44	371.44	471.44	571.44	671.44	771.44	871.44
	v	0.02517	0.2028		0.2274	0.2687	0.3004	0.3275	0.3521	0.3749	0.4171	0.4565	0.4940	0.5303	0.5656	0.6002
	h	660.36	1145.6		1185.7	1251.3	1300.2	1341.4	1378.4	1412.9	1477.4	1538.8	1599.1	1658.8	1718.6	1778.4
	s	0.8522	1.2981		1.3346	1.3925	1.4338	1.4672	1.4960	1.5219	1.5677	1.6084	1.6458	1.6808	1.7138	1.7451
2000 (635.80)	Sh				14.20	64.20	114.20	164.20	214.20	264.20	364.20	464.20	564.20	664.20	764.20	864.20
	v	0.02565	0.1883		0.2056	0.2488	0.2805	0.3072	0.3312	0.3534	0.3942	0.4320	0.4680	0.5027	0.5365	0.5695
	h	672.11	1138.3		1168.3	1240.9	1292.6	1335.4	1373.5	1408.7	1474.1	1536.2	1596.9	1657.0	1717.0	1777.1
	s	0.8625	1.2881		1.3154	1.3794	1.4231	1.4578	1.4874	1.5138	1.5603	1.6014	1.6391	1.6743	1.7075	1.7389
2100 (642.76)	Sh				7.24	57.24	107.24	157.24	207.24	257.24	357.24	457.24	557.24	657.24	757.24	857.24
	v	0.02615	0.1750		0.1847	0.2304	0.2624	0.2888	0.3123	0.3339	0.3734	0.4099	0.4445	0.4778	0.5101	0.5418
	h	683.79	1130.5		1148.5	1229.8	1284.9	1329.3	1368.4	1404.4	1470.9	1533.6	1594.7	1655.2	1715.4	1775.7
	s	0.8727	1.2780		1.2942	1.3661	1.4125	1.4486	1.4790	1.5060	1.5532	1.5948	1.6327	1.6681	1.7014	1.7330
2200 (649.45)	Sh				0.55	50.55	100.55	150.55	200.55	250.55	350.55	450.55	550.55	650.55	750.55	850.55
	v	0.02669	0.1627		0.1636	0.2134	0.2458	0.2720	0.2950	0.3161	0.3545	0.3897	0.4231	0.4551	0.4862	0.5165
	h	695.46	1122.2		1123.9	1218.0	1276.8	1323.1	1363.3	1400.0	1467.6	1530.9	1592.5	1653.3	1713.9	1774.4
	s	0.8828	1.2676		1.2691	1.3523	1.4020	1.4395	1.4708	1.4984	1.5463	1.5883	1.6266	1.6622	1.6956	1.7273
2300 (655.89)	Sh					44.11	94.11	144.11	194.11	244.11	344.11	444.11	544.11	644.11	744.11	844.11
	v	0.02727	0.1513			0.1975	0.2305	0.2566	0.2793	0.2999	0.3372	0.3714	0.4035	0.4344	0.4643	0.4935
	h	707.18	1113.2			1205.3	1268.4	1316.7	1358.1	1395.7	1464.2	1528.3	1590.3	1651.5	1712.3	1773.1
	s	0.8929	1.2569			1.3381	1.3914	1.4305	1.4628	1.4910	1.5397	1.5821	1.6207	1.6565	1.6901	1.7219

Table 16-15. (Continued)

Abs Press. Lb/Sq In. (Sat Temp)		Sat Water	Sat Steam	700	750	800	850	900	950	1000	1050	1100	1150	1200	1300	1400	1500
							Temperature – Degrees Fahrenheit										
2400 (662.11)	Sh			37.89	87.89	137.89	187.89	237.89	287.89	337.89	387.89	437.89	487.89	537.89	637.89	737.89	837.89
	v	0.02790	0.1408	0.1824	0.2164	0.2424	0.2648	0.2850	0.3037	0.3214	0.3382	0.3545	0.3703	0.3856	0.4155	0.4443	0.4724
	h	718.95	1103.7	1191.6	1259.7	1310.1	1352.8	1391.2	1426.9	1460.9	1493.7	1525.6	1557.0	1588.1	1649.6	1710.8	1771.8
	s	0.9031	1.2460	1.3232	1.3808	1.4217	1.4549	1.4837	1.5095	1.5332	1.5553	1.5761	1.5959	1.6149	1.6509	1.6847	1.7167
2500 (668.11)	Sh			31.89	81.89	131.89	181.89	231.89	281.89	331.89	381.89	431.89	481.89	531.89	631.89	731.89	831.89
	v	0.02859	0.1307	0.1681	0.2032	0.2293	0.2514	0.2712	0.2896	0.3068	0.3232	0.3390	0.3543	0.3692	0.3980	0.4259	0.4529
	h	731.71	1093.3	1176.7	1250.6	1303.4	1347.4	1386.7	1423.1	1457.5	1490.7	1522.9	1554.6	1585.9	1647.8	1709.2	1770.4
	s	0.9139	1.2345	1.3076	1.3701	1.4129	1.4472	1.4766	1.5029	1.5269	1.5492	1.5703	1.5903	1.6094	1.6456	1.6796	1.7116
2600 (673.91)	Sh			26.09	76.09	126.09	176.09	226.09	276.09	326.09	376.09	426.09	476.09	526.09	626.09	726.09	826.09
	v	0.02938	0.1211	0.1544	0.1909	0.2171	0.2390	0.2585	0.2765	0.2933	0.3093	0.3247	0.3395	0.3540	0.3819	0.4088	0.4350
	h	744.47	1082.0	1160.2	1241.1	1296.5	1341.9	1382.1	1419.2	1454.1	1487.7	1520.2	1552.2	1583.7	1646.0	1707.7	1769.1
	s	0.9247	1.2225	1.2908	1.3592	1.4042	1.4395	1.4696	1.4964	1.5208	1.5434	1.5646	1.5848	1.6040	1.6405	1.6746	1.7068
2700 (679.53)	Sh			20.47	70.47	120.47	170.47	220.47	270.47	320.47	370.47	420.47	470.47	520.47	620.47	720.47	820.47
	v	0.03029	0.1119	0.1411	0.1794	0.2058	0.2275	0.2468	0.2644	0.2809	0.2965	0.3114	0.3259	0.3399	0.3670	0.3931	0.4184
	h	757.34	1069.7	1142.0	1231.1	1289.5	1336.3	1377.5	1415.2	1450.7	1484.6	1517.5	1549.8	1581.5	1644.1	1706.1	1767.8
	s	0.9356	1.2097	1.2727	1.3481	1.3954	1.4319	1.4628	1.4900	1.5148	1.5376	1.5591	1.5794	1.5988	1.6355	1.6697	1.7021
2800 (684.96)	Sh			15.04	65.04	115.04	165.04	215.04	265.04	315.04	365.04	415.04	465.04	515.04	615.04	715.04	815.04
	v	0.03134	0.1030	0.1278	0.1685	0.1952	0.2168	0.2358	0.2531	0.2693	0.2845	0.2991	0.3132	0.3268	0.3532	0.3785	0.4030
	h	770.69	1055.8	1121.2	1220.6	1282.2	1330.7	1372.8	1411.2	1447.2	1481.6	1514.8	1547.3	1579.3	1642.2	1704.5	1766.5
	s	0.9468	1.1958	1.2527	1.3368	1.3867	1.4245	1.4561	1.4838	1.5089	1.5321	1.5537	1.5742	1.5938	1.6306	1.6651	1.6975
2900 (690.22)	Sh			9.78	59.78	109.78	159.78	209.78	259.78	309.78	359.78	409.78	459.78	509.78	609.78	709.78	809.78
	v	0.03262	0.0942	0.1138	0.1581	0.1853	0.2068	0.2256	0.2427	0.2585	0.2734	0.2877	0.3014	0.3147	0.3403	0.3649	0.3887
	h	785.13	1039.8	1095.3	1209.6	1274.7	1324.9	1368.0	1407.2	1443.7	1478.5	1512.1	1544.9	1577.0	1640.4	1703.0	1765.2
	s	0.9588	1.1803	1.2283	1.3251	1.3780	1.4171	1.4494	1.4777	1.5032	1.5266	1.5485	1.5692	1.5889	1.6259	1.6605	1.6931
3000 (695.33)	Sh			4.67	54.67	104.67	154.67	204.67	254.67	304.67	354.67	404.67	454.67	504.67	604.67	704.67	804.67
	v	0.03428	0.0850	0.0982	0.1483	0.1759	0.1975	0.2161	0.2329	0.2484	0.2630	0.2770	0.2904	0.3033	0.3282	0.3522	0.3753
	h	801.84	1020.3	1060.5	1197.9	1267.0	1319.0	1363.2	1403.1	1440.2	1475.4	1509.4	1542.4	1574.8	1638.5	1701.4	1763.8
	s	0.9728	1.1619	1.1966	1.3131	1.3692	1.4097	1.4429	1.4717	1.4976	1.5213	1.5434	1.5642	1.5841	1.6214	1.6561	1.6888
3100 (700.28)	Sh				49.72	99.72	149.72	199.72	249.72	299.72	349.72	399.72	449.72	499.72	599.72	699.72	799.72
	v	0.03681	0.0745		0.1389	0.1671	0.1887	0.2071	0.2237	0.2390	0.2533	0.2670	0.2800	0.2927	0.3170	0.3403	0.3628
	h	823.97	993.3		1185.4	1259.1	1313.0	1358.4	1399.0	1436.7	1472.3	1506.6	1539.9	1572.6	1636.7	1699.8	1762.5
	s	0.9914	1.1373		1.3007	1.3604	1.4024	1.4364	1.4658	1.4920	1.5161	1.5384	1.5594	1.5794	1.6169	1.6518	1.6847

434

Pressure		Sat.	44.92	94.92	144.92	194.92	244.92	294.92	344.92	394.92	444.92	494.92	594.92	694.92	794.92
3200 (705.08)	Sh v	0.04472	0.1300	0.1588	0.1804	0.1987	0.2151	0.2301	0.2442	0.2576	0.2704	0.2827	0.3065	0.3291	0.3510
	h	875.54	1172.3	1250.9	1306.9	1353.4	1394.9	1433.1	1469.2	1503.8	1537.4	1570.3	1634.8	1698.3	1761.2
	s	1.0351	1.2877	1.3515	1.3951	1.4300	1.4600	1.4866	1.5110	1.5335	1.5547	1.5749	1.6126	1.6477	1.6806
		0.0566													
		9316													
		1.0832													
3300	Sh v		0.1213	0.1510	0.1727	0.1908	0.2070	0.2218	0.2357	0.2488	0.2613	0.2734	0.2966	0.3187	0.3400
	h		1158.2	1242.5	1300.7	1348.4	1390.7	1429.5	1466.1	1501.0	1534.9	1568.1	1632.9	1696.7	1759.9
	s		1.2742	1.3425	1.3879	1.4237	1.4542	1.4813	1.5059	1.5287	1.5501	1.5704	1.6084	1.6436	1.6767
3400	Sh v		0.1129	0.1435	0.1653	0.1834	0.1994	0.2140	0.2276	0.2405	0.2528	0.2646	0.2872	0.3088	0.3296
	h		1143.2	1233.7	1294.3	1343.4	1386.4	1425.9	1462.9	1498.3	1532.4	1565.8	1631.1	1695.1	1758.5
	s		1.2600	1.3334	1.3807	1.4174	1.4486	1.4761	1.5010	1.5240	1.5456	1.5660	1.6042	1.6396	1.6728
3500	Sh v		0.1048	0.1364	0.1583	0.1764	0.1922	0.2066	0.2200	0.2326	0.2447	0.2563	0.2784	0.2995	0.3198
	h		1127.1	1224.6	1287.8	1338.2	1382.2	1422.2	1459.7	1495.5	1529.9	1563.6	1629.2	1693.6	1757.2
	s		1.2450	1.3242	1.3734	1.4112	1.4430	1.4709	1.4962	1.5194	1.5412	1.5618	1.6002	1.6358	1.6691
3600	Sh v		0.0966	0.1296	0.1517	0.1697	0.1854	0.1996	0.2128	0.2252	0.2371	0.2485	0.2702	0.2908	0.3106
	h		1108.6	1215.3	1281.2	1333.0	1377.9	1418.6	1456.5	1492.6	1527.4	1561.3	1627.3	1692.0	1755.9
	s		1.2281	1.3148	1.3662	1.4050	1.4374	1.4658	1.4914	1.5149	1.5369	1.5576	1.5962	1.6320	1.6654
3800	Sh v		0.0799	0.1169	0.1395	0.1574	0.1729	0.1868	0.1996	0.2116	0.2231	0.2340	0.2549	0.2746	0.2936
	h		1064.2	1195.5	1267.6	1322.4	1369.1	1411.2	1450.1	1487.0	1522.4	1556.8	1623.6	1688.9	1753.2
	s		1.1888	1.2955	1.3517	1.3928	1.4265	1.4558	1.4821	1.5061	1.5284	1.5495	1.5886	1.6247	1.6584
4000	Sh v		0.0631	0.1052	0.1284	0.1463	0.1616	0.1752	0.1877	0.1994	0.2105	0.2210	0.2411	0.2601	0.2783
	h		1007.4	1174.3	1253.4	1311.6	1360.2	1403.6	1443.6	1481.3	1517.3	1552.2	1619.8	1685.7	1750.6
	s		1.1396	1.2754	1.3371	1.3807	1.4158	1.4461	1.4730	1.4976	1.5203	1.5417	1.5812	1.6177	1.6516
4200	Sh v		0.0498	0.0945	0.1183	0.1362	0.1513	0.1647	0.1769	0.1883	0.1991	0.2093	0.2287	0.2470	0.2645
	h		950.1	1151.6	1238.6	1300.4	1351.2	1396.0	1437.1	1475.5	1512.2	1547.6	1616.1	1682.6	1748.0
	s		1.0905	1.2544	1.3223	1.3686	1.4053	1.4366	1.4642	1.4893	1.5124	1.5341	1.5742	1.6109	1.6452
4400	Sh v		0.0421	0.0846	0.1090	0.1270	0.1420	0.1552	0.1671	0.1782	0.1887	0.1986	0.2174	0.2351	0.2519
	h		909.5	1127.3	1223.3	1289.0	1342.0	1388.3	1430.4	1469.7	1507.1	1543.0	1612.3	1679.4	1745.3
	s		1.0556	1.2325	1.3073	1.3566	1.3949	1.4272	1.4556	1.4812	1.5048	1.5268	1.5673	1.6044	1.6389

Sh = superheat, F
v = specific volume, cu ft per lb
h = enthalpy, Btu per lb
s = entropy, Btu per F per lb

Table 16-15. (Continued)

Abs Press. Lb/Sq In. (Sat. Temp)		750	800	850	900	950	1000	1050	1100	1150	1200	1250	1300	1400	1500
4600	v	0.0380	0.0751	0.1005	0.1186	0.1335	0.1465	0.1582	0.1691	0.1792	0.1889	0.1982	0.2071	0.2242	0.2404
	h	883.8	1100.0	1207.3	1277.2	1332.6	1380.5	1423.7	1463.9	1501.9	1538.4	1573.8	1608.5	1676.3	1742.7
	s	1.0331	1.2084	1.2922	1.3446	1.3847	1.4181	1.4472	1.4734	1.4974	1.5197	1.5407	1.5607	1.5982	1.6330
4800	v	0.0355	0.0665	0.0927	0.1109	0.1257	0.1385	0.1500	0.1606	0.1706	0.1800	0.1890	0.1977	0.2142	0.2299
	h	866.9	1071.2	1190.7	1265.2	1323.1	1372.6	1417.0	1458.0	1496.7	1533.8	1569.7	1604.7	1673.1	1740.0
	s	1.0180	1.1835	1.2768	1.3327	1.3745	1.4090	1.4390	1.4657	1.4901	1.5128	1.5341	1.5543	1.5921	1.6272
5000	v	0.0338	0.0591	0.0855	0.1038	0.1185	0.1312	0.1425	0.1529	0.1626	0.1718	0.1806	0.1890	0.2050	0.2203
	h	854.9	1042.9	1173.6	1252.9	1313.5	1364.6	1410.2	1452.1	1491.5	1529.1	1565.5	1600.9	1670.0	1737.4
	s	1.0070	1.1593	1.2612	1.3207	1.3645	1.4001	1.4309	1.4582	1.4831	1.5061	1.5277	1.5481	1.5863	1.6216
5200	v	0.0326	0.0531	0.0789	0.0973	0.1119	0.1244	0.1356	0.1458	0.1553	0.1642	0.1728	0.1810	0.1966	0.2114
	h	845.8	1016.9	1156.0	1240.4	1303.7	1356.6	1403.4	1446.2	1486.3	1524.5	1561.3	1597.2	1666.8	1734.7
	s	0.9985	1.1370	1.2455	1.3088	1.3545	1.3914	1.4229	1.4509	1.4762	1.4995	1.5214	1.5420	1.5806	1.6161
5400	v	0.0317	0.0483	0.0728	0.0912	0.1058	0.1182	0.1292	0.1392	0.1485	0.1572	0.1656	0.1736	0.1888	0.2031
	h	838.5	994.3	1138.1	1227.7	1293.7	1348.4	1396.5	1440.3	1481.1	1519.8	1557.1	1593.4	1663.7	1732.1
	s	0.9915	1.1175	1.2296	1.2969	1.3446	1.3827	1.4151	1.4437	1.4694	1.4931	1.5153	1.5362	1.5750	1.6109
5600	v	0.0309	0.0447	0.0672	0.0856	0.1001	0.1124	0.1232	0.1331	0.1422	0.1508	0.1589	0.1667	0.1815	0.1954
	h	832.4	975.0	1119.9	1214.8	1283.7	1340.2	1389.6	1434.3	1475.9	1515.2	1552.9	1589.6	1660.5	1729.5
	s	0.9855	1.1008	1.2137	1.2850	1.3348	1.3742	1.4075	1.4366	1.4628	1.4869	1.5093	1.5304	1.5697	1.6058
5800	v	0.0303	0.0419	0.0622	0.0805	0.0949	0.1070	0.1177	0.1274	0.1363	0.1447	0.1527	0.1603	0.1747	0.1883
	h	827.3	958.8	1101.8	1201.8	1273.6	1332.0	1382.6	1428.3	1470.6	1510.5	1548.7	1585.8	1657.4	1726.8
	s	0.9803	1.0867	1.1981	1.2732	1.3250	1.3658	1.3999	1.4297	1.4564	1.4808	1.5035	1.5248	1.5644	1.6008
6000	v	0.0298	0.0397	0.0579	0.0757	0.0900	0.1020	0.1126	0.1221	0.1309	0.1391	0.1469	0.1544	0.1684	0.1817
	h	822.9	945.1	1084.6	1188.8	1263.4	1323.6	1375.7	1422.3	1465.4	1505.9	1544.6	1582.0	1654.2	1724.2
	s	0.9758	1.0746	1.1833	1.2615	1.3154	1.3574	1.3925	1.4229	1.4500	1.4748	1.4978	1.5194	1.5593	1.5960

Temperature — Degrees Fahrenheit

Sat Water   Sat Steam

436

Press.															
6500	Sh														
	v	0.0287	0.0358	0.0495	0.0655	0.0793	0.0909	0.1012	0.1104	0.1188	0.1266	0.1340	0.1411	0.1544	0.1669
	h	813.9	919.5	1046.7	1156.3	1237.8	1302.7	1358.1	1407.3	1452.2	1494.2	1534.1	1572.5	1646.4	1717.6
	s	0.9661	1.0515	1.1506	1.2328	1.2917	1.3370	1.3743	1.4064	1.4347	1.4604	1.4841	1.5062	1.5471	1.5844
7000	Sh														
	v	0.0279	0.0334	0.0438	0.0573	0.0704	0.0816	0.0915	0.1004	0.1085	0.1160	0.1231	0.1298	0.1424	0.1542
	h	806.9	901.8	1016.5	1124.9	1212.6	1281.7	1340.5	1392.2	1439.1	1482.6	1523.7	1563.1	1638.6	1711.1
	s	0.9582	1.0350	1.1243	1.2055	1.2689	1.3171	1.3567	1.3904	1.4200	1.4466	1.4710	1.4938	1.5355	1.5735
7500	Sh														
	v	0.0272	0.0318	0.0399	0.0512	0.0631	0.0737	0.0833	0.0918	0.0996	0.1068	0.1136	0.1200	0.1321	0.1433
	h	801.3	889.0	992.9	1097.7	1188.3	1261.0	1322.9	1377.2	1426.0	1471.0	1513.3	1553.7	1630.8	1704.6
	s	0.9514	1.0224	1.1033	1.1818	1.2473	1.2980	1.3397	1.3751	1.4059	1.4335	1.4586	1.4819	1.5245	1.5632
8000	Sh														
	v	0.0267	0.0306	0.0371	0.0465	0.0571	0.0671	0.0762	0.0845	0.0920	0.0989	0.1054	0.1115	0.1230	0.1338
	h	796.6	879.1	974.4	1074.3	1165.4	1241.0	1305.5	1362.2	1413.0	1459.6	1503.1	1544.5	1623.1	1698.1
	s	0.9455	1.0122	1.0864	1.1613	1.2271	1.2798	1.3233	1.3603	1.3924	1.4208	1.4467	1.4705	1.5140	1.5533
8500	Sh														
	v	0.0262	0.0296	0.0350	0.0429	0.0522	0.0615	0.0701	0.0780	0.0853	0.0919	0.0982	0.1041	0.1151	0.1254
	h	792.7	871.2	959.8	1054.5	1144.0	1221.9	1288.5	1347.5	1400.2	1448.2	1492.9	1535.3	1615.4	1691.7
	s	0.9402	1.0037	1.0727	1.1437	1.2084	1.2627	1.3076	1.3460	1.3793	1.4087	1.4352	1.4597	1.5040	1.5439
9000	Sh														
	v	0.0258	0.0288	0.0335	0.0402	0.0483	0.0568	0.0649	0.0724	0.0794	0.0858	0.0918	0.0975	0.1081	0.1179
	h	789.3	864.7	948.0	1037.6	1125.4	1204.1	1272.1	1333.0	1387.5	1437.1	1482.9	1526.3	1607.9	1685.3
	s	0.9354	0.9964	1.0613	1.1285	1.1918	1.2468	1.2926	1.3323	1.3667	1.3970	1.4243	1.4492	1.4944	1.5349
9500	Sh														
	v	0.0254	0.0282	0.0322	0.0380	0.0451	0.0528	0.0603	0.0675	0.0742	0.0804	0.0862	0.0917	0.1019	0.1113
	h	786.4	859.2	938.3	1023.4	1108.9	1187.7	1256.6	1318.9	1375.1	1426.1	1473.1	1517.3	1600.4	1679.0
	s	0.9310	0.9900	1.0516	1.1153	1.1771	1.2320	1.2785	1.3191	1.3546	1.3858	1.4137	1.4392	1.4851	1.5263
10000	Sh														
	v	0.0251	0.0276	0.0312	0.0362	0.0425	0.0495	0.0565	0.0633	0.0697	0.0757	0.0812	0.0865	0.0963	0.1054
	h	783.8	854.5	930.2	1011.3	1094.2	1172.6	1242.0	1305.3	1362.9	1415.3	1463.4	1508.6	1593.1	1672.8
	s	0.9270	0.9842	1.0432	1.1039	1.1638	1.2185	1.2652	1.3065	1.3429	1.3749	1.4035	1.4295	1.4763	1.5180
10500.	Sh														
	v	0.0248	0.0271	0.0303	0.0347	0.0404	0.0467	0.0532	0.0595	0.0656	0.0714	0.0768	0.0818	0.0913	0.1001
	h	781.5	850.5	923.4	1001.0	1081.3	1158.9	1228.4	1292.4	1351.1	1404.7	1453.9	1500.0	1585.8	1666.7
	s	0.9232	0.9790	1.0358	1.0939	1.1519	1.2060	1.2529	1.2946	1.3371	1.3644	1.3937	1.4202	1.4677	1.5100

Sh = superheat, F
v = specific volume, cu ft per lb

h = enthalpy, Btu per lb
s = entropy, Btu per F per lb

Table 16-15. (Concluded)

Abs Press. Lb/Sq In. (Sat. Temp)		Sat Water	Sat Steam	750	800	850	900	950	1000	1050	1100	1150	1200	1250	1300	1400	1500
11000	v			0.0245	0.0267	0.0296	0.0335	0.0386	0.0443	0.0503	0.0562	0.0620	0.0676	0.0727	0.0776	0.0868	0.0952
	h			779.5	846.9	917.5	992.1	1069.9	1146.3	1215.9	1280.2	1339.7	1394.4	1444.6	1491.5	1578.7	1660.6
	s			0.9196	0.9742	1.0292	1.0851	1.1412	1.1945	1.2414	1.2833	1.3209	1.3544	1.3842	1.4112	1.4595	1.5023
11500	v			0.0243	0.0263	0.0290	0.0325	0.0370	0.0423	0.0478	0.0534	0.0588	0.0641	0.0691	0.0739	0.0827	0.0909
	h			777.7	843.8	912.4	984.5	1059.8	1134.9	1204.3	1268.7	1328.8	1384.4	1435.5	1483.2	1571.8	1654.7
	s			0.9163	0.9698	1.0232	1.0772	1.1316	1.1840	1.2308	1.2727	1.3107	1.3446	1.3750	1.4025	1.4515	1.4949
12000	v			0.0241	0.0260	0.0284	0.0317	0.0357	0.0405	0.0456	0.0508	0.0560	0.0610	0.0659	0.0704	0.0790	0.0869
	h			776.1	841.0	907.9	977.8	1050.9	1124.5	1193.7	1258.0	1318.5	1374.7	1426.6	1475.1	1564.9	1648.8
	s			0.9131	0.9657	1.0177	1.0701	1.1229	1.1742	1.2209	1.2627	1.3010	1.3353	1.3662	1.3941	1.4438	1.4877
12500	v			0.0238	0.0256	0.0279	0.0309	0.0346	0.0390	0.0437	0.0486	0.0535	0.0583	0.0629	0.0673	0.0756	0.0832
	h			774.7	838.6	903.9	971.9	1043.1	1115.2	1184.1	1247.9	1308.8	1365.4	1418.0	1467.2	1558.2	1643.1
	s			0.9101	0.9618	1.0127	1.0637	1.1151	1.1653	1.2117	1.2534	1.2918	1.3264	1.3576	1.3860	1.4363	1.4808
13000	v			0.0236	0.0253	0.0275	0.0302	0.0336	0.0376	0.0420	0.0466	0.0512	0.0558	0.0602	0.0645	0.0725	0.0799
	h			773.5	836.3	900.4	966.8	1036.2	1106.7	1174.8	1238.5	1299.6	1356.5	1409.6	1459.4	1551.6	1637.4
	s			0.9073	0.9582	1.0080	1.0578	1.1079	1.1571	1.2030	1.2445	1.2831	1.3179	1.3494	1.3781	1.4291	1.4741
13500	v			0.0235	0.0251	0.0271	0.0297	0.0328	0.0364	0.0405	0.0448	0.0492	0.0535	0.0577	0.0619	0.0696	0.0768
	h			772.3	834.4	897.2	962.2	1030.0	1099.1	1166.3	1229.7	1291.0	1348.1	1401.5	1451.8	1545.2	1631.9
	s			0.9045	0.9548	1.0037	1.0524	1.1014	1.1495	1.1948	1.2361	1.2749	1.3098	1.3415	1.3705	1.4221	1.4675
14000	v			0.0233	0.0248	0.0267	0.0291	0.0320	0.0354	0.0392	0.0432	0.0474	0.0515	0.0555	0.0595	0.0670	0.0740
	h			771.3	832.6	894.3	958.0	1024.5	1092.3	1158.5	1221.4	1283.0	1340.2	1393.8	1444.4	1538.8	1626.5
	s			0.9019	0.9515	0.9996	1.0473	1.0953	1.1426	1.1872	1.2282	1.2671	1.3021	1.3339	1.3631	1.4153	1.4612
14500	v			0.0231	0.0246	0.0264	0.0287	0.0314	0.0345	0.0380	0.0418	0.0458	0.0496	0.0534	0.0573	0.0646	0.0714
	h			770.4	831.0	891.7	954.3	1019.6	1086.2	1151.4	1213.8	1275.4	1332.9	1386.4	1437.3	1532.6	1621.1
	s			0.8994	0.9484	0.9957	1.0426	1.0897	1.1362	1.1801	1.2208	1.2597	1.2949	1.3266	1.3560	1.4087	1.4551
15000	v			0.0230	0.0244	0.0261	0.0282	0.0308	0.0337	0.0369	0.0405	0.0443	0.0479	0.0516	0.0552	0.0624	0.0690
	h			769.6	829.5	889.3	950.9	1015.1	1080.6	1144.9	1206.8	1268.1	1326.0	1379.4	1430.3	1526.4	1615.9
	s			0.8970	0.9455	0.9920	1.0382	1.0846	1.1302	1.1735	1.2139	1.2525	1.2880	1.3197	1.3491	1.4022	1.4491
15500	v			0.0228	0.0242	0.0258	0.0278	0.0302	0.0329	0.0360	0.0393	0.0429	0.0464	0.0499	0.0534	0.0603	0.0668
	h			768.9	828.2	887.2	947.8	1011.1	1075.7	1139.0	1200.3	1261.1	1319.6	1372.8	1423.6	1520.4	1610.8
	s			0.8946	0.9427	0.9886	1.0340	1.0797	1.1247	1.1674	1.2073	1.2457	1.2815	1.3131	1.3424	1.3959	1.4433

Temperature Degrees Fahrenheit

Sh = superheat, F
v = specific volume, cu ft per lb
h = enthalpy, Btu per lb
s = entropy, Btu per F per lb

References

1. Buiness Makes Room for the Energy Executive, J. R. Stock and D. M. Lambert, *Business,* November–December, 1980.
2. 1972 Statistical Review of the World Oil Industry, British Petroleum Company.
3. Patterns of Energy Consumption in the United States, Office of Science and Technology, Executive Office of the President, January, 1972.
4. The Technical Aspects of the Conservation of Energy for Industrial Processes, J. Burroughs, Federal Power Commission National Power Survey Technical Advisory Committee on Conservation of Energy, Position Paper No. 17.
5. Solar Energy: It's Time is Near, W. Morrow, Jr., *Technology Review,* December, 1973.
6. Rankine–Cycle Systems for Waste Heat Recovery, R. Barber, *Chemical Engineering,* 1974.
7. Heat Recovery Combined Cycle Boosts Electric Power with No Added Fuel, *Electrical Consultant,* November, 1974.
8. Energy Conservation in Existing Plants, J. C. Robertson, *Chemical Engineering,* January 21, 1974.
9. Solar Energy System Reduces Life Costs, Louis C. Sullivan, *Specifying Engineer,* August 1974.
10. Energy Self-Sufficiency; An Economic Evaluation, *Technology Review,* May 1974.
11. More Cooling at Less Cost, W. Farwell, *Plant Engineering,* August 22, 1974.

439

12. The Heat Pump: Performance Factor, Possible Improvements, E. Ambrose, *Heating/Pump/Air Conditioning*, May 1974.

13. Energy, Economy and Industrial HVAC Systems, A. Hallstrom, *Specifying Engineer*, December 1974.

14. A New Approach to Factory Air Conditioning, F. Dean, *Heating/Piping/Air Conditioning*, June 1974.

15. Insulation Saves Energy, R. Hughes and V. Deumaga, *Chemical Engineering*, May 27, 1974.

16. Energy Conservation in New Plant Designs, J. Fleming, J. Lambrix and M. Smith, *Chemical Engineering*, January 21, 1974.

17. Fuel Savings Through No-Vent Systems, M. Stout, *Chemical Engineering*, September 30, 1974.

18. Heating Fuel Conservation, R. Meister, *Plant Engineering*, July 11, 1974.

19. Conserving Utilities—Energy is New Construction, C. Schumacher, B. Girgis, *Chemical Engineering*, February 18, 1974.

20. How About A Steam Turbine Drive?, J. Behl, *Plant Engineering*, July 26, 1974.

21. How to Conserve Energy in the Design of Plants, G. Moor, *Specifying Engineer*, September, 1973.

22. Power Demand Control Uses Energy Better, B. Murphy and R. Putman, *Specifying Engineer*, December 1974.

23. Ways to Save Energy in Existing HVAC Systems, W. Landman, *Specifying Engineer*, January 1975.

24. Application Guide Weathertron Heat Pump, General Electric Publication No. 23-3047-2.

25. ASHRAE 1972 Handbook of Fundamentals
    Chapter 2—Heat Transfer
    Chapter 11—Applied Heat Pump Systems
    Chapter 20—Design Heat Transmission Coefficients
    Chapter 21—Heating Load
    Chapter 22—Air Conditioning Cooling Load

26. ASHRAE 1974 Application Handbook
    Chapter 59—Solar Energy Utilization for Heating and Cooling

27. ASHRAE 1973 Systems Handbook
    Chapter 7—Heat Recovery Systems
    Chapter 43—Energy Estimating Methods

28. Trane Air Conditioning Manual

29. How to Save Refinery-Furnace Fuel, D. Cherrington, H. Michelson, *Oil and Gas Journal*, September 2, 1974.

30. Conserve Energy by Design, Applications Engineering Manual, Trane Company.

31. Organizing for Energy Conservation, R. Kulweic, *Plant Engineering*, February 20, 1975.

32. Energy Conservation Program Guide for Industry and Commerce, NBS Handbook 115.

33. ASHRAE Standard 90P, Energy Conservation in New Building Design, 1975.

34. Thermodynamics of Engineering Science, S. L. Soo, Englewood Cliffs, N.J., Prentice Hall, Inc. 1959.

35. Economic Analysis for Engineering and Managerial Decision Making, N. Barish, New York, McGraw–Hill Book Company, Inc., 1962.

36. Thermodynamics, New York, F. Sears, Reading, Mass., Addison-Wesley Publishing Company, Inc., 1952.

37. Which Air Conditioning System *Really* Saves Energy?, I. Naman, *Specifying Engineer*, August 1974.

38. Keeping That Electric Bill Under Control, A. Wright, *Plant Engineering*, June 19, 1974.

39. Steam—Its Generation and Use, The Babcock–Wilcox Company, 1963.

40. Energy—Crisis and Conservation, A. Field, *Heating/Piping/Air Conditioning*, February 1974.

41. Saving Energy in Department Store M/E Systems, H. Argintar, *Electrical Consultant*, October 1974.

42. An Efficient Selection of Modern Energy—Saving Light Sources Can Mean Savings of 10% to 30% in Power Consumption, H. Anderson, *Electrical Consultant*, April 1974.

43. Electric Power Distribution for Industrial Plants, The Institute of Electrical Engineers, 1964.

44. Flow of Fluids Through Valves, Fittings and Pipe, Crane Technical Paper No. 410, 1970.

45. Life Cycle Costing in an Energy Crisis Era, Joseph P. Keefe, *Professional Engineer*, July 1974.

46. The Sad State of Maintenance Management, Albert B. Drui and Paul T. Suul, *Plant Engineering*, June 12, 1975.

47. Evaluating the Effects of Dirt on HID Luminaries, A. C. McNamara and Andy Willingham, *Plant Engineering*, April 17, 1975.

48. Electrical Consulting: Engineering and Design, A. Thumann, Atlanta, Ga., Fairmont Press, 1972.

49. Secrets of Noise Control, A. Thumann and R. K. Miller, Atlanta, Ga., Fairmont Press, 1974.

50. Lighting Handbook, 1971, Westinghouse Electric Company.

51. Design Concepts for Optimum Energy Use in HVAC Systems, Electric Energy Association.

52. Square D Bulletin SM-451-CPG-1-73, Demand Control.

53. The Engineering Basics of Power Factor Improvement, *Specifying Engineer,* February 1975; May 1975.

54. A New Look at Load Shedding, A. Thumann, *Electrical Consultant,* August, 1974.

55. Improving Plant Power Factor, A. Thumann, *Electrical Consultant,* July 1974.

56. The International Solar Energy Society Conference of 1970, Paper No. 4135, R. V. Dunkle and E. T. Davey.

57. Solar Energy Utilization for Heating and Cooling, National Science Foundation, NSF 74–41.

58. Low Temperature Engineering Application of Solar Energy, ASHRAE 1967.

59. Climatic Atlas of the U.S., reprinted under the title: Weather Atlas of the United States, Gale Research Co., 1975.

60. Plant Engineers and Managers Guide to Energy Conservation—2nd Edition, Albert Thumann, Van Nostrand Reinhold, 1983.

61. Handbook of Energy Audits—2nd Edition, Albert Thumann, Fairmont Press, 1983.

62. Waste Heat Management Guidebook, Kreider and McNeil, NBS Handbook 121.

63. The Emerging Synthetic Fuel Industry, Albert Thumann, Fairmont Press, 1981.

64. The Residential Energy Audit Manual, Fairmont Press, 1981.

65. Foundations For The Solar Future, Richard Koral, Fairmont Press, 1981.

Index